This book is to be returned on or before
the last date

04. FEB 01

04. MAR

12.

BSC24

edition r pub.

5064

GEOGRAPHY OF WORLD AFFAIRS

Geography of World Affairs

Sixth edition

J. P. Cole

BA, MA, PhD
Professor of Regional Geography

and

Holder of Rotating Headship
in the Geography Department
Nottingham University

Butterworths

London   Boston   Durban   Singapore   Sydney   Toronto   Wellington

First published as a Penguin Special 1959
Reprinted 1960
Reissued in Pelican Books 1963
Reprinted 1964
Third edition 1964
Reprinted with a Postscript 1965
Reprinted 1966
Fourth edition 1972
Reprinted with revisions, new concluding chapter 1974
Fifth edition 1979
Sixth edition published by Butterworths 1983

© J. P. Cole 1983

British Library Cataloguing in Publication Data

Cole, J. P.
  Geography of world affairs.—6th ed.
  1. Social history—1960–1970
  2. Social history—1970–
  I. Title
  909.82'6   HN16

  ISBN 0-408-10842-8

Library of Congress Cataloguing in Publication Data

Cole, J. P. (John Paul), 1928–
  Geography of world affairs.
  Includes bibliographies and index.
  1. Geography.   I. Title.
  G116.C64   1983   910   83–7600

  ISBN 0-408-10842-8

Photoset by Butterworths Litho Preparation Department
Printed in England by Page Bros Ltd, Norwich, Norfolk.

Preface

The first edition of *Geography of World Affairs* was written in 1957. At that time world affairs seemed to be dominated by a struggle between the communist bloc, led by the Soviet Union, and the capitalist world, strongly influenced by the United States. In the 1960s the world situation apparently became more complex as other powers such as China, Japan and Germany re-emerged, and as economic groupings came into being. Since the late 1960s increasing publicity has been given to the demographic, economic and ecological problems facing mankind, and serious attempts have been made to map out possible futures.

The structure of the present sixth edition of *Geography of World Affairs* differs considerably from that of the first. World problems seem more numerous but less clearcut and world affairs more complex. It is hoped, however, that the present book will serve as a useful geographical background to the study of world affairs, a work of reference and a source of ideas for discussion and speculation.

In view of the difficulty of obtaining complete sets of very recent information for large numbers of countries much of the numerical data will be a little out of date when the book is read. The broad conclusions drawn from the data should, however, be broadly true for some years.

The author is very grateful to an anonymous referee appointed by Butterworths for his (or her) comments on the original text and to Mr C. Lewis for drawing the maps and diagrams.

J. P. C.
Nottingham University

Contents

Abbreviations
(See Table 2.2 for countries)

CMEA	Council for Mutual Economic Assistance (also COMECON)
FAO	Food and Agriculture Organization (Rome)
FAOPY	Food and Agriculture Production Yearbook
GNP	Gross National Product
UNDYB	United Nations Demographic Yearbook
UNSYB	United Nations Statistical Yearbook
UNYITS	United Nations Yearbook of International Trade Statistics
UK	United Kingdom (of Great Britain and Northern Ireland)
USA	United States of America
USSR	Union of Soviet Socialist Republics

Conversions and Equivalents

Length (1 kilometre = 1000 metres = 100 000 centimetres)

1 centimetre	= 0.39 inches		1 inch	= 2.54 centimetres
1 metre	= 3.28 feet		1 yard	= 0.91 metres
1 kilometre	= 0.62 miles		1 mile	= 1.61 kilometres

Area (1 square kilometre = 100 hectares)

1 hectare	= 2.47 acres
1 square kilometre	= 0.386 square miles
1 acre	= 0.405 hectares
1 square mile	= 2.59 square kilometres

Weight (1 tonne = 1000 kilograms)

1 kilogram	= 2.205 pounds
1 tonne (metric)	= 0.984 long tons and 1.102 short tons

Temperature

Centigrade	$-10°$	$0°$	$+10°$	$+20°$	$+30°$
Fahrenheit	$+14°$	$+32°$	$+50°$	$+68°$	$+86°$

Oil (petroleum)

Oil is measured more precisely by volume than by weight.
1 barrel = 42 US gallons, about 35 Imperial gallons, or 159 litres.
One tonne of oil is equivalent approximately to 7.5 barrels (for example in the USA 7.77, Middle East 7.35).

Introduction

1.1 Introduction

The subject of world affairs includes aspects of many traditional disciplines. The same total complex of objects and events in the world is seen from a different angle by each discipline. All are seeking some pattern in the complex picture, but none has a monopoly of the field. What special contributions can the geographer make to the study of world affairs?

The word 'geography' has two distinct meanings in general use. It may refer to the physical and topographical conditions in a given area or it may refer to location or distribution in the world. Such a use as: 'the north of Canada is difficult to develop on account of its geographical conditions' is common, but so also is the following use: 'because of its geographical location, Gibraltar is of great strategic significance'. The two above uses of the word geography cover only part of the field of study accepted by contemporary professional geographers.

Although the scope of geography has broadened considerably in recent decades it is possible, according to the American geographer W. D. Pattison (1964), to identify four broad, overlapping traditions. Each tradition noted below has been fashionable in particular periods:

(1) The *earth sciences* involve the study of the land, the oceans, the atmosphere and the biosphere. Knowledge about the physical background is essential in the study of world affairs since it reveals many limitations and constraints to the decisions and actions of individuals and of whole countries and is a guide to the availability of natural resources in different regions. Unless, for example, the very frail qualities of the soil beneath the dense tropical rain forests of the Amazon region are properly appreciated it is not easy to see why there is a great risk in clearing them for cultivation or grazing. The development of tectonic plate theory has led to a greater appreciation of the occurrence of earthquakes. Weather satellites show quickly and comprehensively the weather situation on a global and continental scale.

(2) The study of *man-environment relationships* has been a strong tradition in British geography in the last hundred years, especially with reference to the effect of the physical environment on man. In the northern parts of Canada and the USSR climatic conditions are so harsh that people cannot easily be persuaded to settle there permanently. Recently increasing attention has been devoted to the effect, often adverse, that man has on the environment.

(3) *Area Studies* concentrate on particular parts of the earth's surface. Regional geography is largely based on the area studies tradition. In world affairs each sovereign country is a clearly defined, organized region. How it works (or does not) and how it is related to other regions is a matter of particular concern in the study of world affairs.

(4) The study of *spatial situations* became fashionable in the 1960s. One of the main functions of geography and cartography has been to map places according to their position on the 'space' of the earth's surface. Many situations and problems in world affairs can only be fully appreciated when the relative position of places to one another is made clear. The landlocked location of Switzerland, for example, has forced that country to seek agreement with its neighbours with regard to access to the coast.

World affairs may be viewed by the geographer in various ways. One way of thinking of the world is to see it as a mosaic or jigsaw puzzle, the main pieces being the countries of the world. Since countries are related to one another through various transactions, however, world affairs may also be thought of as taking place in a complex system and the countries as elements in this system. An event in one place may have repercussions anywhere else in the system. The occurrence of bad grain harvests in the USSR may enable US farmers to sell more grain. The decision of Middle East countries to cut down oil exports may lead to an increase in the extraction of coal in western USA or in Queensland, Australia.

The main purpose of this edition of *Geography of World Affairs* is to describe 'what is where' in the world and 'where somewhere is in relation to somewhere else'. The study of the distribution of phenomena over area is one of the central traditions of geography and the theme has been approached here through an appraisal of the technical, historical, topical and regional features of the study of world affairs on the finite but unbounded surface of the globe. It is hoped that the information given in the book will provide the reader with a background into which specific situations and events can be fitted.

The European culture region has had a great influence on world affairs during the last few centuries. The European concern about the way society works and about why the world is there at all, has produced a number of interpretations that partly overlap. Such interpretations of the world must be taken into account in geographical studies. Examples of general views of the world widely held in the West include the following:

(1) A Christian view in which, strictly, the earth is a testing ground to determine the future of individuals for all time. There definitely is a purpose in life.
(2) An evolutionary view in which *Homo sapiens* is one of many species and should be occupying a niche, now a big one, in competition with other species. Eventually the human species could be superseded, reduced or even eliminated by the emergence of other species.
(3) The tough socio-economic 'market economy' view whereby competition *within* the species is to be expected and not hindered (too much), following the idea that the ablest members of a given species should be the ones to flourish and reproduce. The hidden hand guides, but not the hand of government.
(4) The tender socio-economic Marxian model of the inevitable progress of societies to conditions where everyone is well off and there is no competition.

Approaches 3 and 4 both contain ideas from 1 and 2. Non-Western regions of the world have received the above essentially European approaches to society and the economy with mixtures of scepticism and enthusiasm. Such views of the world are generally impossible to test since they are vague about what should happen when, but they may greatly influence the decisions and actions of influential individuals or of political parties and, therefore, affect world affairs.

1.2 Geometrical aspects of the earth's surface

The construction of maps that cover large parts of the earth's surface causes technical problems. These problems·will now be discussed since the cartographical problems of map projections are not adequately appreciated even by some geographers. Reference is not often enough made to a globe to show the true positions of places relative to one another.

Whether or not the earth is flat was a matter of speculation in Europe some centuries ago. Everyday experience would lead one to assume that it is indeed flat. Laws of physics that prevent people from falling over somewhere on a curved surface were not appreciated. On the other hand, if the earth was flat, where would its surface end and what might lie beyond the edge?

Few people now disagree that the earth is actually a sphere. Pictures from satellites and moon probes have revealed the curvature and spherical form convincingly. Actually the earth is not quite a perfect sphere but the fact that it is slightly flattened at the poles need not concern us in this book. World affairs, then, take place on a surface that is finite but unbounded, two features that are of considerable significance.

European geographers and cartographers have long been concerned with examining and recording the results of exploration. Several centuries passed before European cartographers finally got even the main details of major land and sea areas correct. In mapping the location of places on the earth's surface cartographers have constantly faced a major problem. There is little distortion of shapes and areas on a flat map of a fairly small part of the earth's surface, but when a substantial area is mapped, distortion is bound to occur. This problem must be appreciated in the study of world affairs because the way a particular situation is perceived and interpreted may depend on the way it is mapped.

The surface of a model globe is the only place to reproduce a map that represents the world accurately. Globes of many different sizes, prices and themes are commercially available. One produced by the National Geographic Society of the USA is particularly useful. It is illustrated in *Figure 1.1*. The globe rests in a holder but is not fixed to rotate on an axis. A separate transparent 'cap' can be centred on any point on the earth's surface and distances can be measured out from it. The adjoining Soviet map of part of the world in *Figure 1.1* brings out nicely the spherical nature of the world but it 'cheats' because it shows more of the earth's surface than could actually be seen on a globe from one viewpoint.

Trying to map the earth's surface on a flat piece of paper is similar to trying to flatten a large piece of orange peel without tearing it or pulling it out of

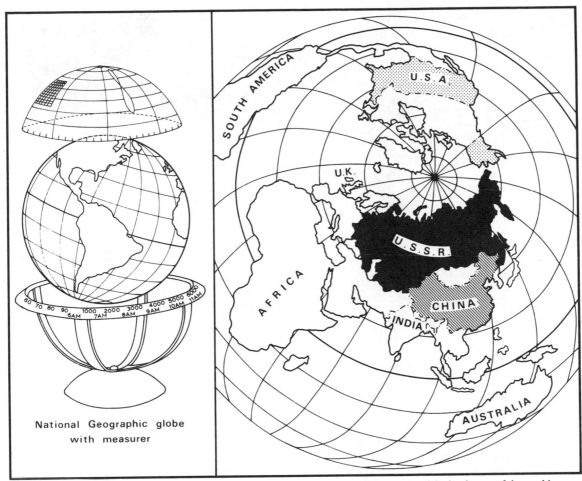

Figure 1.1   *National Geographic globe, and a Soviet Atlas map that attempts to show most of the land areas of the world*

shape. When represented on the flat surface of a map, any sizeable part of the earth's surface becomes distorted, whatever projection is used. Area, distance, shape, and direction cannot all be correctly shown simultaneously. Distortion is negligible when an area the size of a British county is shown on a map, but it becomes appreciable for an area the size of the USA or Europe.

On some projections area is shown correctly (projections of this kind are called equal area) but for this to be done shape must be distorted. On other projections shape is preserved (orthomorphic projections) but correct area must be sacrificed. Some areas come out much larger than they should and a scale will be reasonably true only for part of the map. When a large proportion of the earth's surface is mapped on a flat piece of paper great distortions occur. It is impossible to include every place in the world on one flat map. Some places that are close to one another in reality (and on the globe) end up far apart.

Among the most familiar world maps are those with the equator running left to right across the map and with north at the 'top'. Versions of this kind of projection are shown in *Figures 1.2* and *1.3*. Such a projection is described as 'cylindrical' because the resulting map starts as a 'cylinder' of paper touching the globe round the equator to be subsequently unrolled. Mercator's projection in *Figure 1.2* shows the world in a familiar way. On this projection, direction and shape are represented correctly but area is greatly exaggerated towards the poles because the scale increases away from the equator. Although Greenland appears to be as large as South America, in reality it is not much more than one-tenth the size. *Figure 1.3* shows the world on Interrupted Mollweide's Homolographic projection, with area correct but shape distorted, especially towards the poles. The oceans have been broken into so that the shape of the land areas is not too unrecognizable, but the oceans themselves are sacrificed.

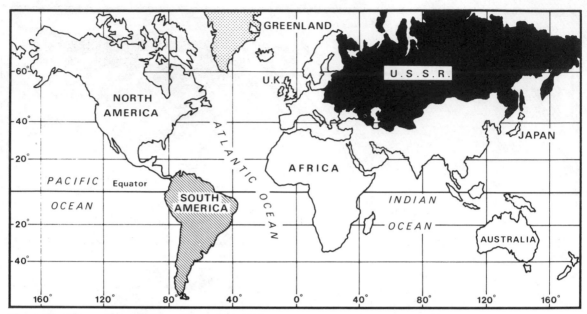

Figure 1.2   *The world portrayed on Mercator's projection*

On the two projections shown so far the equator is a straight line running across the map from one side to the other. A world map with the equator in this position is useful when the purpose of the map is to show how temperatures decrease away from it towards the poles. It is also satisfactory for showing how climate, vegetation, soils, agriculture and even types of building and clothing are related to latitude, because these are all related to temperature. In the study of world affairs it is often more necessary to know how various places are located in relation to one another than to know how they are located in relation to the equator. Other projections can be made.

Since it is not possible to bore a tunnel straight through the earth, the shortest practical distance between two places on the earth's surface is a 'great circle'. A great circle line between two places is one that you would find on a globe by holding one end of a piece of string over the first place and then pulling the string taut to its shortest length over the other place. In *Figure 1.4(a)* the ten black dots are towns and the lines joining some of them are the paths that ships, aircraft or missiles would follow assuming no obstacles such as land in the way of the ships or hostile airspace to deflect the aircraft. On the globe the lines of longitude (meridians), which all pass through both poles, are great circles, but of the lines of latitude (parallels) only the equator is a great circle.

Some special properties of the earth's surface are shown in *Figures 1.4(b)* and *(c)*. The centre of map *(b)* is the North Pole and the centre of map *(c)* is the

South Pole. The straight lines radiating from the poles are meridians and also great circles. The equator is not a straight line on maps *(b)* and *(c)*. The unbounded but finite nature of the earth's surface is illustrated by a hypothetical air journey made by an eccentric American millionaire flying round the world from New York. He flies to London and then somewhat deviously to Delhi to avoid the USSR, continuing to Singapore. At Singapore he leaves the northern hemisphere (ie half sphere) and crosses the southern hemisphere via Wellington, New Zealand, Antarctica and Brazil, stopping at Belem near the mouth of the Amazon. At Belem he is back in the northern hemisphere and he continues to New York, where he started.

Maps *(b)* and *(c)* in *Figure 1.4* are called zenithal projections. They may be thought of as flat maps each touching the globe at a chosen point, in this case the poles. The two maps show that the whole of the earth's surface can be represented successfully on *two* flat maps. These both show the equator and together include everywhere in the world, while because they overlap some places are included on both maps.

A zenithal projection need not necessarily be centred on one of the poles. A New Zealand atlas shows the lonely position of that country in a mainly oceanic hemisphere (*Figure 1.5*). The circle shown by a broken line bounds the hemisphere of Wellington. Scale and area increase towards the edge of the map and shape is greatly distorted. Countries on the opposite side of the globe from New Zealand (its antipodes) are stretched out enormously. The air

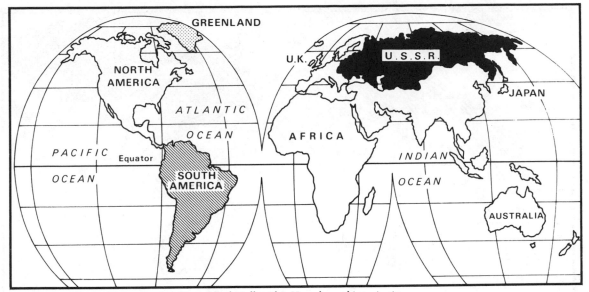

Figure 1.3 *The world portrayed on Interrupted Mollweide's Homolographic projection*

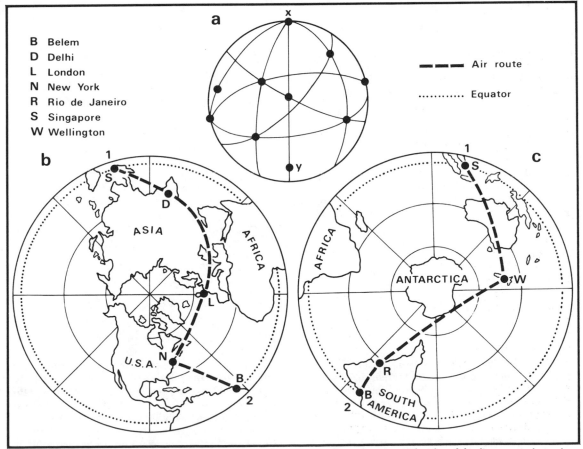

B Belem
D Delhi
L London
N New York
R Rio de Janeiro
S Singapore
W Wellington

— — — Air route

········· Equator

Figure 1.4 *Some geometrical properties of the earth's surface. (See text for explanation.) The idea of the diagrams is derived from Callahan (1976)*

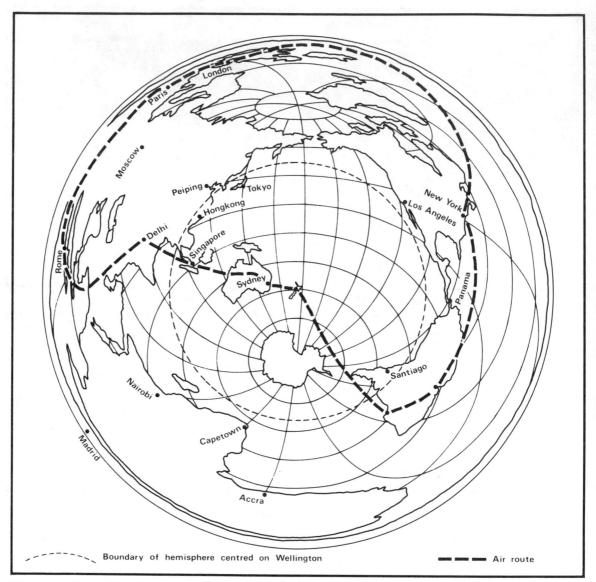

Figure 1.5   *The world in relation to New Zealand. The millionaire's journey in Figure 1.4 is represented on this map, based on A. H. McLintock's A Descriptive Atlas of New Zealand, R. E. Owen, Wellington, 1960, Plate I*

route of the eccentric American can be represented only with difficulty on the map.

In the Middle Ages it was assumed that the earth was flat. *Figure 1.6* shows the Hereford Map of AD 1285. On this map Jerusalem was taken to be the centre of the world. The cartographer of the time had problems. As the accompanying correct map of the same area shows, he made bad errors with area and shape.

It will be noted by the reader that the zenithal maps centred on the poles, on Wellington and on Jerusalem do not have north 'at the top'. It may be useful to adhere to the convention of having north at the top of the map when there is no advantage to be gained by rotating the map or having an unfamiliar projection. It should be remembered, however, that there is no right way up of looking at things and no top and bottom to the Universe (as far as we know). The north face of the earth is not necessarily its 'top' and sometimes a new view of relationships on the earth's surface can be gained if a map is rotated.

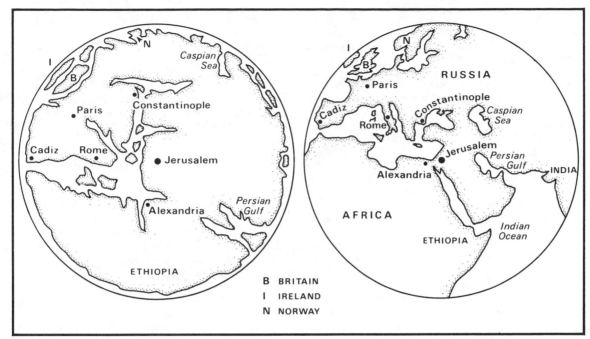

Figure 1.6   *The world according to the Hereford Map of 1285 (left) with correct map (right) for comparison*

1.3 Physical aspects of the earth's surface

With their cultural roots in Christian ideas, and their territorial origins in Europe, Westerners are still divided with regard to the relationship between the natural (or physical) environment and man. Some firmly believe that the world was created as a home for humans. Some equally firmly believe that the human species is basically similar to any other species, but is one that has developed (or evolved) special attitudes and skills. Many people, perhaps, either keep an open mind or privately hope that the two apparently mutually exclusive views of the world can be reconciled.

The human species has increasingly come to dominate and manipulate other species, both plant and animal. Other animal species are consumed as food and raw materials and are used as work animals, for experiments and for entertainment. The assumption of Christian cultures that the world, together with its plant and animal populations, has been created specifically as the home of man has provided us with a licence to exploit other species at will. The Christian view is seen in 2 Esdras 53: 'On the sixth day thou didst command the earth to bring forth before thee cattle, beasts and creeping things: and over these thou didst place Adam, as ruler over all the works which thou hadst made . . .' *The Oxford Annotated Apocrypha*, Oxford University Press, New York, 1965.

It is not the purpose of this section to debate the above interpretations of the origin and purpose of the world as we know it, or perceive it, but in the study of world affairs there are several reasons why it is increasingly necessary to take into account the physical background to man's activities. Since the 1940s it has been possible for the human species to be totally wiped out in a nuclear war. Large parts of the natural environment could be gradually destroyed by pollution. Some natural resources could be used up in a few decades. The population of the world more than doubled between 1940 and 1980 and it could double again within a few decades.

In various ways, increasing pressure is being put by man on the natural environment. On a long time scale going back to the origins of the earth man has very suddenly become a factor in its processes, even if only at and near the surface. Calder (1972) compared the estimated 4600 million years since the earth's surface ceased to be molten to 46 years in a person's life. Against 46 years of earth's existence, 'substantial animal life' has only been present for six years, the great reptiles flourished less than one year ago and man-like apes have been around only a few days. On this 46 year scale, the birth of Christ took place about ten minutes ago and the Industrial Revolution started a minute ago.

Until the launching of the first earth satellite by the USSR in 1957 almost all of man's activities took place in a space extending about ten kilometres above and ten kilometres below sea level. The total area of the earth's surface is about 510 million square kilometers. The total space defined above in cubic kilometres is thus about 20×510 million or 10 200 million cubic kilometres. The space includes the upper part of the earth's crust, the lowest part of the atmosphere and the water or 'hydrosphere', of which at any given time about 97 percent is in the oceans.

Changes are constantly taking place in the physical content of the space. Some changes are gradual, some rapid. Some changes take place smoothly while others result in abrupt alterations from one state to another. Very gradual changes in the earth's crust occur as 'tectonic plates' are formed, move against each other, or are destroyed. Very abrupt events such as earthquakes and the eruption of volcanoes occur unpredictably. Broad patterns of oceanic currents and climatic features occur with some regularity, but day to day changes in the hydrosphere and atmosphere are again difficult to anticipate.

With regard to the influence of the physical background on world affairs one should consider both the limits set by existing physical conditions to man's activities and possible major fluctuations and changes in the next few decades. Particularly since the early 19th century with a great increase in the use of fossil fuels, and a drastic attack on soil and natural vegetation, man's activities have already greatly modified the natural environment. Some kinds of influence and effect may be noted:

(1) The natural vegetation and soil have been removed in many places. Often the natural vegetation of a particular region took millions of years to reach its present form. Once destroyed it cannot be quickly recreated in that form. Soil may be washed or blown away through ploughing and the clearance of protective vegetation. A hundred years may be needed for 2–3 centimetres of soil to form naturally.
(2) Non-renewable fossil fuels are extracted in enormous quantities. Oil that took perhaps 100 million years to accumulate in the earth's rocks has been used up in a hundred years. The burning of hydrocarbons, especially in thermal electric power stations, increases carbon dioxide in the atmosphere. As a result, in a few more decades the atmosphere could become sufficiently hotter to melt the ice in the Arctic Ocean and on part of Antarctica. The resulting rise of at least several metres in world sea level would submerge parts of many well-known large cities as well as some agricultural land.
(3) Though climatic changes tend to occur fairly gradually, it is possible, especially with the help of man's activities, that temperature and precipitation could change substantially in a few decades. Conditions for agriculture could de- teriorate in some traditional farming regions and improve in others.

The above few examples of the nature of the physical environment should suffice to draw the attention of the reader to the complex relationship between man and the physical environment. Further aspects of the physical environment and its natural resources will be discussed in chapter 4.

1.4 Territorial division of the world

Throughout the last several thousand years various regions of the earth's surface have been marked off by man for organizational purposes. In world affairs today the basic units of territorial organization are the set of political units defined technically as sovereign (or independent) states. Almost every place on the earth's land surface is now in the territory of one and only one sovereign state though some disputed areas remain as well as a number of small colonial units. For convenience sovereign states and colonial units will henceforth be referred to as countries.

In 1978, 216 separate political units in the world, most independent, were listed in the Statistical Yearbook of the United Nations. They were distributed by continents as follows:

Africa	56	Asia	43
North America	37	Europe and USSR	38
South America	14	Oceania	28

Many of the countries are small. For example, 12 of the 38 in Europe had fewer than one million people and they included Gibraltar, the Holy See (Vatican), the Isle of Man and Monaco. Of the 28 in Oceania only three had more than one million people and most are groups of small islands. In 1980 there were 146 actual members of the United Nations but the total does not include a number of sizeable countries, among them Switzerland, Taiwan and the two Koreas. The USSR has *three* of the 146 seats.

The countries of the world vary greatly in size. The four largest in population have nearly half of the inhabitants of the world. The six largest in territorial extent contain about half of the world's land area. The four largest producers of goods and services account for about half of the world's production. *Figure 1.7* shows the above countries. With so much concentrated in so few large countries and so many small units at the other end of the scale it is important to give due weight to each unit in order to appreciate its likely influence in world affairs.

Although countries are the major 'players' in the world affairs game they only occupy a particular rank in a hierarchical system in the world of politico- administrative units. Many countries belong to

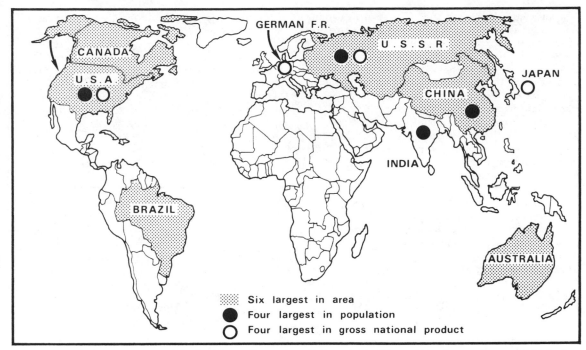

Figure 1.7 *The largest countries of the world according to three distinct criteria. (Here and in other world maps in this book the former Spanish Colony of Spanish Sahara has been represented as part of Morocco)*

supranational units. By 1981, for example, the ten members of the European Economic Community had already renounced some of their individual sovereignty to the new unit. Almost all the independent countries of the world belong to the United Nations though here usually only matters relating to external affairs are debated.

All but the smallest countries are subdivided for administrative purposes. Thus France is divided into 95 Departments, the USA into 50 States (plus one District). Such divisions are usually referred to as *major civil divisions*. In federal countries, including among others the USA, USSR, Canada and Switzerland, varying degrees of political power belong to the federal divisions, though the central government takes care of matters with international significance such as foreign policy and defence. In some countries, as in France, the major civil divisions are largely administrative units serving as convenient areas in which central government policy and measures are applied on the ground. In many countries at least one further level of subdivisions exists as, for example, the counties in the USA and the communes in France. These are often referred to as *minor civil divisions*.

Major and minor subdivisions are primarily of administrative significance and are therefore not of great interest in world affairs. Sometimes, however, they reflect the presence of cultural minorities; such groups may be aspiring to greater autonomy within a country or even complete independence. Scotland and Wales both contain considerable minorities wishing for independence from the UK (or England). In the USSR the Soviet Socialist Republics (SSRs) have consciously been formed round distinct non-Russian peoples such as the Latvians and Uzbeks. Selected examples of internal problems within countries will be given in chapter 6.

It has already been noted that countries vary greatly in size. A country the size of the UK appears very small on a world map. Since it is not customary in atlas maps of separate continents or countries to show other parts of the world on the same scale for comparison it is useful here to draw attention to some contrasts in size. In *Figure 1.8* England and Wales are drawn on three different scales and in each of the sections in the diagram the other regions are drawn to the same scale.

1.5 Distance in world affairs

The distance round the equator is 40 000 kilometres. If you were to travel at 100 kilometres per hour, the speed of a fairly fast train, you would need 400 hours

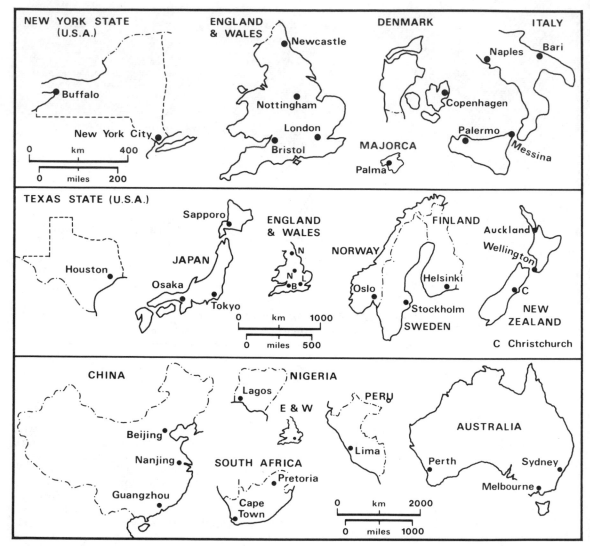

Figure 1.8   *The comparative sizes of selected countries and other regions*

or 16–17 days non-stop to go round the equator once. If you were to string out two-metre-tall humans head to foot round the equator you would need 20 million individuals.

Since the earth is spherical it is helpful in the study of world affairs to have a globe available for reference and as a means of measuring distances. The UK is a comparatively small country in area and the British reader studying world affairs is advised to check up on the scale of other areas in the world from time to time. *Table 1.1* shows some great circle distances on the earth's surface in kilometres. It is not always possible to travel along the great circle distances discussed and exemplified in this section. Other kinds of distance will now be described.

Route or travel distances

These are often much longer than direct distances. While London and Edinburgh are about 530 kilometres apart in a straight line, the road distance is about 600 kilometres by one of the shortest of many possible road routes. The Caucasus Range in the USSR is not crossed by railway. The formidable detour in *Figure 1.9(a)* between Nalchik and Tbilisi or Poti is an example of a route distance greatly in excess of direct distance. Even flights by air rarely go precisely along the shortest (great circle) route from airport to airport. They tend to fly in 'straight' sections from one control point to another. Many journeys get 'bent', as the one in *Figure 1.9(a)*,

Table 1.1 *Selected distances in kilometres*

Places	Distance in km	Places	Distance in km
Cornwall end to end	120	London to New York	5 500
California end to end	1300	London to Tokyo	9 500
London to Manchester	260	Los Angeles to Sydney	12 300
Washington to Boston	630	London to Sydney	17 100
London to Edinburgh	530	North Pole to South Pole	20 000
New York to Los Angeles	4000	Round the equator	40 000

whether by a physical obstacle like a mountain range, an unfriendly country or simply a gap in the transportation system.

Time distance

This may be a more useful measure of distance between a pair of places than route distance. The speed at which one can travel has increased dramatically with railways in the 19th century and air travel in the 20th century. Telegraph in the 19th century allowed information to flow quickly and news can now travel instantaneously so that with the help of communication satellites and television one can be a spectator at an event anywhere in the world being covered by cameras. When it comes to planning a journey, time may be synonymous with distance rather than miles or kilometres. *Figure 1.9(b)* shows miles and estimated car travel time in Florida, USA. Such information is rarely provided on maps. Its usefulness is limited by the fact that there is no accepted travel speed for a vehicle. Where quality of road or railway, steepness of gradient and other such conditions vary, however, time distance does bring out great contrasts.

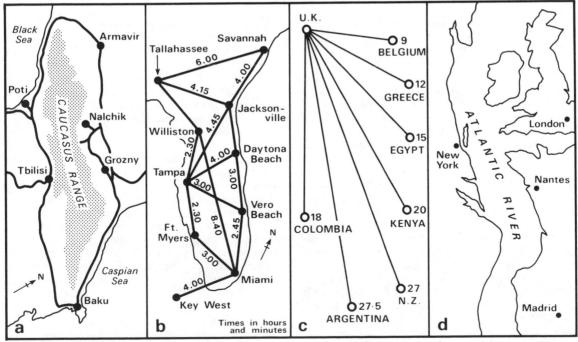

Figure 1.9 *Examples of different types of distance. a – route distance; b – time distance; c – cost distance; d – TransWorld Airline advertisement showing effect of air travel on distance*

Cost distance

Cost distance, like time distance, may not correspond closely to direct or even route distance. It is well known that many goods can be carried over long distances much more cheaply by sea than by road or air. Passenger fares between given pairs of places may also vary according to means of transport. One feature of cost distance is that the charge per unit of distance may diminish (or taper) as the length of journey increases. Such a situation is illustrated by the data in *Table 1.2* and by *Figure 1.9(c)*. Royal Mail

Table 1.2 *Selected cost distances*

From UK to	Distance in km	Charge in £ for 5 kg parcel	Km per £
Belgium	300	8,90	34
Switzerland	800	7.40	108
Portugal	1 500	9.05	166
Greece	2 500	11.95	209
Egypt	3 500	15.10	232
Nigeria	5 000	14.15	353
Kenya	7 000	19.75	354
Colombia	8 500	17.55	484
Argentina	11 300	27.50	411
New Zealand	19 000	26.80	709

Source: Royal Mail advertisement, The Times (1981)

in 1981 advertised a charge for mailing a five kilogram parcel about 300 kilometres from the UK to Belgium as £8.90 and about 19 000 kilometres to New Zealand as £26.60.

Perceived social, economic, ideological and cultural distances should not be overlooked in the study of world affairs even if they cannot readily be expressed numerically. Australia and New Zealand are culturally much closer to the UK than are, for example, Afghanistan or Ethiopia, yet the latter are much closer to the UK in space distance. Such distances may be incorrectly perceived but they reflect a kind of proximity that should be taken into account in world affairs. So numerous and rapid are jet flights between North America and Western Europe that one can think of the Atlantic Ocean being greatly narrowed as in the Trans-world Airlines advertisement in *Figure 1.9(d)* compared with the times when it could only be crossed by ship.

In view of the complexity of constructing maps of large areas and of interpreting spatial information it is not surprising that people perceive places and geographical situations differently. The theme of the next section will be the way situations are conceived or misconceived and the way in which they may even be distorted for propaganda purposes.

1.6 Perception and prejudice

It seems unlikely that a single individual or even a team of people working together could form a completely impartial or objective view of world affairs. An individual sees the world from a particular viewpoint with a specific set of assumptions. The citizens of a given country tend all to have access to the same particular range of educational material and news media and they therefore also share a broadly similar viewpoint.

The way people see things is affected by a number of principles, themselves resulting from personal limitations and mental attitudes:

(1) It is virtually impossible for an individual to receive all the possible information relevant to a particular situation or problem. In world affairs, many key decisions may indeed be made in secret discussions not divulged to the public.
(2) No two individuals have exactly the same information. Each person has a particular life space, a point in place and time, which determines the information received. One sees the world through the limits set by a particular 'window' or even 'keyhole'.
(3) People tend to be selective in the interpretation and use of the information they do receive. They see what is convenient, what they expect or want to see, what fits in with an existing preconceived framework. They tend to be slow and unwilling to change the framework even after receiving convincing evidence that they should. Watson (1967) has given many instances of how explorers crossing North America westwards in search of the Pacific Ocean and a route to China assumed each big river and lake they came upon in turn to be the beginning of the sea leading across to Asia. Such information was passed on to cartographers and often used long after it had been shown to be incorrect.

People are not only limited with regard to the information they receive, and biased and selective in their interpretation of it, but they also tend to think that what they are accustomed to in their own district or country is normal and that everywhere else in the world there are only varying degrees of abnormality. There is little justification for thinking that conditions in one's own country are any more normal than those in other parts of the world. Indeed, such highly industrialized and urbanized countries as the USA and UK are unusual even now and certainly would have been in the past.

People tend to think not only that what they are used to is normal, but also that it is superior. They may base their claim to superiority on their military strength, on the level of consumption they enjoy, on

Figure 1.10 *Places mentioned frequently in a sample of front pages of the Peruvian newspaper El Comercio in the 1970s*

the colour of their skin, or on the fact that they belong to a certain religious faith or have a particular type of political organization.

The idea that one's own group is not merely different from all others, but also better, is widespread. It is reflected in the rivalry between individual communities and between tribes in simple societies, between towns and regions in urbanized countries, and between countries and even continents on a world scale. It is expressed implicitly if not explicitly in many school history books. Perhaps there would be no such thing as world affairs without it. Illustrations of the points made so far in this section may now be given.

The news media reporting international events may attempt to achieve a complete coverage of everything of significance that happens in the world. They may, on the other hand, have to make omissions either because lack of time and space requires selection or, as in the USSR, because a state controlled press is required to do so. The bias of national news media in their coverage of events whether intentional or unintentional is illustrated in *Figure 1.10* by news reported in the Peruvian daily newspaper *El Comercio*, a reputable newspaper actually established as long ago as 1839. In a sample of a hundred front pages taken over selected periods in the 1970s the author found the following 12 places in *Table 1.3* to

Table 1.3 *Places mentioned most frequently on the front pages of one hundred editions of the Peruvian newspaper, El Comercio, during the 1970s*

Place	Number of mentions	Place	Number of mentions
Lima (national capital)	91	Argentina	36
USA	82	USSR (or Russia)	34
Peru	79	Brazil (neighbour)	33
Washington	48	Mexico	31
Callao (port of Lima)	39	Latin America	30
Chile (neighbour)	39	Great Britain (or England)	30

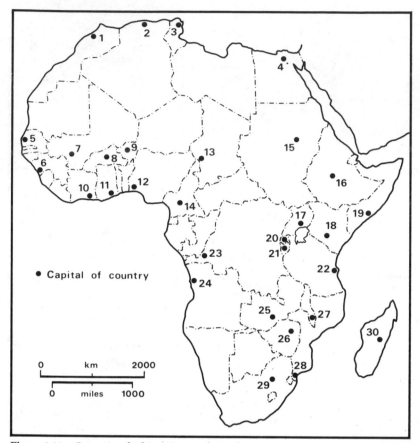

Figure 1.11   *Countries of Africa. (See text for explanation)*

be the ones most frequently mentioned. One half of all places mentioned were in Peru itself. At the other extreme, only about three percent of all places mentioned were in southern and eastern Asia although this region has half of the world's population.

News in the Melbourne newspaper *The Age* was also monitored by the author for a time and an equivalent bias was evident, with Australia and southeast Asia dominant and Africa and Latin America receiving very little attention. Of places mentioned on the front page in the Soviet daily newspaper *Pravda* about two-thirds are within the large, inward looking host country. For the whole of 1968 not one reference was found to the USSR's neighbour China, a country Soviet leaders preferred to forget at that time.

Just as the frequency with which places and regions are mentioned in newspapers and in radio and television broadcasts can be expected to influence people's views on the world, so also certain places are more conspicuous than others. The shape of Italy, the fact that South Africa is where its name suggests or the large size and isolation of Australia seem to make these places more familiar than many others. The reader is invited to identify the numbered countries in *Figure 1.11* on the map of Africa. The correct names can be found on the maps in *Figures 12.4* and *12.6* in chapter 12.

In view of the general lack of knowledge of the location and size of places it is not surprising that confusion occurs even in the minds of political leaders. In 1938 the Prime Minister of Britain, Neville Chamberlain, is credited with having referred to Czechoslovakia as a distant country about which we know little.

According to Pankhurst (1965) a 16th century Portugese visitor to Ethiopia showed Portugal on a map of the world to Ethiopian ministers. They were far from impressed by its small size and doubted the strength of the country. One wonders if Nazi leaders underestimated or even gave a thought to the large size of European USSR before they launched their invasion in 1941. They certainly failed to give due importance to the surface condition of the roads or the severity of the winter.

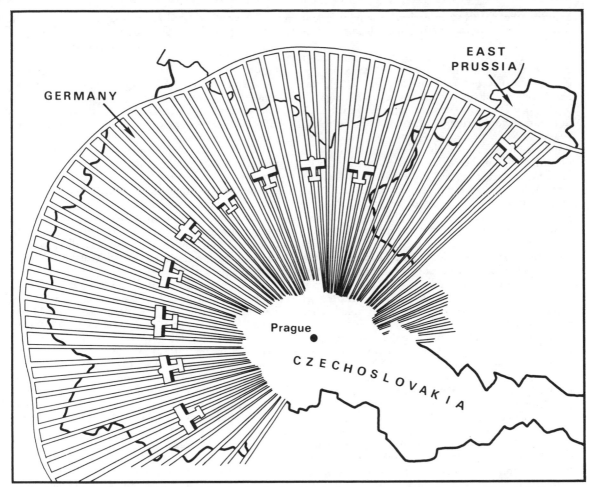

Figure 1.12   *A German map of the 1930s showing Czechoslovakia as a threat to German security. Source: Zeitschrift für Geopolitik (1934)*

Sometimes maps have been used for purposes of propaganda. The Nazis were encouraged in their early plans for expansion in Europe by messages conveniently revealed in maps. The map in *Figure 1.12* was published in the 1930s in a German periodical, *Zeitschrift für Geopolitik*. One of the main figures in the geopolitical institute of Nazi Germany was a geographer, Karl Haushofer. The place names have been added by the author and are not on the original. The message is clear: virtually all of Nazi Germany was vulnerable to air attack from Czechoslovakia, a pretext (if one was needed) for occupying the country (as was done) in the interests of German security.

In the study of world affairs it is difficult to be impartial, even if one wishes to be. It seems virtually impossible for anyone in the USA or the UK to take a balanced view of the Soviet Union (and *vice versa*).

Either nothing the Soviet leaders do is good or everything they do is good. It is remarkable, however, how quickly the public view can be changed. Superficially, at least, the USSR was portrayed in Britain as a heroic ally during 1941–45.

During the period of Mao's leadership of China, and the Cultural Revolution, communist China was visited by a number of eminent Western personalities, including Montgomery, Heath, Nixon and Kissinger, all un- if not anti-socialist in their views. They professed to admire some of the achievements of the system, if anything more 'leftist' than that of the USSR. Yet by the late 1970s the new leaders of China were *themselves* condemning many of the features of the later years of Mao's leadership, referring to serious economic and political problems. The overriding political and strategic consideration of consolidating China's break with its fellow socialist country the USSR has enabled many Western observers to overlook ideological features they normally condemn.

In studying world affairs it is well to distinguish between fact, speculation and opinion. It is a fact, for example, that according to the 1981 census, England and Wales had 49 001 417 inhabitants on *the night of 5/6 April*. The exact number actually varied even during the short period in question, with births and deaths. A few people may have been missed or double-counted. The number, however, is not worth disputing. Speculation is about facts that cannot be verified. It is a fact that the USSR manufactures aircraft, but it is a matter of speculation in the West as to exactly where they are made and how many are made in each place.

When, however, one asks whether the USA is a better country to live in than the USSR then opinion takes over. The conclusion reached depends on the factors taken into consideration in the assessment of 'goodness' in this context. If one took personal liberty to express views publicly or to travel freely as a major consideration, then the USA would score over the USSR. If one gave great weight to the provision of full employment or the equal distribution of wealth (as opposed to income) then the USSR would score over the USA.

There is a danger in practising what Koestler referred to as reductionism. You can reduce all problems in world affairs to single influences such as the ideological conflict between capitalism and communism, the arms race, transnational companies, postcolonialism or modern technology. People like to fit world events into a particular framework or model. Within the limitations and prejudices of the author it is intended in the present book to express views held in various countries and to avoid using one particular framework.

1.7 Numerical information

The reader may not be greatly interested in numerical data and may consider quality to be more important than quantity. Statistics are often unreliable and can easily be misinterpreted or falsely presented. Many situations in world affairs cannot, however, be assessed without some reference to quantity, whether it refers to area, distance, number of people, amount of rainfall or production of goods. Moreover, non-quantitative information can still be represented numerically for some purposes. Lord Kelvin, a 19th century physicist, is quoted as saying:

'When you can measure what you are speaking about, and express it in numbers, you know something about it; when you cannot measure it, when you cannot express it in numbers, your knowledge is of a meagre and unsatisfactory kind; it may be the beginning of knowledge, but you have scarcely, in your thoughts, advanced to the stage of science.'

The words of Lord Kelvin may be too drastic and sweeping to please some social scientists. For those who do expect to use numbers it is disappointing to find that even in the 1980s, when one might have hoped to find a regular flow of numerical data about aspects of population and production, censuses and other data-collecting procedures are still carried out relatively infrequently. Data often contain degrees of error that are far in excess of what might popularly be expected or hoped.

Examples will be given of the inaccuracy of numerical data on population in some 'official sources'. According to Hauser (1971) the 1960 census of the United States is estimated to have undercounted the true population of the country by about 3.1 percent. The discrepancy was discovered by a careful postcensus check of a sample of enumeration districts. Non-whites were particularly prone to omission as, presumably, were illegal Mexican and other immigrants.

Far worse is the case of the population of a former French African colony, now the Central African Republic, which was not even estimated in the *United Nations Statistical Yearbook*, 1974. An estimate for 1959–60 gave 1 202 910, while another (recensement instantané) in 1968 was 2 255 536. The estimated population for 1970 was 1 612 000 or 2 370 000. Such data are hardly a basis for planning the food, educational and other needs of the country.

In between the two examples given is the intriguing case of Nigeria. What follows shows how adaptable one may have to be from year to year in using such works as the *United Nations Statistical Yearbook* (UNSYB). A count of the population of Nigeria in the period July 1952–June 1953 gave a figure of 30.4 million. This figure formed the basis for subsequent estimates. Thus UNSYB 1963 gave 33.8 million for 1958 and 36.5 million for 1962.

When UNSYB 1964 appeared it recorded 33.8 million for 1958 (loyal to previous estimates) but 55.6 million for 1963 according to the census of 4 November 1963. The 1958 estimate was accompanied by the following footnote: 'Appears to be underestimated in relation to results of latest census (or sample survey) shown adjacent'.

UNSYB 1965 contained an adjusted estimate for 1958 of 50.0 million and a figure of 56.4 million for 1964. The 1963 census was taken to be accurate, though it was later accepted that for political reasons some regions of Nigeria had greatly overstated the number of inhabitants they had.

Later still UNSYB 1974 had adjusted to the situation with a compromise. It gave a 1970 population of 55.1 million (close to that for 1963 in the census) and 59.6 million for 1973. Both estimates were tactfully footnoted as follows with regard to data disparities: 'There is a possibility that the 1963 census overstated the population. The size of this

Table 1.4 *Estimates of the population of Nigeria*

Year	UNDYB	UNSYB	Year	UNDYB	UNSYB
1952	34.7	30.4	1963	46.0	55.6
1954	36.4	30.3	1964	47.3	56.4
1958	40.2	33.8 or 50.0	1970	56.3	55.1
1961	43.5	35.7	1973	61.7	59.6
1962	44.7	36.5	1979	74.6	74.6

possible overstatement may be judged from the mid-year estimates for 1970 and 1973 provided by the United Nations Population Division.'

The *United Nations Demographic Yearbook Special Issue* (UNDYB) 1979 seems to have got things satisfactorily sorted out with the help of substantial smoothing. *Table 1.4* gives its estimates for the population of Nigeria in selected years and also those quoted above in earlier publications.

The US Population Reference Bureau gave an estimate of 79.7 million for the population of Nigeria in 1981. In the 23 years since 1958, when all this was going on, the population of Nigeria roughly doubled, in spite of its devastating civil war over Biafra. How little then can we know about the populations of Ethiopia and Afghanistan, neither of which has ever had a census at all, or of China, for which the last proper census in 1953 gave 590 million people until the advent of the 1982 census, giving 1008 million.

Some production figures are likely to be more accurate than the demographic examples given above. It is not difficult to count the number of cars produced in a factory or the barrels of oil extracted from an oil-well. On the other hand many agricultural figures are probably wildly inaccurate even when they are not being falsified for purposes of propaganda.

When you have a set of numerical data, suspect or not, does it necessarily mean anything? The first thing to do, often, is to round the data greatly to get rid of spurious precision and to make the numbers more easily readable. The next thing is to make sure the data have the right number of digits. It is common to find a few zeros missing on a very big number even in the more prestigious newspapers.

Was Soviet electricity production in 1979 1 238 000 kilowatt hours, 1 238 000 000 kilowatt hours or 1 238 000 000 000 kilowatt hours? How many kilowatt hours might be needed per inhabitant in a year in a fairly advanced industrial country? There seems no easy way of appreciating such large numbers and many people seem to have quite low numeracy ceilings. Anything more than a few thousand is just large. Various sets of numerical data will be presented in this book but an attempt will be made to put the data in a context and to make comparisons.

Data can be on a continuous scale or simply the answer, yes or no, to a specific question. For example if data are accurate enough the proportion of the total population of country X under a specific age on a given date can be expressed as a percentage such as 43 percent, even more precisely as 432 per thousand or more approximately as just four-tenths. In contrast, the answer to the question 'is country X landlocked?' is simply yes or no. A yes or no answer is likewise expected to the question: is French the official language of France? Life is never as straightforward as it seems and there would be complications if one asked: is French the official language of Switzerland because the answer would be 'yes', qualified by 'one of them'. Even when a question refers to a quality or attribute (such as: is the bus red?) the answer can be coded as 1 for yes and 0 for no and if required such data can then be handled numerically.

Data about a country can often be expressed in one of two different ways, either of which gives a valid piece of information. For example, Belgium produced about 12.6 million tonnes of steel in 1978, the *absolute* amount. That works out to be about 1280 kilograms *per inhabitant*. The second value allows Belgian steel production to be compared directly with that (also per inhabitant) of another country without having to allow for differences in population size. The difference may be illustrated here by comparing Belgium with India. The latter produced an *absolute* amount of about 10.2 million tonnes of steel. India however had about 640 million people in that year, Belgium only 10 million. The amount produced *per inhabitant* was therefore much larger in Belgium than in India, 1280 kilograms compared with a mere 16.

Another aspect of relative and absolute values may also be noted. Suppose there are two countries, A and B, which in 1955 had steel consumption per inhabitant of 10 and 100 kilograms respectively and in 1980 of 20 and 150 kilograms per inhabitant. During this period, steel consumption per inhabitant increased more *relatively* in A than in B (100 percent compared with 50 percent) but by more *per inhabitant* in B than in A, by 10 and 50 kilograms respectively.

Data are more readily and widely available and more accurate for some topics than for others. For this reason some topics tend to be covered more thoroughly than others and greater emphasis is given to some variables than to others. In Soviet Statistical Yearbooks data are given in great detail for such industries as the manufacture of plywood and knitwear but for security reasons virtually nothing is revealed on engineering or non-ferrous metals.

1.8 Models of world affairs

Space allows the inclusion of only two examples out of many possible ways in which situations in world affairs can be viewed.

The board game

World affairs have often been compared with a game of chess. The analogy is thought-provoking but has several defects. Wrong analogies are made because many people do not apparently actually play chess. Thus the word 'stalemate' is used to mean impasse when it really refers to a technical situation in which one player's King, though *not* checked, cannot move without moving into check. Secondly, chess is played on a board that is finite and bounded. The rules of play depend on a precise layout of rows, columns and moves. There are only two sides.

'Go' (Ko) gives a better though still only partially satisfactory analogy with world affairs. Like chess it is played on a board that is finite and bounded (*see Figure 1.13*). A comparable board could, however, be fitted (topologically) to the earth's surface. Although the game is for two players it would be possible to have more than two. Each player starts with a large set of stones, black or white. The game of 'Go' starts with no pieces on the board. Each player in turn places a piece (stone). Gradually territory is occupied and some territory subsequently changes hands, sometimes with the capture of surrounded pieces and their removal from the board. *Figure 1.13* shows the game applied to the Middle East.

In answer to the question 'who makes the moves?' one might say the leaders or governments of the big powers. The Russians are notoriously good chess players. The Japanese, however, provide the world champions of 'Go'.

Game theory

This is a branch of mathematics that studies the possible strategies and outcomes in games of a certain kind. In noughts and crosses (tick-tack-toe) it is possible to work out the best reply to each move according to the opponent's previous move. In chess there are so many possible moves at any given moment and a huge number of eventual combina-

tions of moves that all possible runs of the game cannot be worked out.

One adaptation of game theory is its use to assess the preferences for different outcomes by two or more participants in a situation of rivalry or in a conflict. In economics one might study the pricing of a product by competing companies. In world affairs the tabulation of preferences may at least help one to appreciate possible or likely outcomes of complex situations. An example is given in *Table 1.5*. There are four active participants in the 'game' and at least two interested observers. For simplicity there are two participants within Northern Ireland, the Protestants and the Catholics. There are also two directly concerned national governments, those of the UK and of Ireland. Assuming (and hoping) that lethal conflict will not continue for ever in Northern Ireland then there are at least five possible mutually exclusive outcomes. The preferences in the table have been estimated by the author and may not necessarily reflect the results of a referendum, if one were held.

The results do not reveal anything earth-shattering, but they do help to clarify some aspects of the situation. The combined preferences of the four interested parties for the five possible outcomes are A 10, B 9, C 15, D 10, E 16. The outcome that might produce the greatest overall satisfaction is B (join the Republic of Ireland) while the least attractive solutions are C and E (become completely sovereign or become a UN trust territory). The problem is, however, that outcomes A and B, though the most attractive collectively, are the solutions *least* attractive to the local Catholics and Protestants respectively. Ironically the US Catholics and the USSR (Communists) both support the IRA. Perhaps in the end an EEC view will be sought.

Near the end of this book a model borrowed from biology will be used to portray the developed countries as predators in the food chain, deriving sustenance through the developing ones, which digest and process materials from their own patches. Catastrophe theory is another way of looking at some aspects of world affairs, situations where a sudden change of policy or attitude may occur. Q-analysis is a fairly recent development for looking at multidimensional situations and could be particularly valuable in illustrating and studying the relations at different hierarchical levels between countries and within countries. A feeling for situations in world affairs can also be gained from using operational games. The Avalon Hill games of Stalingrad and D-Day from the 1960s give excellent operational board portrayals of the German invasion of the USSR (1941–43) and of the allied invasion of France (1944–45). These and other types of technique, model and simulation can be used in the study of the geography of world affairs.

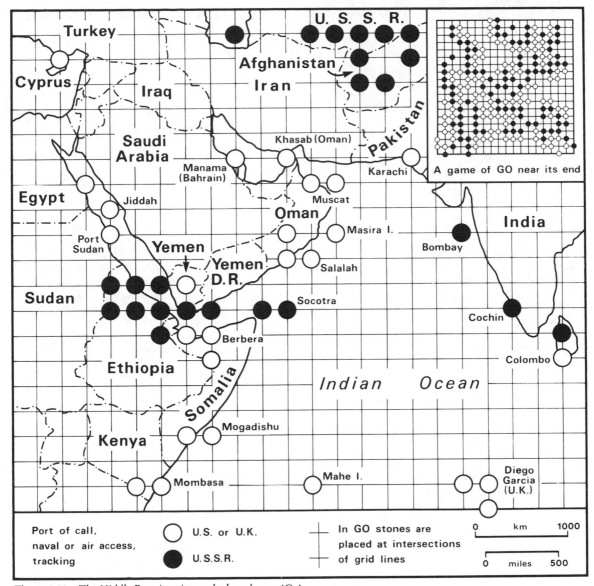

Figure 1.13 *The Middle East situation as the board game 'Go'*

Table 1.5 *Game theory applied to the Ulster situation*

Preferences	Prots NI	Caths NI	Ireland	UK	USA Caths	USSR
A Remain part of the UK	1	5	3	1	4	5
B Become part of the Republic of Ireland	5	1	1	2	1	3
C Become completely sovereign	2	3	5	5	5	1
D The territory partitioned to separate Catholics and Protestants into Ireland and the UK respectively	3	2	2	3	2	4
E Become a UN trust territory	4	4	4	4	3	2

CHAPTER TWO

The Emergence of the Present Countries of the World

2.1 The world around 1500

In 1500 there were roughly a tenth as many people in the world as now. The economies of the time depended very heavily on bioclimatic resources. These were cultivated land, or natural pasture and forest used by herders, by food gatherers or hunters. For the most part the food gatherers and pastoralists were very few in number but used extensive areas. Cultivation was carried out either by hand or by plough and with fertile soils and in places irrigation could support quite high densities of population. The main areas in the world of food gathering, pastoralism and cultivation are indicated in *Figure 2.1*.

Plough cultivation was extensively used throughout Europe, northern Africa and southern and eastern Asia.

Five centuries ago most of the population lived in loosely organized communities. *Figure 2.1* shows the areas in which a more rigorous political control extended over large numbers of people, though still in countries or empires that were less highly integrated than now. The two main civilizations of the 'New World' of the Americas were largely based on hand cultivation, with the hoe, and livestock had only a limited economic function. The main 'Old World' civilizations of Europe, Africa and Asia were based on plough cultivation, with livestock making a

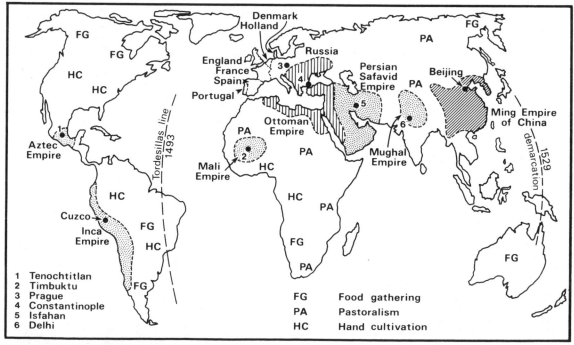

1 Tenochtitlan
2 Timbuktu
3 Prague
4 Constantinople
5 Isfahan
6 Delhi

FG — Food gathering
PA — Pastoralism
HC — Hand cultivation

Figure 2.1 *The world before the Columbian period. (Columbian refers to the explorer Christopher Columbus)*

major contribution to the economy in providing work animals, food and raw materials. Many of the more densely populated areas of cultivation depended on irrigation.

Late in the 15th century, the Ming Empire of China was the most populous single unit of organization in the world. Technologically there was not much difference in 1500 between Europe and China. The two New World empires were in some respects far behind those of the Old World. The Ottoman Empire, centred on what is now Turkey, and with its capital in Constantinople, had been growing for some time before 1500 and occupied extensive areas in Africa and Asia as well as roughly

the southeast quarter of Europe, reaching close to Prague and Vienna.

In the Middle Ages there had been frequent if tenuous trade links between Europe and China. By late in the 15th Century the Ottoman Empire and further north the Khanate of Kazan and other hostile states lay across trade routes. The rivalry between Christian Europe and Islam meant that the trade routes were at risk and could be blocked. The period of European exploration that started in the 15th Century was to some extent a response to the above situation, a result of the attempt to find other trade routes between Europe and southern and eastern Asia. Thus for several centuries during the later

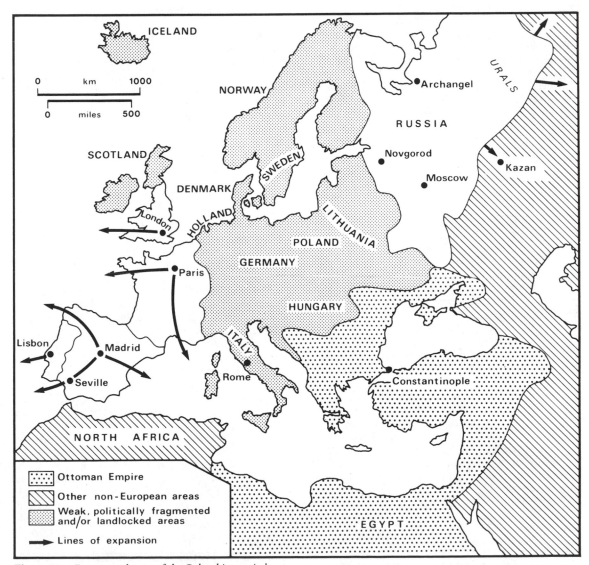

Figure 2.2 *Europe on the eve of the Columbian period*

Middle Ages the Christian culture of Europe had been confined to a small part of the world's area, but in the later part of the 15th century Europeans started to extend their influence across the oceans and across northern Asia. The present form and distribution of countries, religions, languages and economic systems throughout the world can only be understood with reference to the so-called Columbian era, the period following the 'discovery' of the Americas by Europeans.

Around 1500 Europe (*see Figure 2.2*) could be subdivided into three parts. Its western-facing periphery had access to the Atlantic Ocean while the eastern-facing periphery bordered the nearly empty lands of Siberia. In contrast the central part, hemmed in by France in the west, Russia in the east and the Ottoman Empire in the southeast, was less well placed to send explorers and settlers out of Europe. From the central area, indicated by fine dotted shading in *Figure 2.2*, many countries had access to the sea, but only to the Baltic, North Sea or Mediterranean, not directly to the Atlantic Ocean. Even Holland, when it emerged as an influential oceanic power in the 17th century, had access to the Atlantic via the English Channel or North Sea, both adjacent to England.

The distribution of countries on the map in *Figure 2.2* shows that it would have been very difficult for a country such as Hungary or Poland to undertake the exploration and conquest of lands outside Europe, because other European countries could block the way. Indeed when Sweden and more recently Germany attempted to conquer new territories they tended to do so in Europe itself rather than elsewhere. Some areas like Iceland, Scotland, Italy and Germany were either too small in population or politically too fragmented to take much part in the process of European expansion that was starting around 1500.

2.2 Exploration and conquest by European powers from the 1940s to the 1770s.

As explained in chapter 1 (section 1.2), the view became widely held in Europe in the 15th century that the earth is spherical, not flat, and the first globe was made in 1492. On a spherical surface it would be possible to travel from Europe to East Asia either eastwards or westwards. The easterly land route passed through the hostile world of Islam and was difficult to use. The immediate routes east from the Mediterranean were blocked by the Ottoman Empire. It was therefore reasonable to look for a westerly route. Columbus set out in search of such a route to East Asia in 1492. Although his view of a spherical world was correct, the interpretation of its size was not. It was stated in the *Book of Esdras* in the

Apocrypha that six-sevenths of the earth's surface consists of land (actually nearly three-quarters is sea):

'On the third day thou didst command the waters to be gathered together in the seventh part of the earth; six parts thou didst dry up and keep so that some of them might be planted and cultivated and be of service before thee.'

2 Esdras 42

Figure 2.3 shows Fra Mauro's map of the world dated 1459. According to the view that six-sevenths of the world is land, East Asia could not be far to the west across the ocean from Europe.

It is thought by some that Scandinavians may have reached the Americas before Columbus did. It is highly probable that some other European explorer would have done so not long after Columbus, if he himself had not been successful (or lucky). Columbus himself died believing that he *had* reached East Asia rather than a completely unknown new land mass. The land he discovered was thus called the Indies, and the name America was only given later in tribute to the explorer Amerigo Vespucci. The discovery and exploration of the New World by Columbus and his contemporary navigators was a key event in world history. It gave West Europe the incentive to be the first cultural area of the world to become influential in every continent.

Most of the voyages of exploration in the 16th century originated in Spain or Portugal. The world had already been divided by the Pope on behalf of the two Iberian powers in 1493 by the Treaty of Tordesillas. Lands in the tropics west of a meridian passing 370 leagues west of the Cape Verde Islands (about 48 degrees west of Greenwich) were allocated to Spain to colonize, those to the east to Portugal. As the meridian actually passes close to the mouth of the Amazon, eastern South America fell in the Portuguese sphere. The Tordesillas line allocated a 'western' hemisphere to Spain and an 'eastern' one to Portugal; the limit is indicated in *Figure 2.1*. The division was completed in 1529 by the Treaty of Saragossa whereby another meridian, off East Asia, gave the Philippines to Spain but left the rest of Asia and all of Africa as a Portuguese preserve.

As a result of the division of the tropics between Spain and Portugal the latter initially concentrated on the establishment of a route round Southern Africa to gain access to the spices and manufactures of the East. In seeking a route to Asia in a westerly direction from Europe, the energies of Spain were diverted by the discovery of the Americas. Although a link was eventually established between Spain and its Asian colony, the Philippines, the precious metals of Mexico and Peru became its main interest outside Europe. In the 16th and 17th centuries both Portugal and Spain also began to establish plantations to grow such crops as sugar, cotton and coffee in the

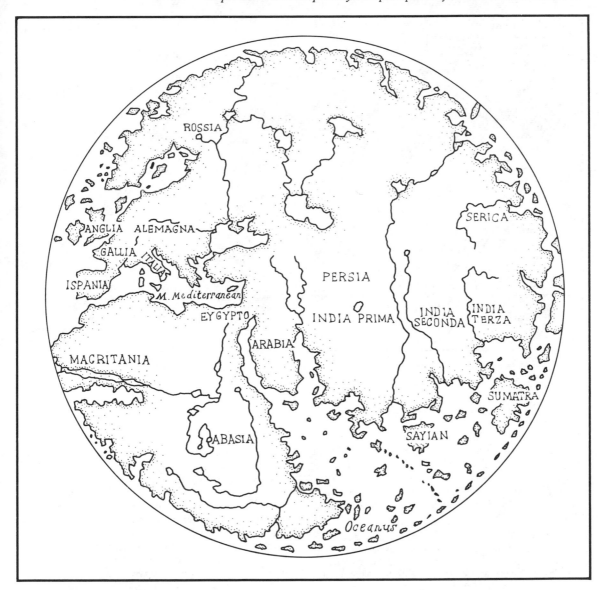

Figure 2.3 *Fra Mauro's map of the world dated 1459*

Americas, especially in Northeast Brazil and the Caribbean, using slaves brought in from Africa.

Following the lead of two Iberian powers, England, France and Holland sent out explorers from Europe but for a long time they failed to gain control of large territories, useful subject populations and extensive natural resources. Not until the 19th century did the British and French Empires reach a very great territorial extent, by which time nearly every Spanish colony had already become independent.

England and France competed extensively in the 18th century both in North America and in India. *Figure 2.4* shows the comparatively limited extent of the English colonies of North America before the War of American Independence (1775–83). They were in danger of being outflanked by French conquests in Canada, Louisiana and the Mississippi lowlands. For both England and France, their Caribbean colonies were of enormous importance. Haiti, in particular, was intensively developed by France in the 18th century with the help of slaves

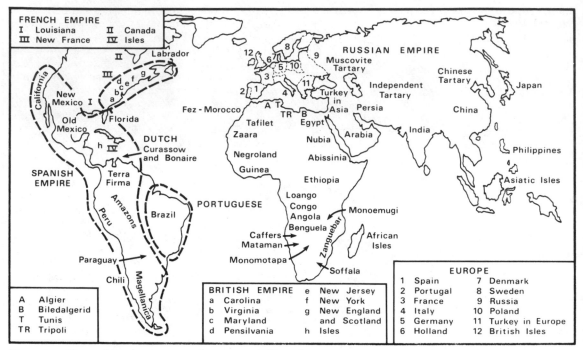

Figure 2.4   *The world in the 1760s according to maps and lists of countries in Encyclopaedia Britannica, vol 2, Edinburgh, 1771, pp 682–4*

tropical crops. The British likewise developed Jamaica and other islands. The importance attached to tropical colonies and the belief in the great fertility of tropical soils, now in some respects discredited, are expressed in the following lines from the article on Britain in *Encyclopaedia Britannica* (fourth edition 1800–1810):

'It is in vain, therefore, that France possesses a superior European population. In the state of things now described, it is impossible for her to support the same number of European soldiers that Britain may do. Every soldier France sends out must be maintained and clothed by the industry of Frenchmen, exerted upon a European soil, of far inferior fertility to that which is cherished by tropical rains, or the periodical floods of the Ganges. Whereas the British soldier is not supported merely by British industry, but by the industry of the natives of Hindustan, or of the labourers of Jamaica.'

Through expansion to the north in the 15th century, Russia had potential access to the Atlantic Ocean via Archangel and the White Sea, but this sea is frozen for several months of the year. Russia did not use this dubious outlet to send expeditions to the New World or to tropical Africa and Asia, but it did push eastwards across the land mass of northern Asia in the 16th and 17th centuries, absorbing a very large territory. The conquest of Siberia was carried out

almost entirely by land, with penetration along rivers. By the mid-17th century, Russia had outlets on the Pacific coast and in the 18th century was establishing settlements in what is now Alaska. Russia thus acquired a vast territory, at the time virtually uninhabited, but providing natural resources of enormous potential in the 20th century. The lands conquered by Russia differed from those acquired by West European powers in extending over a compact area, cut off from the rest of the world by mountains, deserts and frozen seas.

The map in *Figure 2.4* was published shortly before the British colonies in North America became the new independent USA and only a few decades before Spanish control of its Empire in the Americas was broken. Over two centuries ago many of the well-known names in world affairs were already there. Only the names of Japan and the Philippines have been added by the author to the 1771 map, but they also existed at the time. Africa is the area that changed most in the 19th century, being carved up into some 50 colonial units by European powers to give an entirely different structure from that based on tribes and loose kingdoms shown in vague terms on the 1771 map. Here, therefore, is evidence of the durability of countries, if so many can be identified so long ago.

2.3 European influence from the 1770s to the Second World War

Between the period 1775–83 and the 1820s the British colonies in North America, the key French colony of Haiti in the Caribbean and almost all of the vast Spanish and Portuguese empires in what is now Latin America achieved independence. The history of the War of American Independence is well known in the USA and UK. The original 13 new states formed the nucleus of what has become one of the two superstates of the late 20th century. After a slow start, the USA in the 19th century expanded westwards to the Pacific, taking in areas of former French and Spanish colonies and purchasing Alaska from Russia in 1867. Canada remained a British colony.

The history of the emergence of new countries in Latin America is not well known outside the Iberian culture area. By 1770 the Spanish Empire had become subdivided for administrative purposes. Much of the area had originally been administered under the viceroyalties of New Spain (Mexico) and Peru (Lima). Already in the 18th century, however, smaller colonial subdivisions could be recognized, as for example, Chile, Colombia and Guatemala. When the Spanish-American Empire finally broke up in the period following the Napoleonic Wars in Europe, ten separate countries emerged. These later split further to give 18 separate countries of Spanish origin early in the 20th century. Spain continued to hold Cuba, Puerto Rico and the Philippines as colonies until the end of the 19th century.

Portugal's most important colony was Brazil, the territory of which was gradually pushed west, at the expense of Spain, across the Tordesillas line (*see Figure 2.1*) into Amazonia, the interior of South America. The various Portuguese settlements in Brazil stayed together as one political unit to give a new independent state early in the 19th century. Portugal also controlled stretches of coast in Africa, territories behind which the colonies of Angola and Portuguese East Africa (now Mozambique) spread late in the 19th century.

The period from the 1770s to the 1830s saw for the first time the emergence of new sovereign states outside Europe. In addition to the USA itself, by 1830 the following countries had formed from the European colonies of the Americas: Mexico, Peru, Bolivia, Chile, Argentina, Brazil, Central America (which later split into five separate countries), Santo Domingo (and Haiti later), Colombia (and Ecuador, Venezuela and Panama later). Almost all the countries named consisted of a mix of the original indigenous population (American Indians), Europeans whose origin depended on the colonial power, and Africans, brought in after about 1550 as slaves especially to serve in areas of tropical and subtropical agriculture.

While European colonial influence in the Americas was challenged and thrown off in the way described above, European powers were becoming more involved in Asia. In the 18th century the power of Spain declined while Portuguese activities in the tropics were increasingly carried out in association with Holland. France and Britain were the major maritime colonial powers of the late 18th and 19th centuries and Russia continued the expansion of its empire by land.

South Asia (especially India) and Southeast Asia (especially the East Indies, now Indonesia) were the areas most extensively colonized by the European maritime powers in Asia in the 19th century. The contemporary Russian Empire extended southwards in three main areas, Transcaucasia, Middle Asia and the Far East.

Several European powers had gained control over convenient stretches of the coast of the continent of Africa during the 16th–18th centuries. Such territories were used for the supply of provisions to ships passing between Europe and Asia, to collect slaves for the Americas and to trade in a limited way in such products as ivory. In the late 19th century, especially in the 1880s, European powers pushed into the interior of Africa, each aspiring to acquire as much territory as possible behind its coastal footholds. *Figure 2.5* shows Africa in 1898. This map may be compared with the portrayal of Africa in the 1771 map in *Figure 2.4*. By the end of the 19th century most of Africa had been partitioned by mutual agreement among seven European colonial powers.

During the 19th century two main regions of the world largely escaped direct colonization by European powers, Southwest Asia and the Far East, but they also came under pressure from Europe. Persia (now Iran) and Afghanistan might easily have ended up as Russian or British colonies. In the 1850s Japan was forced by the USA and Europe to trade with the outside world and China was under great pressure in the later part of the 19th century from Russia, Europe and the USA to do likewise. Late in the 18th century and early in the 19th, Australia and New Zealand were unobtrusively added to the British Empire. To complete the picture, the building of a substantial empire by Japan should be noted. Though never a European colony, Japan began to adopt Western technology in the 1860s and was later able to conquer Formosa, Korea, and in the 1930s Manchuria and other parts of China.

During the 19th and early 20th century virtual independence was given to certain parts of the British Empire, though these remained subsequently within the British Commonwealth: Canada, Australia, New Zealand and South Africa. By the time of their independence the first three had majorities of people of European origin while the Europeans in South Africa were the dominant minority, politically and

economically speaking. A second phase of decolonization could be said to have started, then, with the
granting of dominion status to Canada in 1867
through the British North America Act.

As a result of the First World War Germany lost its
colonies, most of the territory of which was in Africa
(*see Figure 2.5*). In 1935 Italy attacked Abyssinia and
gained control of the last sizeable part of Africa to
escape colonization. By 1900, however, the acquisi-

tion of extensive territories by European powers
largely came to an end. There were not many places
left that could be easily annexed.

Almost all the colonies of Europe were established
on the initiative of the colonizing power and at it own
invitation. Superior military strength and organization was all that was needed. As the centuries went
by, various reasons were given to justify the
establishment of colonies. One motive was the

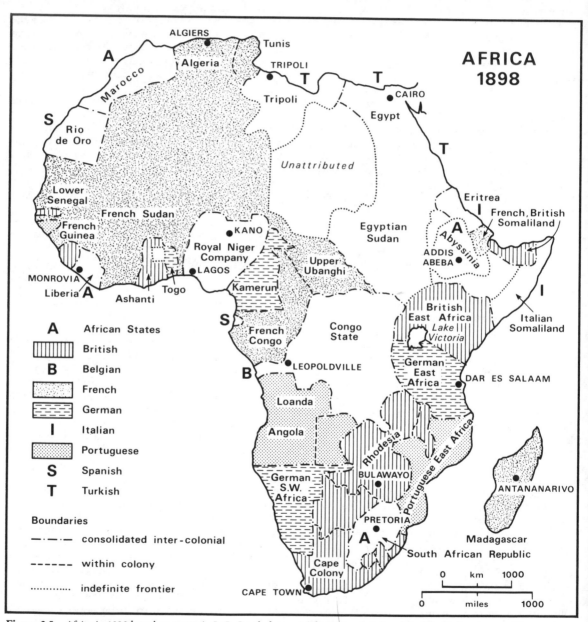

Figure 2.5   *Africa in 1898 based on a map in J. G. Bartholomew's The Citizen's Atlas of the World, London, 1898, pp 83–84.
The original spelling of the place-names is used here*

assumed need and duty to Christianize people in other parts of the world. One pretext for colonizing Africa in the late 19th century was to stop the slave trade. Colonial powers generally brought peace, stability and sound administration to their colonies.

In practice colonies were sources of raw materials and of tropical foods and beverages that could not be produced in Europe. They gave opportunities for some pressure of population to be relieved in Europe but in reality comparatively few people left until the development of the steamship starting in the middle of the 19th century. Altogether about 60 million people left Europe for other continents during 1850–1930, most to the countries of the Americas *after* their independence.

It was common for colonial powers to discourage or prevent the development of manufacturing in their colonies, except for branches needed to satisfy local requirements in less sophisticated goods or to process materials for exports. The colonies were captive markets for the manufactures of Europe. In the case of India, Britain even prevented the manufacture of cotton goods in its colony so that cotton grown in India could be taken to Britain for manufacture there and some of the products sent back.

Whatever is said for and against the process of colonization by European powers there is no doubt that at the height of European influence in the world Europeans never doubted their superiority over other 'races' and their self-confidence, if not arrogance, was impressive. The following extract from the entry, Colony, in *Encyclopaedia Britannica* (eighth edition, 1853–60) illustrates the attitude:

'There is no doubt that increasing and multiplying the British stock increases the producing and trading populations over the earth. Hence, if we can conveniently spare a portion of our people to inhabit distant regions, it may be politic to aid them by a government staff and other provisions in making a settlement. It happens that in general this country has not only been able to spare a portion of her citizens, but has found a relief in seeing them depart from her shores. To those indeed who have fallen into an unproductive and unhappy position in the mother country, the fresh emigration field has had the same effect as a supply of capital; for the occupation and use of a piece of productive land which has not previously been productively applied is to those who take it equivalent to the obtaining of so much capital. It is for the purpose of protecting if not creating emigration fields, that our colonial system has thus extended itself, and hence the more important considerations connected with it came under the head of emigration. Wherever we can plant a numerous and prosperous body of the British people, we create a trading population which will increase our own commerce; and if it be said that they take capital out of the country to employ it elsewhere, after the secondary answer that colonists do not generally remove much capital, the main answer is, that for a nation like the British, in the full employment of free trade, the best part of the world for their capital to be placed in, is that where it is most productive.'

The europeanization of most of the rest of the world has been described at some length. Even so, many important details have been left out. In the view of the author the number, distribution and composition of the countries of the world cannot be fully appreciated without some knowledge of their origins. Each country bears traces of its history, some at least through several centuries. Sometimes the present attitude or 'behaviour' in a given country may apparently be accounted for by some past experience.

The impact of improved communications on world affairs in the 19th century and the further impact expected in the 20th century are expressed in a paper given in 1904 and published later in that year, by Sir Halford Mackinder, one of the first modern British geographers:

'Broadly speaking, we may contrast the Columbian epoch with the age which preceded it, by describing its essential characteristic as the expansion of Europe against almost negligible resistances, whereas medieval Christendom was pent into a narrow region and threatened by external barbarism. From the present time forth, in the post-Columbian age we shall again have to deal with a closed political system, and none the less that it will be one of world-wide scope. Every explosion of social forces, instead of being dissipated in a surrounding circuit of unknown space and barbaric chaos, will be sharply re-echoed from the far side of the globe, and weak elements in the political and economic organism of the world will be shattered in consequence. There is a vast difference of effect in the fall of a shell into an earthwork and its fall amid the closed spaces and rigid structure of a great building or ship. Probably some half-consciousness of this fact is at last diverting much of the attention of statesmen in all parts of the world from territorial expansion to the struggle for relative efficiency.'

Mackinder expected the world of the 20th century to be a single closed system.

In the BBC Reith Lectures given in 1953 the British historian Arnold Toynbee (1953) spoke on the theme of the World and the West. One point strongly argued in the series was that the day of reckoning had come for Europe. Those peoples of the world that it had pushed around for so long would retaliate. Looking back at the decades since the Second World War the response of the colonies to their ex-masters has been restrained on the whole and rather muted. You have to mention the USA to a Mexican, not Spain, to make his blood pressure rise. Italy has been one of Ethiopia's main trading partners since 1945 in spite of quite recent unpleasant memories. The UK is highly respected in India in some ways and the English language is still widely spoken. The non-Russian Soviet Republics of Middle Asia have accepted the Soviet Communist Party and Uzbeks and Tadjiks there apparently sit back and let Russians get some of the best jobs in their part of the world.

2.4 The period since the Second World War

The world that emerged from the Second World War was clearly one that was subdivided into independent and dependent countries. The location of the 51 countries that in 1945 originally formed the new United Nations is given in *Figure 2.6*. They included countries that had been victors in the Second World War or 'respectable' countries, such as those of Latin America, Southwest Asia and Africa or German occupied Europe that had been anti-German or at least neutral. For no obvious reason the USSR managed to gain three seats in the United Nations.

Between 1945 and the late 1970s the number of members of the United Nations increased about three times. Some countries that were already independent in 1945 such as Sweden, Finland and Japan, subsequently joined. Nearly all the additional members were ex-colonies of varying shapes and sizes, about half from Africa alone.

Several types of country could be distinguished at the end of the Second World War, based on the study of the origins of modern countries given in this chapter. The following types may be proposed:

(1) European countries such as France, Hungary, Sweden, Russia.

(2) Former European colonies, independent in 1945, including virtually all the Americas and also Australia. These could be broadly subdivided into:

    (a) Those with a large proportion of people of European origin, including Canada, the USA, Argentina, Uruguay, Australia, New Zealand.

    (b) Those with a large proportion of non-European people, including the rest of Latin America with regional exceptions (such as southeast Brazil), and South Africa.

(3) Countries that have never been held directly as European colonies for any length of time. Such countries as Japan and China, Thailand, Afghanistan, most of Southwest Asia, Turkey and Ethiopia qualify, though virtually none has entirely avoided some conflict with European powers and most have come to terms with the need in some way to change their culture and technology to exist in a europeanized world.

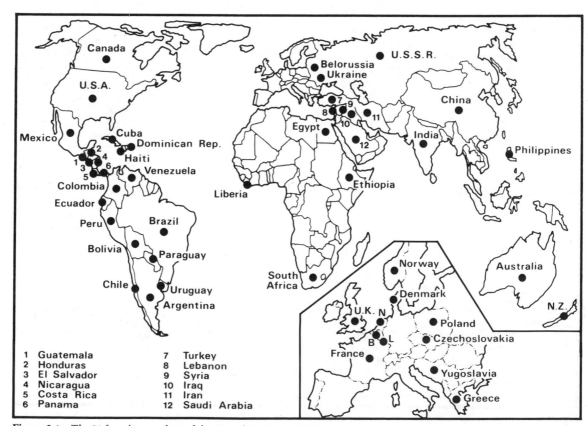

1	Guatemala	7	Turkey
2	Honduras	8	Lebanon
3	El Salvador	9	Syria
4	Nicaragua	10	Iraq
5	Costa Rica	11	Iran
6	Panama	12	Saudi Arabia

Figure 2.6   *The 51 founder members of the United Nations in 1945*

(4) European colonies, including almost all of Africa and much of South and Southeast Asia as well as many small units in the Caribbean, Indian and Pacific Oceans. All the non-Russian parts of the USSR are included in the above definition. In the early communist period various non-Russian peoples were given the status of Soviet Socialist Republics, each with a boundary on the edge of the USSR and with an option (theoretically) to secede. Soviet thinking has not found it difficult to justify the continuing inclusion of former colonies within the new multinational Soviet state.

The Japanese occupation of much of Southeast Asia during the Second World War helped to loosen the hold of European powers there and made the return of Netherlands control to its East Indies and of French control to Indo-China virtually impossible. In Africa the European powers could probably have kept their colonies longer but through the initiatives of de Gaulle and of Macmillan with his 'winds of change' attitude many new countries suddenly appeared around 1960, a situation that prompted the cartoons by David Langdon and Larry in *Figure 2.7*.

The theme of this chapter so far has been the spread of European political, cultural and technological influence over much of the rest of the world during four centuries, 1500–1900. A counter process started during the period 1770–1830 when most of the European colonies of the Americas became new independent countries. It continued with the granting of dominion status to British possessions largely settled (Canada, Australia, New Zealand) or dominated (South Africa) by Europeans. The concluding stage has been the achievement of independence by virtually all European colonies of Asia and Africa during 1945–75.

At least through the eyes of a European it could be said that since 1500 world affairs have increasingly been a result of two struggles, one between national states within Europe itself, the other between some of these states to achieve and maintain control of areas outside Europe. One should make at least three reservations about the above generalization. Firstly, Russia did not come into conflict to any great extent *outside* Europe with other European powers during its territorial expansion. Secondly, during the last hundred years, Japan built up a 'regional' empire without conflicting to any great extent with European powers except Russia, at least until the showdown in 1941 when Japan overran various American, French, British and Netherlands colonies. Thirdly, in 1945 the USSR and USA emerged as the dominant powers in the world from a military point of view.

Since 1945 the main struggle in the world has been between the USA and the USSR and it can be said that Europe, apart from Russia, has lost the initiative it previously had in determining the course of world affairs. Both the USA and the USSR pressed, at the end of the Second World War, for the then weak European colonial powers to give up their remaining colonies. The USA set an example by giving independence to the Philippines, which it had 'inherited' from Spain in 1898 but the USSR made no signs of offering independence to its own colonies.

Figure 2.8 is a view of the world as seen in 1957 by the present author at the time of writing the 1959 edition of *Geography of World Affairs* and representing the situation about a decade after the end of the war. Around 1955 European powers still controlled virtually all of Africa. With a ring of allies and bases in West Europe, South and East Asia, the USA was attempting to contain the Soviet bloc, which at that time already included the East European countries (such as Poland and Bulgaria) and mainland China. China itself depended on the USSR for technical assistance and was not a power of world influence.

The USSR had some influence within the 'zone of friction' shown in *Figure 2.8* but little influence

Figure 2.7 *David Langdon illustrates the rush around 1960 to accommodate many new African countries in the United Nations General Assembly. Reproduced with the permission of Punch. Larry's drawing reflects the same theme*

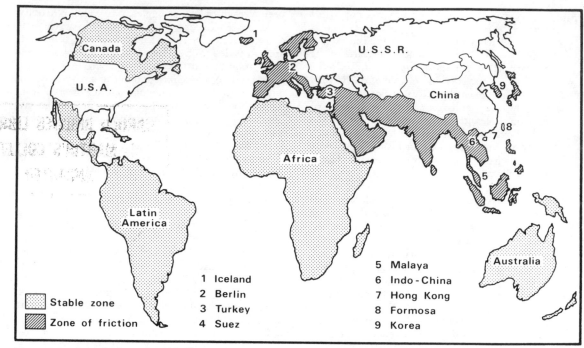

Figure 2.8   *A simple view of world affairs as seen in the mid-1950s*

beyond. Most of the world's trouble spots were located in the sensitive ring of places close to the Soviet bloc. Some are indicated by the numbering on the map. Whatever the defects of the model of world affairs in *Figure 2.8* it was a simple framework in which many happenings, whether political or spatial, seemed to fit easily.

Since 1955 several new trends have appeared:

(1) By the early 1960s it was evident from the withdrawal of Soviet aid and technicians from China and a drastic reduction in trade between the two countries that relations had become much cooler than they had been in the 1950s. China would no longer be ready to fall in with Soviet policy.

(2) Soviet influence began to be felt more strongly than previously in parts of the world beyond the 'zone of friction' in *Figure 2.8*. In Cuba the USSR found an ally in Fidel Castro and it was involved in the Congo when that colony became independent from Belgium in 1960. In the 1970s, with the help of Cuban forces, it has gained considerable influence in Angola, Ethiopia and other parts of Africa.

(3) Weaker countries began to form groupings of an economic nature, The Treaty of Rome (1957) started a Common Market for six West European countries while the Latin American Free Trade

Association (1961) brought together most of the larger countries of that region. Japan began to regain economic influence in the Pacific, and China, with such a large population, could hardly be ignored in world affairs. In other words, new powers were emerging to challenge, though not to equal the dominant position of the USA and USSR.

(4) In the immediate postwar years world affairs could be interpreted to a considerable extent as an ideological conflict between communism and capitalism. Any country in the world might be expected to be committed to one side or the other. Various countries refused, however, to be clearly 'committed' and the concept of the Third World developed, its members unwilling to be aligned with either the USA or the USSR. In addition to such 'neutral' countries in Europe as Sweden, Switzerland and Austria, countries such as Yugoslavia, Egypt and Indonesia belonged. There was nothing initially in the concept of the Third World to imply that its members were poor or developing countries since it was a political rather than an economic concept. Increasing awareness and concern over the gap between rich and poor countries led to a new division of the world as developed and developing. This has been referred to as the North-South problem since most of the richer countries are to

Table 2.1 *Development and ideological dimensions*

	Capitalist (market economy)	Intermediate (mixed)	Communist (centrally planned)
Rich	USA Australia Switzerland	UK Italy	Czechoslovakia USSR
Intermediate	Brazil	Yugoslavia	Cuba China Korean DPR
Poor	Nigeria	Tanzania India	Viet Nam

the north of the poorer ones. Ideology has little to do with the gap. The communist-capitalist and rich-poor 'dimensions' or dichotomies are shown in *Table 2.1*.

It was noted in chapter 1 that the concept of a sovereign state is fundamental in world affairs. International relations, whether through trade, aid or conflict, are between such countries. As made clear in this chapter, almost all the direct effects of European and other colonialism have been removed from the world, politically speaking, by the creation of new independent countries. The origins of the present countries of the world have now been briefly described. The countries may be regarded as 'players' in a complex game. Their strengths, locations and aspirations will be assessed in the chapters that follow. It remains to identify the present players. Some with new or unfamiliar names are noted in *Table 2.2*.

2.5 The present countries of the world

At any given moment it is difficult to put an exact number to the sovereign states of the world. Some countries, like Hong Kong and Puerto Rico, are strictly dependent. Some independent units like the Pitcairn Islands and San Marino are so small that they can hardly be counted as sovereign states. One, Taiwan, does not officially exist for it is not now listed in United Nations publications. Taiwan (formerly Formosa) has been turned out to make way for China, which regards it as part of itself. The North/South subdivision of Korea is not recognized although for practical purposes two separate sovereign states exist. Germany, on the other hand, is represented as two separate states.

The set of sovereign states in existence in the 1980s is one arrived at through the rather haphazard events

Table 2.2 *Countries with new or unfamiliar names*

Name used in this book	Full and/or former name
Cameroon	United Republic of Cameroon
Central African Rep.	Republic, formerly Empire
German DR	German Democratic Republic (East Germany)
German FR	German Federal Republic (West Germany)
Kampuchea	Democratic Kampuchea, formerly Cambodia
Korean DPR	Democratic People's Republic (North Korea)
Korean R	Republic of Korea (South Korea)
Laos	Lao People's Democratic Republic, formerly Laos
Libya	Libyan Arab Jamahiriya
Namibia	Formerly South West Africa
South Africa	Republic of South Africa
Syria	Syrian Arab Republic
Tanzania	United Republic of Tanzania, formerly Tanganyika and Zanzibar
Viet Nam	Also Vietnam, formerly North and South
Western Sahara	For simplicity this former colony of Spain has been entirely attributed to Morocco
Yemen AR	Yemen Arab Republic (North Yemen)
South Yemen	Democratic Yemen (South Yemen), formerly Aden Protectorate
Zimbabwe	Formerly (Southern) Rhodesia

of the last few hundred years. Many countries were arbitrarily formed colonies of European powers. The system was not, therefore, designed, as many *internal* administrative systems are, to form a balanced set of regions, roughly comparable in area, in population or in some other way. *Table 2.3* shows that area, population and production are all highly concentrated in a few out of more than 200 countries. The four largest in population have nearly half of the world's total and the 15 largest contain about 70 percent. The six largest in area have half of the world's total and the 15 largest 65 percent. The four largest in gross national product have half and the 15 largest 75 percent.

For reference the fifty largest countries of the world in population are given in *Table 2.4* and the next fifty in *Table 2.5*. The location of each set of countries is shown in *Figures 2.9* and *2.10*. The striking contrasts in area size are shown in these maps. Contrasts in population size may be shown in various ways. One way is to represent countries diagrammatically, proportional in area to their population size, as has been done in *Figure 2.11*.

Table 2.3 *Fifteen largest countries of the world in population, area and gross national product*

Population	Area	Gross national product
1 China	1 USSR	1 USA
2 India	2 Canada	2 USSR
3 USSR	3 China	3 Japan
4 USA	4 USA	4 German FR
5 Indonesia	5 Brazil	5 France
6 Brazil	6 Australia	6 UK
7 Japan	7 India	7 Italy
8 Bangladesh	8 Argentina	8 Canada
9 Pakistan	9 Sudan	9 China
10 Nigeria	10 Algeria	10 Brazil
11 Mexico	11 Zaire	11 Spain
12 German FR	12 Saudi Arabia	12 Netherlands
13 Italy	13 Mexico	13 Poland
14 UK	14 Libya	14 Australia
15 Viet Nam	15 Iran	15 India

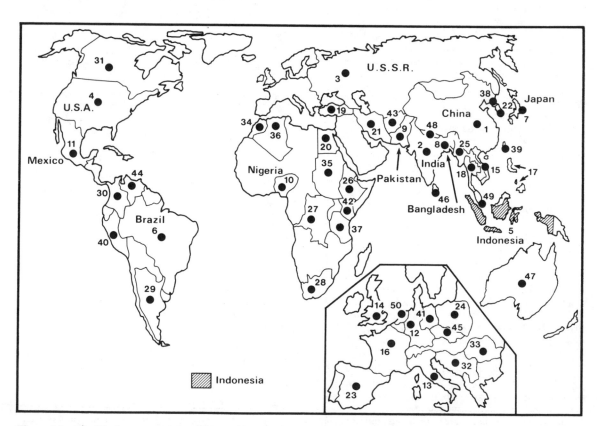

Figure 2.9   *The fifty largest countries of the world in population in 1981. (See Table 2.4)*

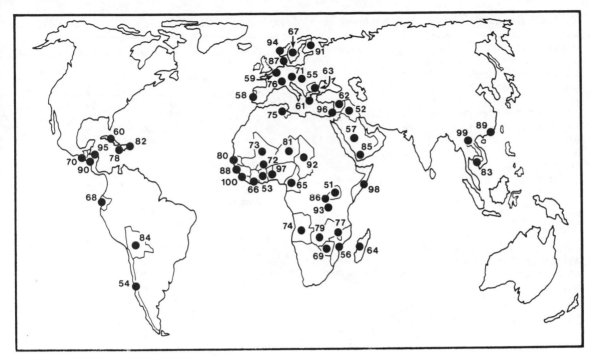

Figure 2.10 *The countries ranked 51 to 100 in population size in the world in 1981. (See Table 2.5)*

Table 2.4 *Fifty largest countries of the world in population in 1981*

Country	Popn 1	Area 2	GNP 3	Country	Popn 1	Area 2	GNP 3
1 China	985	9 597	219	26 Ethiopia	34	1222	4
2 India	689	3 288	126	27 Zaire	30	2345	7
3 USSR	268	22 402	1085	28 South Africa	29	1221	49
4 USA	230	9 363	2384	29 Argentina	28	2767	61
5 Indonesia	149	1 904	54	30 Colombia	28	1139	26
6 Brazil	121	8 512	201	31 Canada	24	9976	229
7 Japan	118	372	1020	32 Yugoslavia	23	256	54
8 Bangladesh	93	144	9	33 Romania	22	238	42
9 Pakistan	89	804	22	34 Morocco	22	447	14
10 Nigeria	80	924	50	35 Sudan	20	2506	7
11 Mexico	69	1 973	108	36 Algeria	19	2382	30
12 German FR	61	249	718	37 Tanzania	19	945	5
13 Italy	57	301	298	38 Korean DPR	18	121	20
14 UK	56	244	354	39 Taiwan	18	36	24
15 Viet Nam	55	330	9	40 Peru	18	1285	13
16 France	54	547	531	41 German DR	17	108	107
17 Philippines	49	300	28	42 Kenya	17	583	6
18 Thailand	49	514	27	43 Afghanistan	16	647	3
19 Turkey	46	781	59	44 Venezuela	16	912	42
20 Egypt	44	1 001	19	45 Czechoslovakia	15	128	82
21 Iran	40	1 648	70	46 Sri Lanka	15	66	3
22 Korean R	39	98	56	47 Australia	15	7687	131
23 Spain	38	505	163	48 Nepal	14	141	2
24 Poland	36	313	136	49 Malaysia	14	330	18
25 Burma	35	677	5	50 Netherlands	14	41	143

1 = population in millions, 1981; *2* = area in thousands of square kilometres; *3* = gross national product in thousands of millions of US dollars in late 1970s.

Table 2.5 *The countries ranked fifty-one to one hundred in population size in 1981*

Country	Popn 1	Area 2	GNP 3	Country	Popn 1	Area 2	GNP 3
51 Uganda	14.1	236	3.8	76 Switzerland	6.3	41	89.7
52 Iraq	13.6	435	31.1	77 Malawi	6.2	118	1.2
53 Ghana	12.0	239	4.5	78 Haiti	6.0	28	1.5
54 Chile	11.2	757	18.6	79 Zambia	6.0	753	2.9
55 Hungary	10.7	93	41.2	80 Senegal	5.8	197	2.4
56 Mozambique	10.7	783	2.6	81 Niger	5.7	1267	1.4
57 Saudi Arabia	10.4	2150	59.7	82 Dominican Rep.	5.6	49	5.2
58 Portugal	10.0	92	21.6	83 Kampuchea	5.5	181	0.9
59 Belgium	9.9	31	106.7	84 Bolivia	5.5	1099	2.9
60 Cuba	9.8	115	14.0	85 Yemen	5.4	195	2.4
61 Greece	9.6	132	37.0	86 Rwanda	5.3	26	1.0
62 Syria	9.3	185	8.9	87 Denmark	5.1	43	60.7
63 Bulgaria	8.9	111	32.8	88 Guinea	5.1	246	1.3
64 Madagascar	8.8	587	2.5	89 Hong Kong	5.0	1	18.8
65 Cameroon	8.7	475	4.6	90 El Salvador	4.9	21	3.0
66 Ivory Coast	8.5	322	8.2	91 Finland	4.8	337	39.6
67 Sweden	8.3	450	98.9	92 Chad	4.6	1284	0.5
68 Ecuador	8.2	284	8.4	93 Burundi	4.2	28	0.8
69 Zimbabwe	7.6	391	3.4	94 Norway	4.1	324	43.9
70 Guatemala	7.5	109	6.9	95 Honduras	3.9	112	1.6
71 Austria	7.5	84	52.7	96 Israel	3.9	21	15.8
72 Upper Volta	7.1	274	1.2	97 Benin	3.8	113	0.9
73 Mali	6.8	1240	0.9	98 Somalia	3.8	638	0.5
74 Angola	6.7	1247	3.0	99 Laos	3.6	237	0.3
75 Tunisia	6.6	164	7.2	100 Sierra Leone	3.6	72	0.9

1 = population in millions to nearest 100 000 mid-year 1981; *2* = area in thousands of square kilometres; *3* = gross national product in thousands of millions of US dollars to the nearest 100 million.

The size and strength of countries, present or potential, can be measured in other ways in addition to those described so far in this chapter. An estimate of the destructive power in megatons might show 90 percent or more concentrated in the hands of the USA and USSR. Any assessment of natural resources would show that certain countries, some with quite small populations, like Canada and Australia, have impressive shares of the world total. Military strength will be discussed in chapter 3 and natural resources in chapter 4. The influence of a country in world affairs is not only related to its size, measured as shown in various ways, but also to its geographical position in the world and to its leadership and ideology. These aspects will also be discussed in chapter 3.

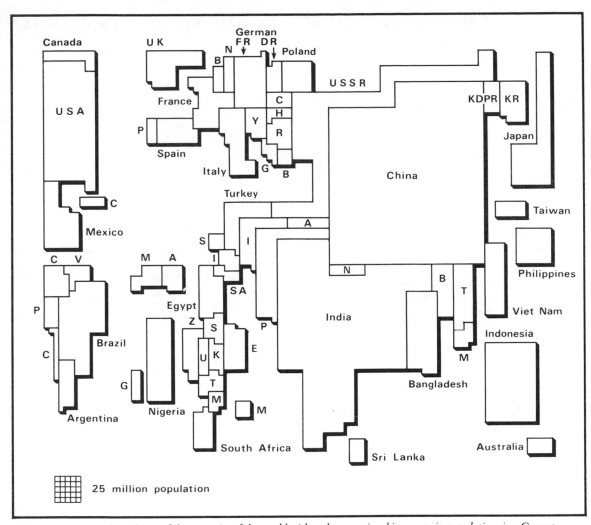

Figure 2.11 *A topological map of the countries of the world with each proportional in area to its population size. Correct contiguity is not conserved throughout*

CHAPTER THREE

The Players in the Game of World Affairs

3.1 Spatial aspects

The countries of the world are like the pieces of a giant jig-saw puzzle, and they occupy the land on a finite but unbounded spherical surface. Each country of the world fits into a slot with a unique shape and with a unique position in relation to all other countries. In this section some spatial features of the countries will be noted:

Shape (see Figure 3.1)

One might expect a country that is greatly elongated or fragmented to be more difficult to administer politically and to organize economically than a compact country of comparable area. Such a hypothsis is, however, difficult to test because there is no single measure of shape and also because it is not easy to decide with much precision whether or not a political unit is functioning well or badly.

Chile is a very long narrow country. It has a north-south extent of some 4000 kilometres but over four-fifths of its population live between Valparaiso and Puerto Montt (*see Figure 3.1*). The desert to the north and the rugged forests to the south can be reached by road and by shipping services from the central area. Chile's long boundary with neighbouring Argentina runs through the high, uninhabited Southern Andes. In spite of its inconvenient shape Chile apparently has fewer problems of internal organization than some countries that are much more compact, but have their population living round the periphery. Norway is probably the second most elongated country in the world. Here population tends to gravitate towards the south, the part of the country with the best economic prospects and the least unpleasant climatic conditions.

Compared with many countries the UK is lacking in compactness. It is, however, very small in area for a country with 56 million inhabitants. In Japan most of the population lives on the four main islands named in *Figure 3.1*. Long tunnels link Hokkaido and Kyushu to Honshu and in spite of the very rugged nature of most of the terrain, road and rail links ensure reasonable access between the main towns, while nowhere is far from shipping services. Of the five countries shown in *Figure 3.1* perhaps the Philippines is the most inconvenient to integrate since inter-regional links must largely depend on ferries. There is not, however, a great deal of goods traffic to be carried between the islands. Indonesia (*see Figure 11.4*) is even larger and more fragmented than the Philippines.

Position

Two aspects of the position of countries will be referred to here briefly: landlocked countries and number of neighbours.

(1) *Landlocked countries* are particularly sensitive politically because their links with the rest of the world pass through the territory of other countries to reach the coast and the oceans. Although international agreements protect access to the coast such links can in practice be interrupted or made difficult to use. Of the hundred largest countries of the world in population, 17 are entirely without a sea coast.

Table 3.1 *Largest landlocked countries with population in millions*

Country	Population in millions	Country	Population in millions
Afghanistan	16	Switzerland	6
Czechoslovakia	15	Bolivia	6
Nepal	14	Malawi	6
Uganda	14	Zambia	6
Hungary	11	Niger	6
Austria	8	Rwanda	5
Zimbabwe	8	Chad	5
Upper Volta	7	Burundi	4
		Laos	4

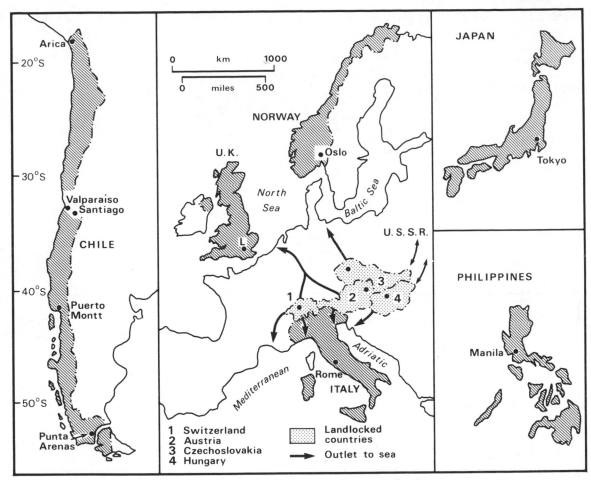

Figure 3.1   *The shape and position of selected countries. All maps are on the same scale*

They are listed in *Table 3.1*. All are comparatively small in population and together they only have about three percent of the population of the world, without counting several that are still smaller in population, including Paraguay and the Mongolian People's Republic.

The four largest landlocked countries in Europe are shown in *Figure 3.1*. All four depend heavily on foreign trade but Hungary and Czechoslovakia trade mainly with the USSR and other fellow-communist countries and their main outlets are therefore by land, direct to these countries. For Austria the shortest route to the coast is through the Italian ports of Venice and Trieste. Swiss outlets are to the Mediterranean through France or Italy and to the North Sea via the River Rhine.

Nine of the 17 largest landlocked countries are in Africa. After its unilateral declaration of independence in 1964 and until the resolution of its disputed status in 1979, Southern Rhodesia, now Zimbabwe, had all but one of its outlets to the coast blocked by its neighbours. Southern Rhodesia in its turn hindered the flow of Zambian copper to ports in Mozambique, and a special line, the Tanzam, was built by the Chinese through Tanzania as a new outlet. In South America, Bolivia is particularly sensitive about its links to the Pacific coast since all but one pass through provinces now held by Chile, one of them actually taken from Bolivia in the Pacific War of 1879–84.

(2) *How many neighbours?* The position of a country on the earth's surface greatly affects its prospects and role in world affairs as may be seen in a comparison of two small countries, Israel and New Zealand. Within 1000 kilometres of Israel there are all or part of ten countries, while within the same distance of New Zealand there is not one. Israel is more likely to be closely involved

Table 3.2 *The number of neighbours of selected countries*

Number of countries within 1000 km		Number of contiguous countries			
USSR	28	USSR	12	Mexico	3
Poland	25	Canada	1	Libya	6
France	24	China	12	Iran	5
Italy	23	USA	2	Indonesia	3
Turkey	22	Brazil	10	Japan	0
		Australia	0	Bangladesh	2
Japan	4	India	6	Pakistan	4
Australia	2	Argentina	5	Nigeria	4
Canada	1	Sudan	8	German FR	9
Sri Lanka	1	Algeria	6	Italy	4
New Zealand	0	Zaire	9	UK	1
		Saudi Arabia	7	Viet Nam	3

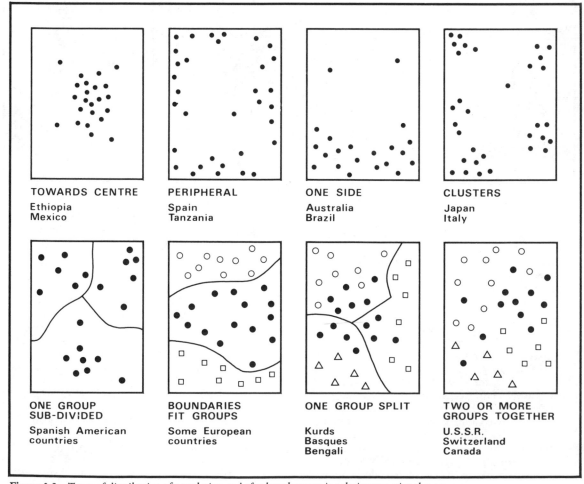

TOWARDS CENTRE

Ethiopia
Mexico

PERIPHERAL

Spain
Tanzania

ONE SIDE

Australia
Brazil

CLUSTERS

Japan
Italy

ONE GROUP SUB-DIVIDED

Spanish American
countries

BOUNDARIES FIT GROUPS

Some European
countries

ONE GROUP SPLIT

Kurds
Basques
Bengali

TWO OR MORE GROUPS TOGETHER

U.S.S.R.
Switzerland
Canada

Figure 3.2 *Types of distribution of population and of cultural groups in relation to national areas*

with other countries than is New Zealand. Column *1* in *Table 3.2* shows the number of countries within 1000 kilometres of selected countries.

The 24 countries of the world among the top 15 either in area size or in populaton size (or both) are shown in *Table 3.2*, columns *2* and *3*, with the number of countries contiguous to each. For example, ten other South American countries (including the three 'Guianas') touch Brazil, whereas Australia, like smaller island countries such as Japan, Sri Lanka, Jamaica and New Zealand, does not share an international boundary with any other country.

3.2 The internal cohesion of countries

Few countries are as solid and homogeneous as they appear to be, each represented in a particular colour, on conventional political maps. The purpose of this

section is to illustrate variations within countries that cause friction and conflicts, with possible repercussions on other countries and on world affairs in general.

Population and natural resources are not spread uniformly over the national area of any country. Very few countries contain people all with the same cultural background and outlook. Some hypothetical distributions of population in *Figure 3.2* exemplify different possible situations. No actual country exactly fits any simply described spatial pattern but some that roughly compare with the examples are listed.

The four lower diagrams in *Figure 3.2* show four possible ways in which different cultural groups can be related to countries. Such groups could be people with the same language or religion or of the same 'race', as defined by skin colour or by some other characteristic.

Case 1 shows members of the same group found in different countries. There are German speakers in the

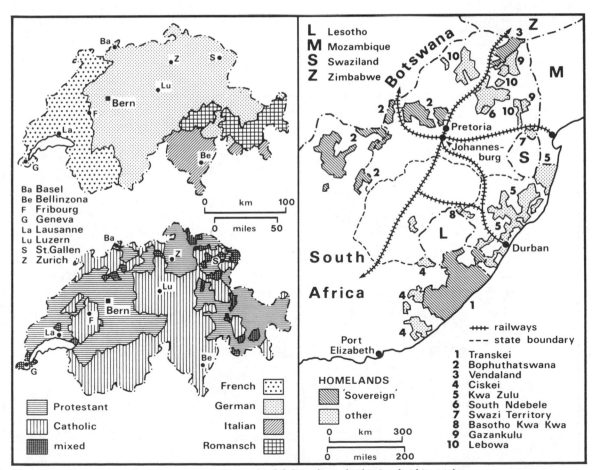

Figure 3.3 *Linguistic and religious groups in Switzerland (left) and apartheid in South Africa (right)*

German FR and DR and also in Austria and Switzerland. Spanish is the only or main language in nine south American countries.

Case 2 illustrates a theoretically desirable situation in which countries and cultural groups coincide exactly. Attempts have been made at various times to match European countries with European national groups but rarely to the complete satisfaction of everyone.

Case 3 is a national group (black dots) not forming a majority in any country, but forming a minority in two or more. The Kurds of Southwest Asia are an example. On numerical grounds they could certainly argue the case for separate existence as an independent (though landlocked) country larger than many existing countries.

Case 4 is found widely in the world, the multinational country, often with a large majority group and several or many small peoples. Examples include Spain, the UK, Yugoslavia, the USSR, India and China.

In much of western Europe the Christian Church became fragmented in the 16th century. Roughly half the population of Switzerland is Roman Catholic and half Protestant (*Figure 3.3*). Some areas, including the larger towns, having mixed populations. Four languages are spoken in Switzerland. About two-

thirds of the population is German speaking, but French and Italian are also used. Only Romansch, a minor 'language' of Latin origin (like French and Italian) could be referred to as distinctly 'Swiss'. In other words, Switzerland exists as a cohesive and successful sovereign state in spite of depending on the languages of other countries and being divided between two potentially conflicting views of Christianity. Fortunately, perhaps, the linguistic and religious divisions do not coincide.

The distribution of two national groups in Southwest Asia, the Kurds and the Azerbaijanis, is shown in *Figure 3.4*. Azerbaijanis only inhabit two countries, the USSR and Iran. They do not have a country of their own though they number several million but they do have the status of Soviet Socialist Republic (capital Baku) in the USSR. The Azerbaijan SSR includes many peoples of other national groups of the USSR, including some Russians. Unification of Azerbaijanis might come through a merger of those living in northwest Iran with the existing Soviet Republic.

Most of the Kurds of Southwest Asia are in Turkey, Iran or Iraq (*see Figure 3.4*) but there are some in Syria and a few tens of thousands in the USSR in the Soviet Socialist Republics of Armenia and Azerbaijan. Though the lands inhabited by the

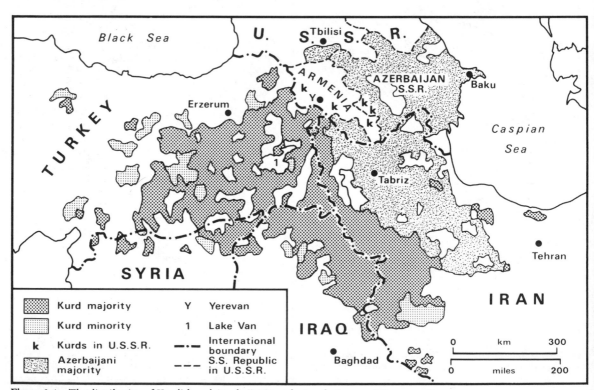

Figure 3.4 *The distribution of Kurdish and Azerbaijani peoples in relation to existing sovereign states*

Kurds are mostly rugged and dry, with limited natural resources, it is hardly likely that any one of the host countries would give complete independence to their own Kurds unless the other countries did so too.

The spatial appearance of apartheid, the gradual progress of relegating the African part of the population of South Africa to certain parts of the national territory is shown in *Figure 3.3*. Most of the ten homeland territories are fragmented. It seems to be accepted also that on the whole they are carefully located to avoid including good agricultural land, important mineral deposits, or the courses of major roads and railways. The international significance of the Bantu homelands is considerable since according to South Africa three are already sovereign units, entitled to form direct links with other countries, while the rest are destined to have similar status. The price is that the inhabitants become 'foreigners' in white South Africa. In contrast therefore to most other parts of Africa, African peoples or 'tribes' are being given apparent sovereign status. In most other countries the trend is towards playing down the role of tribal groups and consolidating the ex-colonial territories that form the basis of African states at present.

Few countries are so homogeneous that they are completely exempt from internal regional feelings and potential movements towards autonomy or independence. In European countries there are numerous submerged peoples aspiring to greater independence or at least attempting to keep their traditions and cultural heritage alive. In many Latin American countries relics of pre-Columbian cultures survive while in Africa many countries are formed out of various tribal groups, the integration of which has barely started.

Countries that have received immigrants in the last hundred years or so may also have problems of integration. The USA is a 'melting pot' for peoples from many parts of the world but its policy of integration for all 'races' has not eliminated minority groups.

In some countries the minority group or groups may be from countries that are much poorer than the host country. Such are the Italians (mainly from southern Italy) and Portuguese in Switzerland, the West Indians and Pakistanis in England and the Tamils in Sri Lanka. Such minority groups tend to remain under-represented and under-privileged. In other cases a minority is from a more developed country than the host country. Such are the British in Zimbabwe, the Japanese in southeast Brazil, the Asians in Kenya and the Chinese in parts of Southeast Asia. Such minorities may acquire a share of wealth and influence in the country far larger than their numbers would merit, possibly causing resentment at their presence.

3.3 Transport and communications

In the sense that one could theoretically post a letter from any place in the world to any other it may be said that the world operates as a single system, with all its parts linked. Modern transportation and communication links have brought many places in the world close to a railway or motor road. Air transport gives great speed and flexibility in linking remote places to the system. The diffusion of cheap battery radios has brought poor, remote regions with no electricity supply within reach of broadcasting services. The continuing expansion of mass transportation media is illustrated by the data in *Tables 3.3* and *3.4*. In spite of the 1973–74 rises in oil prices an enormous increase can be seen in facilities even in the decade between the late 1960s and the late 1970s.

While it is assumed to be desirable for a country to have a high level of production of material goods, it is simply an additional cost on production to have to move the goods from one place to another. The increasing specialization of production by regions in the world, and the efficiency that is assumed to go with it, have had to be paid for by the organization of ever more extensive and elaborate transportation systems. The data in *Tables 3.3* and *3.4* cover both internal and international transportation features. The media included are mostly general purpose but the importance of special media for the movement of such items as water, oil and electricity should not be overlooked.

Within large countries such as the USSR and Canada, in which natural resources and centres of production are widely dispersed over the national area, very large quantities of goods have to be moved long distances between regions. In the USSR, for example, the various means of transport available carry on average one tonne of goods some 20 000 kilometres per year for every inhabitant of the country. In small industrialized countries, such as the UK and Japan, distances between the main centres of production are much smaller.

International goods traffic is mainly carried by sea, but oil and gas are sent by pipeline from Canada to the USA and there is also much movement of goods by rail, road and pipeline between the countries of CMEA (USSR and East Europe) and also between West European countries. The trade between the developed and developing countries of the world is mainly by sea. One main function of the internal transportation networks in developing countries has been to link centres of production in the interior with seaports.

It is difficult to assess the relative importance of different forms of transport in the world as a whole since to do so fairly one should take into account the value of goods carried as well as their weight. Although air cargo services move only a minute share

Table 3.3 *Rail traffic and merchant shipping*

Goods traffic handled by railways in thousands of millions of tonne-kilometres				*Merchant fleets in millions of gross registered tonnes of shipping*			
Country	1968	1973	1978	Country	1969	1973	1980
1 USSR	2275	2958	3430	1 Liberia	29	55	80
2 USA	1086	1243	1250	2 Japan	24	39	41
3 Canada	139	191	215	3 Greece	9	22	39
4 India	125	120	150	4 UK	24	32	27
5 Poland	93	116	138	5 Panama	5	11	24
6 South Africa	50	60	80	6 USSR	14	18	23
7 Romania	41	57	74	7 Norway	20	25	22
8 Czechoslovakia	57	65	72	8 USA	20	14	18
9 France	63	74	69	9 France	6	9	12
10 Brazil	22	43	64	10 Italy	7	9	11
11 German DR	45	53	53	11 German FR	7	8	8
12 Japan	60	59	40	12 Spain	3	5	8
13 German FR	59	67	57	13 Singapore	–	2	8
14 Mexico	20	26	36	14 China	1	2	7
15 Australia	20	26	32	15 India	2	3	6
16 Hungary	18	21	24	16 Sweden	5	6	4
17 UK	24	26	20				
Author's estimates for China 300				World	212	290	420

Source: UNSYB 74, 79/80, Table 146 in 79/80   Source: UNSYB 74, 79, 80, Table 148 in 79/80

of the weight of all the goods carried in the world they carry items of high value. Even so within countries there are great differences in emphasis on different transport media. In the USSR and East Europe rail transport is much more important than road traffic except for very short hauls and in certain areas not easily served by rail. In the USA in contrast, the railways have declined in relative importance since the 1930s though they have been regaining goods traffic recently. In many developing countries it has been more convenient to build new roads rather than railways. Since the Second World War, however, there has been some railway construction in developing countries, as in Mexico, Colombia and Tanzania, while in both the USSR and China many new inter-regional lines have been completed and other lines are still under construction. In the transportation of oil, gas and electricity, pipelines and transmission lines take an increasing share of land traffic.

Table 3.3 shows the great amount of goods traffic carried by rail in the USSR and USA compared with that carried in other countries. In relation to population size, however, Canada has one of the busiest systems of any country in the world. Of the total world merchant shipping tonnage of 420 million tonnes in 1980, 175 million consisted of tankers. Liberia and Panama provide flags of convenience for other countries while Norway and Greece in particular serve the needs of other countries as well as

their own. In relation to the size of their populations, the merchant fleets of the USSR and the USA seem small. That of the USSR is growing fast, while the USA has additional shipping based in other countries. Without including the freak cases of Liberia and Panama, developing countries have only a small share of the world's shipping fleet though the fleets of China, Singapore and India have grown fast since the late 1960s. Much of the traffic entering and leaving the ports of developing countries is therefore carried by vessels owned in developed countries.

Although almost exactly the same countries appear in the top nine in *Table 3.4* with regard to the number of passenger cars and commercial vehicles in use, they differ in order. The main anomaly is the USSR in which the passenger car has only restricted use (about three million in the late 1970s) while commercial vehicles play a vital part in transportation. The low position of the German FR compared with France on the list of commercial vehicles may be to some extent a matter of classification and data collection. In the late 1970s the USA had almost 40 percent of all the passenger cars in the world. In the early 1950s, however, it had nearly three-quarters of the much smaller world total of the time. The private car has become virtually a necessity in American life and the two- and even three-car family is common in some states. Gasoline (petrol) used largely in road transport accounts for about a third of all the US consumption of refined oil.

Table 3.4 *Motor vehicles in use in millions*

	Passenger cars				Commercial vehicles		
Country	1968	1973	1978	Country	1969	1973	1978
1 USA	83.3	101.0	117.1	1 USA	16.3	22.3	31.9
2 Japan	5.2	14.5	21.3	2 Japan	7.4	10.2	12.2
3 German FR	11.3	16.6	21.2	3 USSR			8.0 approx.
4 France	11.5	13.9	17.4	4 France	2.5	2.0	2.5
5 Italy	8.2	13.4	17.0	5 Canada	1.4	2.2	2.5
6 UK	10.9	13.7	14.7	6 UK	1.7	1.8	1.8
7 Canada	6.2	7.9	9.6	7 Italy	0.8	1.5	1.8
8 Australia	3.4	4.5	6.8	8 Australia	0.9	1.1	1.7
9 Spain	1.6	3.8	6.5	9 German FR	0.9	1.1	1.4
World	170.5	235.3	296.8	World	44.8	58.4	80.8

Source: UNSYB 74, 79/80, Table 147 in 79/80

3.4 Demographic contrasts between countries

Between 1950 and the late 1980s the population of the world will probably have doubled, growing from 2500 million to 5000 million. During the period in question, however, the population of some 'developed' countries will have changed little in size while that of many 'developing' countries will have more than doubled.

Until the 19th century it was common in most societies throughout history for more babies to be born on average per female than the theoretical number of about 2.2 needed to maintain a stable (or zero growth) population. So many infants and young children died before the females among them reached the reproductive age that extra children were needed even to maintain population size. Furthermore it was considered desirable to have one or two male children alive to support their parents in old age. Thus for social as well as biological reasons several children were born to most women and a limit was set to total population size by high levels of mortality in all age groups, caused by diseases and food shortages.

In the more highly industrialized countries of the world several developments have led to a reduction in fertility, reflected in a drop in birthrates. Briefly they include:

(1) The virtual eradication of contagious diseases that killed off children and young people in large numbers.
(2) Increasing knowledge of means of family planning and of ways of applying it.
(3) Provision of state pensions for the elderly, reducing their dependence on their own children and therefore the incentive to produce children for support later in life.

Other influences seem also to have worked in certain countries. In France the conditions of inheritance of property, particularly land, made it desirable for there to be only one son in a family. In China in the 1970s late marriage became common and in India and other South Asian countries sterilization has been encouraged.

In spite of the acceptance by governments in most developing countries that family planning is now an urgent need the application of policies takes time. To families in rural areas in particular it may not seem realistic simply to cut down the number of births. The means to do so are not readily available and the economic conditions are not appropriate. On the other hand widespread improvements in medicine and hygiene have already cut the death rate sharply in many such areas.

The highest rates of population growth in the world in the early 1980s were recorded in some Latin American countries, in much of Africa and in Southwest Asia. To some extent the attitude to family planning of the Roman Catholic church in Latin America and of Islam in northern Africa and parts of southern Asia has discouraged a possible reduction in fertility. The lowest rates of population growth are in Europe, including the western part of the USSR, while growth is somewhat faster in North America, Japan and Australia. International migration had only a slight effect on population change in most countries in the 1970s.

Some demographic characteristics of the 15 largest countries of the world in population and of other selected countries are given in *Table 3.5*. The annual natural change is the difference between birthrate and deathrate (with rounding effects). Death rate, which is related to total population, is affected by the age structure of a population. In the populations of

Table 3.5 *Demographic features of the largest and other selected countries 1980 or late 1970s*

Largest countries	Nat inc 1	BR 2	DR 3	Inf mort 4	Life exp 5	Other selected countries	Nat inc 1	BR 2	DR 3	Inf mort 4	Life exp 5
Nigeria	3.2	50	18	135	48	Kenya	3.9	53	14	87	54
Pakistan	2.8	44	16	126	51	Libya	3.5	47	13	100	55
Viet Nam	2.8	37	9	100	53	Honduras	3.5	47	12	88	57
Bangladesh	2.8	47	19	136	46	Syria	3.8	46	9	62	64
Mexico	2.5	32	6	56	65	Iraq	3.4	47	13	78	55
Brazil	2.4	32	9	77	62	Algeria	3.2	46	14	118	56
India	2.0	35	15	123	49	Venezuela	2.9	34	5	42	66
Indonesia	1.7	34	16	93	48						
China	1.4	22	7	45	65	World	1.7	29	11	85	60
Japan	0.8	14	6	7	76	Canada	0.8	16	7	11	74
USSR	0.8	18	10	36	69	France	0.5	15	10	10	74
USA	0.7	16	9	12	74	Belgium	0.1	13	12	11	73
Italy	0.2	11	10	14	73	Sweden	0.1	12	11	7	75
UK	0.2	14	12	12	73	German DR	0.0	15	14	12	72
German FR	−0.2	10	12	13	72	Austria	0.0	12	12	14	72

Source: Population Reference Bureau: World Population Data Sheet 1981

1 = percentage rate of annual natural increase; *2* = crude birth rates per 1000 population; *3* = crude death rates per 1000 population; *4* = deaths of infants before first birthday per 1000 live births; *5* = life expectancy at birth in years.

developing countries there are so many children and young people (usually over 40 percent under 15 years of age) that older people at higher risk of dying (only on average three percent over 64 years of age) are very few. The comparison in *Figure 3.5* of the population structure of Sweden in 1978 with that of itself in 1751 and with those of Venezuela (1977) and Ivory Coast (1975) illustrates the contrast between the population of developed and developing countries.

When the population of a country is structured by age in the form of that of Venezuela or Ivory Coast, even if the potential mothers of the present suddenly begin to have fewer children, the rate of increase of population does not immediately change by as much as might be expected. This is because the age groups of women leaving their reproductive ages have fewer members than those of girls and young women entering the reproductive ages. The typical broad based triangular pyramid of a developing country needs perhaps 70 years to be transformed smoothly into the more rectangular shaped pyramid of a developed country. If fertility is cut too drastically, as possibly in China in the early 1980s, where one child families have become the recommended official size, then the future age structure could be very seriously distorted with 'bulges' and 'waists' similar to those still found in the population pyramids of some European countries that suffered a large number of deaths in the two World Wars.

An all out nuclear war, a natural disaster of enormous proportions (a meteorite, the sudden melting of ice caps) or simply widespread food shortages *could* upset the fairly predictable path of population change in the world as a whole. If none of these events does occur then the world seems to be heading for a population of at least 6000 million in the year 2000. *Table 3.6* shows actual population and expected population change by major regions of the world in the second half of the 20th century according to the *United Nations Demographic Yearbook*. An estimate in *Global 2000* (p. 9) of 6350 million for 2000 compares with the UN figure of 6100.

During the period 1950–2000 the population of Africa may increase nearly four times. At the other extreme the population of Europe should only increase by about 30 percent. The distribution of population in the world has indeed changed greatly in recent decades with an ever increasing proportion living in the poorer regions.

3.5 Cultural and ideological variations in the world

The dominant position of sovereign states as the most precisely defined and fiercely defended set of territorial units in world affairs should not prevent one from considering many other influences based on

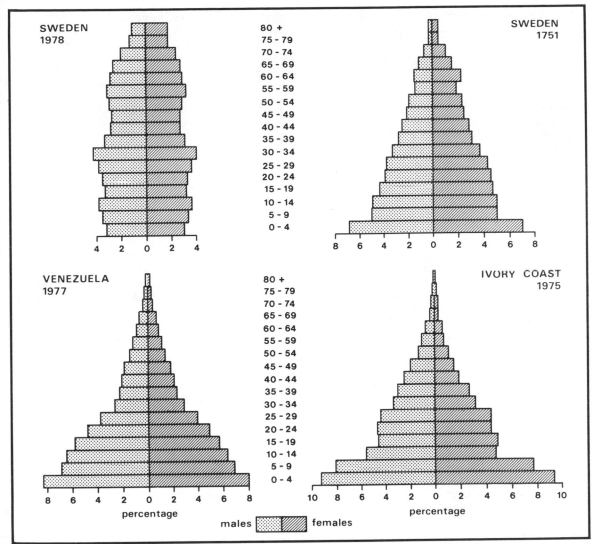

Figure 3.5 *The population structure of Sweden in 1978 and 1751 and of Venezuela in 1977 and Ivory Coast in 1975*

Table 3.6 *The population of major regions of the world*

Region	Popn 1950	Popn 1960	Popn 1970	Popn 1981	Popn 2000	BR 1975–1980	DR 1975–1980	Change 1950–2000	Inf mort
World	2486	2982	3632	4500	6100	28	11	245	97
Africa	217	270	344	486	830	46	17	382	142
North America	166	199	228	254	290	16	9	175	13
Latin America	162	213	283	366	560	32	9	346	75
East Asia	657	780	930	1185	1440	18	6	219	51
South Asia	698	865	1126	1423	2130	37	14	305	120
Europe (excl. USSR)	392	425	462	486	510	14	10	130	17
Oceania	13	16	19	23	30	21	8	230	51
USSR	180	214	243	268	310	18	10	172	36

Popn = population in millions; BR = birth rate; DR = death rate.

human views and endeavours. Four of these features will now be examined: religion, ideology, language and 'race'.

Religion

Although religious movements have played a major part in world affairs they are often difficult to relate to the policy of specific countries. The Revolution against the Shah of Iran in the late 1970s was unusual for recent times because a religious leader became a political one as well.

The boundaries of sovereign states do not often coincide with boundaries between religious groups. Some broad areas of the world contain several or many countries with a particular religious tradition. Latin America, for example, is officially Roman Catholic, though minority religions are now found in most countries and the influence of the Catholic Church itself is much stronger in some countries than in others. Many countries of northern Africa and southern Asia are predominantly Muslim. In some countries of West Europe, Roman Catholics and

Protestants are found in distinct parts of the same national area, but in others they live side by side. India and Pakistan were formed on the basis of majorities of Hindus and Muslims in different parts of British India, though minorities of each religion remain in the territory of the other. Again, Israel could be thought of as a territorial expression of a people bound through history by their religion.

The distribution of the Muslim religion is of particular interest in world affairs at present since it is found in some significant proportion in about 50 countries. As shown in *Figure 3.6* its influence extends from Morocco on the Atlantic to Indonesia and the Philippines on the Pacific. There are minority groups of Muslims in such important countries as the USSR, China and Yugoslavia.

Ideology

As an ideology, if not strictly a religion, communism is one of the major influences in world affairs today. Indeed historians of the future may regard communism as a kind of religion, though one based on

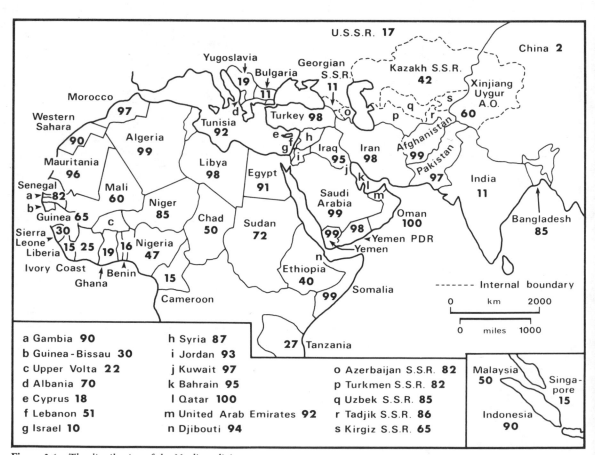

Figure 3.6 *The distribution of the Muslim religion*

material rather than spiritual considerations. The difference could be that conventional religions tend to offer rewards outside this world, where it is impossible to go to test the promises. Communism and other 'ideal' political systems make promises about life in this world and their achievements can be verified and assessed, albeit subjectively.

The Bolshevik Party, later referred to as the Communist Party of the Soviet Union, gained control of Russia in 1917 and was the first Communist Party to hold complete power in any country in the world. Soviet leaders and thinkers on political matters stress that the USSR is 'still' only at a stage of 'advanced socialism'. In view of Soviet influence in world affairs it is important to attempt to see other countries of the world through Soviet eyes on a scale measuring 'level' of communism reached. The degree to which other countries were moving or leaning towards communism might be assessed in various ways. Is a communist party actually in power? If not, what proportion of votes (where applicable) do communist candidates get in elections? What proportion of the means of production of goods and services are in the public sector rather than the private sector? How friendly is the country towards the USSR?

An entirely notional scale of points from 10 to 0 provides a basis for thinking about a possible arrangement of countries, as seen through Soviet eyes. All the countries with a score between 10 and 7 are termed socialist in Soviet publications but they have been differentiated here on grounds of loyalty or allegiance to Soviet 'Communism'. Scores might be USSR 10, Bulgaria and the German DR 9, Romania 8, Yugoslavia 7. China appears to score 7 but Cuba would receive 8 or 9, depending on the fluctuation of Fidel Castro's image in Moscow. Towards the other extreme, France and Italy (perhaps at 3) would have higher scores than the German FR or the UK (at 1) on account of the large Communist vote. The USA (at 0) is at the far end of the scale.

The above example of an ideological variable may seem flippant to the reader but it is a theme that is relevant to world affairs as a result of the apparent conviction in the USSR that communism will prevail. The 10 to 0 scale, or some other such ordering of countries gives a more subtle measure of this particular ideological dimension than simply a 'yes–no' classification of countries. It also serves to illustrate that two countries, like two individuals, may see the same situation differently. Communist China in the 1970s barely discriminated publicly between the USA annd the USSR, describing them as the two most aggressive powers in the world, the USA through its economic imperialism (and its inflexibility over Taiwan), the USSR through 'reactionary socialist imperialism'. The ten point scale

separating the USSR from the USA therefore collapses to nothing in Chinese eyes. In contrast, the USSR in its more aggressive attacks on its fellow socialist country even demotes China to a position alongside the USA and for long periods has tried to ignore its existence altogether.

Language

At first sight language might seem to be a cultural feature of human groups related closely to the countries of the world. A few decades ago there were however about 3000 languages in the world. Many languages are spoken only by very small groups of people (in extreme cases a few thousand) and many are dying out completely. Others, like Arabic, Chinese, English and Spanish, which are all spoken by hundreds of millions of people, are not confined to single countries.

In view of what has been said above, it is difficult to argue that there should be an exact correspondence between countries and languages. In some cases there is a fairly close one-to-one correspondence between a country and a language. Although there are many Poles in the USSR and the USA, modern Poland is the only Polish speaking country in the world and almost all of its population is Polish speaking. Japan and Iceland are the only countries where Japanese and Icelandic respectively are spoken. Several languages are, however, shared by many countries. Spanish and Arabic are each the dominant language in about 20 countries and English in approximately 10, not counting ex-colonies, large and small, such as India and Kenya, in which it is still widely used.

The complex relationship between languages and sovereign countries may be illustrated by the use of English. Between the extremes of Australia at one end and, for example, Chad and Thailand at the other, English is spoken by proportions of the population of different countries ranging virtually from 100 percent to 0 percent. It is widely learned for business, technical or cultural purposes, and may be read or understood, if not spoken, in many countries. Canada, however, is lower on an English speaking scale than Australia on account of the presence of several million French speakers in Quebec. The USA contains many millions of Spanish speaking inhabitants who have recently come permanently or temporarily from Latin America.

In Africa, English is more widely used and known in countries that were British colonies such as Kenya and Nigeria than in former French colonies such as Chad or the Ivory Coast. English seems to be increasing its role as the 'second' language in many countries, to some degree at the expense of French and German. Many countries have two or more widely spoken languages. Athough about 55 percent of the population of the USSR consists of Russians, almost one hundred other languages are spoken in the country and many are recognized by administrative units that fit, as well as possible, the territorial distribution of the speakers. In many of the larger developing countries, in which a former colonial language is only spoken by a limited number of people and no local language is dominant, many languages may be used within the country. Indonesia and the Philippines both contain many small linguistic groups, as does Brazil among the Indians of the Amazon forest.

National groups may feel that their identity and cohesion are based on a common language or on a particular territory. In the UK, the identity of Scotland comes from territorial apartness rather than from a strong cultural distinctiveness from England, whereas the Welsh have perhaps a weaker territorial claim to national status but a living language to support their desire to be culturally different from England. As pointed out in section 3.2, many countries have some minority group that would claim greater regional autonomy if not complete sovereignty, often on linguistic grounds.

Race

Like religion, ideology and language, race may tie people into groups that do not coincide with the territory of particular countries. In 1976 a BBC television interviewer in the streets of Johannesburg was told by a (white) 'man in the street' that the Africans have just come down from the trees. While such a 'well-informed' South African might claim to know the 'truth', scientific evidence regarding the physiological characteristics of the human species seems to indicate otherwise. According to Cavalli-Sforza (1974) the difference between the extremes of human groups is only about one-hundredth as great as that between any human beings and the nearest species of animals, the anthropoid apes. Another sobering thought for the white or Caucasian breeds is that of the three broad 'racial' types, the Caucasians (Europeans) are more similar to Africans and East Asians than these groups are to each other. In other words, on the complex dimension of racial type Europeans are intermediate.

A difficulty in classifying people into races and relating a particular race to a particular country is that there are many different measurable physiological human characteristics. To some extent it has been a matter of opinion and fashion as to which are particularly important or revealing. Cross-section of hair, relationship of width and length of skull, skin colour and many other characteristics have been used for scaling human beings. Once blood groups could be accurately classified the measurement of blood types became a widely used means of sorting people into groups.

This section concludes with some lines from *Race relations in Minas Velhas, a community in the mountain region of Central Brazil*, written by Marvin Harris (1952). The item quoted may at first sight appear offensive to those who would like to see a final scientific proof that skin colour has nothing to do with the intelligence, behaviour or other mental and psychological attributes of an individual. In the view of the author, the important point to appreciate is that whatever scientific evidence there is regarding the abilities and shortcomings of different human groups 'unscientific' views are what affect the running of communities and countries. Thus even in Brazil, which claims to be free from racial prejudice and discrimination, many people genuinely believe the descendants of former slaves from Africa to be inferior to the European settlers.

'In Minas Velhas, the superiority of the white man over the Negro is considered to be a scientific fact as well as the incontrovertible lesson of daily experience. Literacy only serves to reinforce the folk opinion with the usual pseudo scientific re-working into more grammatical and hence more authoritative forms. A school textbook used in Minas Velhas plainly states the case:

> Of all races the white race is the most intelligent, persevering, and the most enterprising . . . The Negro is much more retarded than the others . . .

None of the six urban teachers (who are all, incidentally, white females) could find ground to take exception with this view. They all contended that in their experience the intelligent Negro student was a great rarity. When asked to explain why this should be so, the invariable answer was: *E uma característica da raça negra.* (It is a characteristic of the Negro race.) Only one of the teachers thought that some other factor might be involved, such as the amount of interest which a child's parents took in his schoolwork. But of this she was very uncertain, since the textbook said nothing about it.'

3.6 Defence

The topics covered so far in this chapter illustrate the complexity of world affairs and the lack of precise correspondence between the present countries of the world and various economic, social and cultural features. Even so, each country is very conscious of its territory and limits and will go to great lengths to defend these. Since the Second World War the term defence has become fashionable. The subject of defence will now be discussed but world trade in arms will be referred to in chapter 7.

A country may invest in armed forces either for returns in the form of acquisition of 'foreign' territory with accompanying natural resources and subject population or it may arm to defend itself against other countries should they attack its own territory or that of allies. Nazi Germany armed with the slogan 'guns before butter' and the explicit intention of gaining access to agricultural land, oil fields and other mineral resources in Europe. Japan had even more success in the 1930s in eastern Asia. Both countries were following a precedent long established by European empire building countries. Germany's mistake, or misfortune, was that it had to take on other European countries to achieve its aims. Opposition was much tougher than in Africa half a century earlier and the destruction of war was delivered to neighbours, not to far-away lands.

Since the Second World War military force has rarely been used openly for predatory purposes. Military action has been taken by 'invitation'. The United Nations invited its members to repel the invasion of South Korea by North Korea. The USA became involved in Viet Nam to help a 'democratic' country being attacked by a 'socialist' one. Tanzanian troops entered neighbouring Uganda to rid the country of a dictator, the USSR occupied Afghanistan to help a friendly Communist Party. Iraq attacked Iran to gain an outlet to the sea after attempts to use negotiation had failed, a pretext used also in 1982 by Argentina for occupying the Falkland Islands or Islas Malvinas. Such are the interpretations of the countries involved. The declaration of an 'official' state of war has gone out of fashion.

Since the Second World War the military establishment in each country of the world has generally been referred to respectably as defence. Its existence depends on the production and use of goods and the use of trained manpower, and it provides a service. Goods and services are ordinarily produced to become the means to produce more goods and services, to be consumed directly, or to be exported. Defence is more an investment with a passive role, like an insurance, not, it is hoped, to be needed, but to be drawn upon in an emergency.

The economically active population in the defence business, whether producing arms and equipment or serving directly in the armed forces, produces nothing that others 'consume' but consumes part of the goods and services produced by others. Military equipment may be used for civilian purposes in emergencies as after natural disasters or to replace strikers and even for more conventional purposes such as building roads into uninhabited regions. Research for military purposes may turn out to be useful for civilian activities. The services may even help to absorb unemployed, as in interwar Italy. Even so it is widely argued that manpower and materials used in defence could theoretically be used more productively in other ways. Complex, expensive equipment becomes obsolete and is scrapped and even in peacetime fuel and other materials are consumed to no productive purpose.

Towards the end of the Second World War it was hoped that some peace-keeping world body or force might be enough to save the 'need' for individual countries to arm, thus at once removing the prospect of further armed conflict between countries and saving wasteful production. In reality international law and the resolutions and peace-keeping forces of the United Nations have failed to prevent many conflicts since 1945 and have certainly not been seriously counted upon by the stronger powers of the world. Unfortunately, or perhaps fortunately, there is not a world government with a police force.

Defence today revolves primarily round the USSR and USA, a result perhaps of the way these two countries were attacked respectively by Germany in June and Japan in December of 1941. By 1945 the USSR had suffered far greater material losses than the USA but the message was clear to each never again to be unprepared. As also became obvious after 1945, neither country was seriously in danger any longer of being attacked by a third country. Each had to be prepared for an attack by the other. Such an attack could be indirect, through some other country or countries, or direct. For some years the USA felt capable of deterring a Soviet attack through its possession of nuclear weapons, a monopoly, however, soon threatened when the USSR first tested its own nuclear weapons in 1949. Ever since the Second World War the superpowers have set the pace for defence.

How much is spent on defence?

Before the First World War and in the interwar period about three percent of the total output of the world was spent on defence. After the Second World War the proportion has fluctuated but in 1980 it was about six percent of total world output and was valued in US dollars at more than 500 000 million. This compares with the equivalent, in 1980 dollars, of only 140 000 million in 1950.

The USA and USSR are the two largest single spenders on defence and since the Second World War they have together accounted for around half of total world expenditure. The destructive capacity of their weapons is however much greater than half of the world total.

US defence expenditure in particular and world expenditure in general tended to increase sharply:

(1) In 1949–51 after the lull following the Second World War and with the war in Korea (starting 1950).
(2) In 1960–62 when the Khrushchev–Kennedy rivalry and confrontation grew and the Communist Party gained control in Cuba.
(3) In 1965–68 as a result partly of the Viet Nam conflict.
(4) In 1980–85 it was planned to increase US defence spending by four percent per year in real terms then early in 1982 an increase of as much as 16 percent per year was being considered.

Increases in arms expenditure or the acquisition of some new weapon system in the USA no doubt quickly trigger off a response in the USSR and *vice versa*. In the late 1970s US nerves over Iran and Afghanistan, fear of lagging behind the USSR in nuclear capability, and uncertainty about comparatively new methods of destruction, seem to have revived the US fear of being caught unprepared. Even so the actual US share of total world spending on arms dropped from 36 percent in 1965 to 25 percent in 1979 while the Soviet share rose slightly from 22 to 24 percent.

The military expenditure of NATO and WTO countries is shown in *Table 3.7* for the late 1970s. There are striking differences in the share of gross national product devoted to defence. *Table 3.8* shows that the NATO allies of the USA have increased their share of total NATO spending between 1965 and 1979. Japan is now a major spender in the group of other industrial countries. The data also show that spending has increased greatly in developing world countries, not only absolutely but also relatively. In spite of the 1970s being supposedly the Decade of Disarmament the OPEC countries have lately been increasing defence spending by 15 percent a year in real terms.

What weapons?

For simplicity, weapons may be subdivided into 'conventional' and nuclear. Conventional weapons have been used in all conflicts before and since the atom bombs were dropped on Hiroshima and Nagasaki in 1945 and their presence in virtually every country in the world is critical in world affairs. It is however the nuclear weapons that could produce a war of hitherto unknown destruction. The first nuclear bomb was equivalent to 20000 tonnes of conventional explosive. A single bomb did more damage to Hiroshima than the many bombs dropped during any of the largest single air raids on German and Japanese cities shortly before the end of the Second World War. NATO, WTO and China have the most advanced nuclear weapons and delivery systems and by far the largest quantity. There has been growing concern, however, about the increasing number of countries possessing nuclear weapons. It is feared that these weapons could be used with devastating effect by 'lesser' powers, at least against comparatively small enemies.

In 1954 the US Secretary of State John Foster Dulles enunciated the US strategy of massive retaliation, implying the capability and intention in the event of Soviet aggression of delivering its nuclear warheads to targets in the USSR, to inflict unacceptable damage on the enemy. As Soviet ability to reach targets in the USA improved in the 1950s the USA then had to be able to inflict unacceptable damage on

Table 3.7 *The military expenditures of individual NATO and WTO countries in 1978*

North Atlantic Treaty Organization

Country	Million US dollars	Dollars per head	Percentage of gnp
Belgium	3 140	315	3.5
Canada	3 690	156	1.8
Denmark	1 320	258	2.4
France	15 230	285	3.3
German FR	21 370	347	3.4
Greece	1 520	163	4.7
Italy	6 210	109	2.4
Luxembourg	40	102	1.1
Netherlands	4 320	309	3.3
Norway	1 250	308	3.2
Portugal	540	55	2.8
Turkey	2 030	47	4.5
UK	14 090	252	4.7
USA	105 140	481	5.0

Warsaw Treaty Organization

Country	Million US dollars	Dollars per head	Percentage of gnp
Bulgaria	440	66	2.5
Czechoslovakia	2 340	153	3.8
German DR	4 240	253	5.8
Hungary	810	76	2.4
Poland	3 340	95	3.0
Romania	1 260	58	1.7
USSR	148 000	574	11–14

Source: International Institute for Strategic Studies, The Military Balance, 1979–1980, London, 1979

Table 3.8 *Expenditure on arms in 1965 and 1979*

Country	Percentage of world total	
	1965	1979
USA	36.1	24.7
Other NATO	20.4	18.3
USSR	22.2	23.7
Other WTO	2.2	2.7
Other industrial	4.6	5.4
China	8.5	9.9
Third World	6.3	15.3

Source: Sipri Yearbook 1980

Table 3.9 *Arms available to NATO and WTO in 1981*

Weapons	NATO	WTO
A European 'Theatre' weapons		
1 Medium and short range		
Troops	2.5 million	4 million
Main battle tanks	12 000	40 000
Dual capable aircraft	760	2 350
Nuclear warheads	1 400	1 600
2 Long range		
Bombers	260	400
Warheads (NATO + Cruise and Pershing not yet deployed 1981)	18 (+572)	775
B Inter-Continental Strategic		
Long range bombers	573	156
Submarine launched missiles	704	950
Intercontinental ballistic missiles	1 052	1 398
Warheads	9 000	8 500

Source: Sunday Times 29 November 1981

the USSR even after it had itself been the target of a Soviet nuclear attack. To ensure that its own delivery systems would not be destroyed in a Soviet attack it had to deploy these in aircraft constantly in the air, in (nuclear powered) submarines and in highly protected silos or, possibly, at even greater expense, on mobile launchers capable of being moved randomly around a rail track.

According to Lewis (1979):

'It was calculated by McNamara's system-analysis staff in the 1960s that the reliable delivery of 400 equivalent megatons would kill 30 percent of the population of the USSR and destroy 75 percent of the industrial capacity; more recently the population damage and industrial damage have been estimated to be closer to 35 and 70 percent.'

The USA has some 3200 deliverable equivalent megatons, eight times the amount estimated to be needed to inflict unacceptable damage on the USSR. The Soviet stock of deliverable nuclear warheads is estimated to be even larger than that of the USA.

Although the USSR is more than twice as large in area as the USA and has a somewhat larger population, many of its key industries, including presumably those making arms, are largely concentrated in some 200 cities. While population might be evacuated from cities in anticipation of a nuclear attack, industrial capacity cannot be. Even so, civilian casualties would inevitably be far larger than military ones. The USSR lost about 20 million people, military and civilians, as a result of four years

of war with Germany. It could lose 80 million in four days.

The strength of the military forces of 'west' and 'east' is estimated in *Table 3.9* for NATO (plus France) and for WTO. The data cover forces that are in Europe or could be quickly deployed there. The USA has forces in other parts of the world not counted here, as has the USSR in its eastern regions facing the long border with China. The Soviet side had superiority in 1981 in both conventional forces and Long Range 'Theatre' weapons. In sheer numbers it should 'win' a conflict in Europe fought with conventional weapons. Similarly its own European territory is partly out of range of European based NATO nuclear weapons though not of course of Inter-Continental strategic weapons. Such a gap would be filled by the deployment in the UK, the German FR and other NATO countries of Pershing and Cruise missiles.

Reduce arms?

The USA and no doubt the USSR are both developing weapons that will not be deployed until the 1990s and will be available in the next century. There is talk of the neutron bomb, which kills population but minimizes damage to property, of chemical and biological warfare, and of conflict in space with spacecraft manned with lasers. What are the prospects for disarmament?

Some agreement might have been reached since the Second World War on arms had the USSR agreed with the USA to allow mutual inspection of arms and territory. Typically the USSR is secretive about what is going on inside its boundaries, as anyone hoping to travel there freely will soon find out. In the event, with reconnaissance aircraft and then satellites the USSR has come under close scrutiny from the air by the USA. The only positive progress in limiting nuclear weapons has come in two forms, the Partial Test Ban Treaty of 1963 and the limitation of strategic arms.

In 1963 most but not all of the countries of the world agreed to confine the testing of nuclear weapons to underground sites where the danger of radioactive fall-out would be negligible compared with that from tests in the atmosphere. Neither France nor China signed the treaty but France adhered to it whereas China later carried out tests in the atmosphere. The total number of tests carried out by all countries during 1945–79 is estimated at more than 1200, so underground testing since 1963 has been a positive achievement. In 1980 alone there were some 80 tests.

The USSR and USA have for some time been negotiating the limitation of their strategic arms. While some agreement has been achieved it seems to be of little total importance, since each already has enough warheads to destroy the other's industry and cities many times over. As long as a major nuclear conflict never takes place the two superpowers retain enormous influence in the rest of the world and also the contradiction of security and preparedness vis-a-vis each other. Neither, perhaps, is really willing to lose its powerful position.

In 1968 the Non-Proliferation Treaty was signed by most countries of the world to discourage the spread of nuclear weapons to many new countries. It was supported by the major nuclear powers and mainly by smaller countries unlikely ever to have (or want) nuclear weapons. Among others, the following eight countries did not sign: France, Argentina, Brazil, India, Israel, Pakistan, South Africa and China. In addition to the five countries first to produce nuclear weapons (the USA, USSR, UK, France and China) in 1981 (according to *The Times* 10 June 1981) 13 other countries had nuclear energy programmes capable of producing nuclear weapons in the future and three more were planning to have them:

Argentina	South Africa	India
Brazil	Israel	Taiwan
Chile	Iraq	Philippines
Peru	Iran	
Uruguay	Pakistan	

Planning

Libya	Egypt	Turkey

What kind of damage?

To those who remember the Second World War the atom bombs dropped in 1945 on Hiroshima and Nagasaki in Japan finally put an end to conflict, and nuclear weapons seemed to be an assurance that no-one would risk embarking on another war. The particularly effective conventional air raid on Dresden in Germany early in 1945 was, however, hardly less damaging though it took many aircraft, not just one. Lewis (1979) notes:

'In Dresden, where a firestorm ignited by chemical bombs killed more than 100 000 people, only those inhabitants who had left their shelters before the firestorm began were able to survive the twin threats of noxious gases and shelter heating.'

Most people who remember the Second World War in the UK and certainly in the USA experienced nothing comparable to Dresden. The present author spent the Second World War in Orpington, Kent, southeast of London, the urban district on which the largest number of German bombs fell in the whole of Britain (it was a large district). Life went on in spite of nightly alerts in 1940–41 for several months and continuous ones under V1 and V2 weapons in 1944–45, relieved by frequent cups of tea. One adapted to a particular way of life.

Figure 3.7 shows that things would be very different in London and New York with just a single one-megaton nuclear bomb. It has been estimated that within 4.3 miles of the point where the bomb fell the number of survivors would equal the number of deaths outside the circle but only the first and instant deaths from the initial effects, primarily a shock wave. Depending on how densely the inner circle is populated anything from one to three or four million people could be killed in a very large city. In London the number would be about two million, about 30 times *all* the civilian casualties from air attacks in Britain in six years of the Second World War. Additional energy from the bomb creates 'prompt' nuclear radiation, and 'delayed' thermal and nuclear radiation. Fires and nuclear fallout could take further toll of property and life. The idea that it is better to try to do something rather than nothing by making civil defence preparations seems worthy but naive. Switzerland is reputedly the only country capable of ensuring the survival of a reasonable number of people after a major nuclear attack. One can think of less likeable nations taking over the world after the Third World War.

When the possible impact of one good sized nuclear warhead described is extended to a full scale nuclear war then the prospect is unthinkable. The magazine *Ambio*, environmental journal of the Royal Swedish Academy of Science, reported a calculation of the result (*The Times* 23 June 1982). The consensus of 16 international experts was:

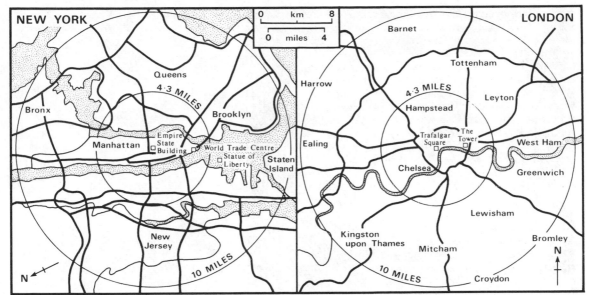

Figure 3.7 *The possible damage in New York and in London from a one-megaton bomb. Based on Lewis (1979)*

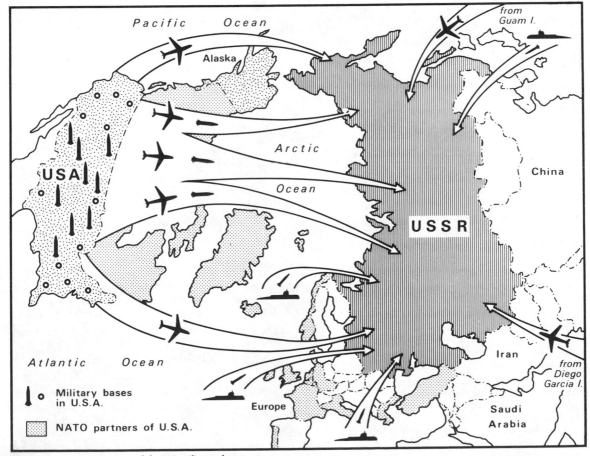

Figure 3.8 *A Soviet view of the US military threat*

54

Caption:
Jenny, did you plant anything on our lawn?
The Englishman suddenly realises there is a
US missile in his garden (with atomic warhead).
Source: Krokadil (Moscow), no 16, June 1981

Figure 3.9 *Soviet cartoons illustrating US military preparations*

Caption:
The mill on which they pour water
The US military and big business sustain a myth
about the Soviet threat in order to keep the
War-Industrial Complex running.
Source: Krokodil (Moscow), no 17, June 1976

'750 million people, roughly half the population of the cities of the northern hemisphere (where the bombing would be concentrated) would die within 24 hours of the exchange, of blast, radiation burns and fire; another 250 million would be destined to die within a few weeks or months from radiation sickness and the expected collapse of organized medical care.

Thus some 1000 million of the total 1300 million population of the urban northern hemisphere would die in the *Ambio* battle or shortly thereafter.'

The rest of the world would be drastically affected in many ways with, for example, food production being greatly reduced through climatic change and loss of fertilizer production capacity.

As argued already, the USA and the USSR have certain advantages in being so powerful militarily, both internationally and internally. Internationally they are able to force or persuade smaller powers to join them in alliances. Internally the arms industry is financially advantageous in the USA to many manufacturers in the private sector and to researchers. The presence of strong armed forces in the USSR provides backing for the Communist Party against any opposition, an asset shown in Poland in December 1981 when the Polish army apparently prevented the collapse of the existing political system.

The world arms situation seems unlikely to change rapidly. Conflicts on a small to moderate scale seem likely to recur indefinitely. Unilateral disarmament by any major country seems improbable. Neutrality implies in some cases arming yourself to the teeth, like Switzerland or Israel. The USA and USSR continue to portray themselves as the bastions of world peace and to point out the threat the other poses to the world. In an unusually glossy and well presented publication in English, the Military Publishing House of the USSR Ministry of Defense produced *Whence the Threat to Peace?* (Moscow 1982) with tables and maps showing the widespread distribution of US forces in the world. *Figure 3.8* is one map, reminiscent of the German map in *Figure 1.12* showing the threat of Czechoslovakia. The Soviet satirical journal *Krokodil* constantly attacks the US industrial and military establishments, as in *Figure 3.9*.

4.1 The assessment of natural resources

The importance of natural resources in the study of world affairs is widely appreciated, but their significance has not been clearly or precisely stated. In this section a viable definition of natural resources will be sought and reasons will be proposed as to why their role in world affairs has not been adequately assessed.

For the purposes of the present book the ingredients of world affairs can mostly be put under six headings:

(1) People themselves (sometimes referred to as human resources).
(2) Natural resources.
(3) Means of production of goods and services (sometimes referred to as capital resources).
(4) Goods and services produced by 3 – means of production.
(5) Transportation links and the movement of goods, people and information.
(6) Organization, the co-ordination of 1–5.

Natural resources mainly fall into two distinct categories:

(1) Bioclimatic resources, the soil, water, temperature, with which or in which plants grow and animals are raised.
(2) Minerals, divisible into fossil fuels and non-fuel minerals.

A particular aspect of the relationship between the six ingredients listed above must be noted. People, natural resources and means of production are combined to produce goods and services. Some of the goods and services (eg food, domestic appliances, education) go back to people as consumer goods and services. Some of the goods (eg tractors, machine tools) become the means of production that turn out more goods and services.

The production of goods and services over a given period such as a year can be assessed in money units. However approximate and debatable the concept of gross national product is in detail, it allows the inclusion together of different products and makes comparison possible between different countries. One can also put a rough monetary value to means of production, expressing for example the construction or replacement cost of a factory or a hospital, though such an assessment is less widely known or used than gross national product. When it comes to assessing the size of natural resources many difficulties arise.

It has been argued by Eyre (1978) that in the two centuries since Adam Smith wrote *The Wealth of Nations* material progress and economic development have usually been associated with the possession of means of production rather than of natural resources. The idea that population growth and large-scale industrialization could at some stage reach a limit has, however, worried various people in the last two centuries. Among these the views of Thomas Malthus, subsequently simplified and even distorted, have been taken to represent the 'pessimistic' view of the limits to growth set by finite natural resources. Not until the 1970s did preoccupation over the limited extent of the earth's natural resources become a matter of widespread investigation and controversy.

One reason, then, for a lack of attempts to assess natural resources has been that globally these have not been considered to have limits, even though locally particular natural resources have in practice been exhausted. Another reason is that it is much more difficult to quantify natural resources than means of production or production itself. Some natural resources such as fossil fuels and fertilizer minerals are exhaustible or non-renewable, while others, such as water, are virtually indestructible; some are intermediate. Inherent soil fertility is ultimately exhaustible if a piece of land is cultivated over a very long period and not replenished with nutrients. Metallic ores are exhaustible, but some metals can be recycled or reused.

The exact extent of natural resources is not calculable because some have not yet been carefully assessed while others, particularly mineral resources, presumably have not even all been discovered yet. Others still are known to exist, as for example a measurable quantity of solar energy falling on the earth's surface, but so far have hardly been directly used at all.

There are considerable difficulties in comparing different types of natural resource and in allowing for those that have still to be discovered or exist but are not yet used. Even so it is evident that some regions of the world are far better endowed with natural resources than others. On average, for example, each inhabitant of Australia has to his name much more cultivable land and many times more fossil fuels and non-fuel minerals than each inhabitant of Bangladesh.

During the last two centuries the highly industrialized countries of the world have either developed largely on the basis of their own natural resources, as the USA has done, or have been able to import food, fuels and raw materials. The UK, France, Russia and Japan have controlled colonies and drawn on the natural resources of these. Germany and Italy were two countries that lacked natural resources at home and only had colonies of limited value. The attempt of Germany to gain control of natural resources in the USSR failed.

In the 1970s natural resources became a major issue. Many developing countries were industrializing and traditional exporters of primary products were increasingly using their own materials and even becoming importers. The special situation in the world oil industry also gave publicity to the matter and the comparatively short life expectancy of proved oil reserves suddenly became appreciated. In a study of the geography of world affairs it is necessary to have at least a rough idea of the nature and size of the natural resources of each major country.

The main types of natural resource will be dealt with in the sections that follow and their world distribution will be given particular attention. One possible subdivision of natural resources is as follows

(1) Cropland
(2) (*a*) Pasture
    (*b*) Forest
(3) (*a*) Water
    (*b*) Direct sun energy
(4) Fossil and nuclear fuels
(5) Mineral raw materials

The above seven items could be broken down and/or grouped differently. Thus they could be related back to their occurrence in the lithosphere, hydrosphere, atmosphere and biosphere, or they could be related forward to their more direct application to the needs of man, as water (fresh and salt), food (from cropland or pasture), energy (from fuels, sun or falling water) and raw materials (derived from cropland, livestock, forest, fossil fuels or mineral raw materials).

4.2 Bioclimatic resources

In the early 1980s about 11 percent of the earth's land surface was defined as being arable or under permanent crops. Another 55 percent consisted of permanent pasture (24 percent) and forest (31 percent). The remaining third of the land was either used for buildings, roads and similar constructions, or was not used at all. The above categories and their areas are only very rough. For example some forest land is also used for grazing, while trees may be planted as 'permanent crops' in areas of cultivation. Further sources of plant and animal products are the lakes and rivers on the land and the oceans, mainly the shallow offshore seas.

The Food and Agriculture Organization publishes estimates of the area under the various land use categories given above for most countries of the world. The areas of each category for ten regions of the world are given in *Table 4.1*. From the table it can be seen that some regions have far more arable, pasture and/or forest per inhabitant (*see* total population in column 6) than others. Oceania, in effect Australia and New Zealand, has about 25 hectares of arable land per inhabitant, China and its neighbouring socialist countries less than one-tenth of a hectare. The disparity of about 1:250 needs qualification because yields tend to be somewhat higher in China than in Australia, while a considerable part of the arable land in China is cropped twice a year and some even three times. Even so the contrast is enormous.

It is the purpose of this section to give a brief description of different types of natural environment in the world. These have been determined by climatic influences, particularly temperature and rainfall, and by underlying rock, on which soil has formed. Various combinations of different plants and animals have evolved together as a result of the above influences and they reflect to a large extent recent and present climatic conditions. Much of the natural environment has, however, been greatly modified by man through clearance for cultivation, grazing or the gathering of wood as a raw material or a fuel. The stability of present land uses and the prospect of transforming nature further in order to bring new areas under cultivation can be better understood if the main features of natural environments are taken into account.

When natural vegetation types are represented on a small map of the whole world, great simplification is needed. *Figure 4.1* shows one of many versions of natural regions to be found in various atlases and textbooks. Although the limits of the distribution of particular types of vegetation are shown by lines on the map, in reality there is usually no abrupt transition from one type to another. The isotherm of mean annual temperature of 18°C is shown on the map rather than the artificial geometrical lines of the two tropics.

(1) The *tropical and temperate rain forest* occurs in the hotter parts of the world, where rainfall is heavy and lasts most if not all the year. The natural forest cover consists often of hundreds of different species of tree in any given small area. There is not necessarily a very dense undergrowth. The soil under the forest is often very limited in fertility and when cleared of plant cover its nutrients are easily washed away. The largest areas of this type of vegetation are in the Amazon region of South America, in Central Africa and in Southeast Asia.

(2) *Savanna vegetation* consists of a low, grassland vegetation with occasional patches of forest or individual trees. Rainfall is seasonal, only moderate and often unreliable. Cultivation is limited and may not be permanent. Africa has extensive areas of savanna, which include many famous wild life reserves.

(3) *Sclerophyllous vegetation and thorn scrub* consist mainly of trees of limited size and of shrubs. Together these form a comparatively dense but low vegetation, able to withstand long periods of drought. They are distinct because the areas of schlerophyllous vegetation are associated with dry conditions during the hot season while the thorn scrub is in areas of transition between savanna and desert and may not have a marked rainy season.

(4) Most areas of *semi-desert and even desert* support some vegetation. This consists either of very hardy perennial dwarf trees and shrubs or of herbaceous plants that flourish for brief spells after rain. Parts of the surface of the deserts consist of shifting sand or bare rock. Little use can be made of semi-desert or desert areas for

Table 4.1 *Major land use types in ten regions of the world around 1980*

Region	Total area 1	Arable 2	Pasture 3	Forest 4	Other 5	Popn 6	% agric 7
North America	1 835	236	266	616	717	242	2.7
West Europe[a]	373	96	72	126	80	369	11.4
Oceania	789	43	460	114	171	17	6.8
Other developed market economies[b]	161	20	83	30	29	146	14.7
USSR and East Europe	2 327	278	390	949	710	372	21.5
Africa[c]	2 330	142	694	540	954	355	70.5
Latin America	2 020	142	534	1021	324	349	35.4
Near East[d]	1 192	81	267	140	704	206	55.4
Far East	809	264	35	315	196	1186	63.5
China and other centrally planned economies[e]	1 149	112	350	165	522	1010	61.8
World	13 074	1414	3151	4057	4452	4258	46.4

Source: FAO Production Yearbook vol 33; FAO Rome 1980 Tables 1 and 2

1, 2, 3, 4, 5 = area in millions of hectares; *6* = population in millions; *7* = percentage of economically active population in agriculture in 1978.

[a] includes Yugoslavia
[b] Israel, Japan, South Africa
[c] excludes South Africa, Egypt, Libya, Sudan
[d] includes Egypt, Libya and Sudan and extends east as far as Iran and Afghanistan
[e] China, Korean DPR, Mongolia, Viet Nam

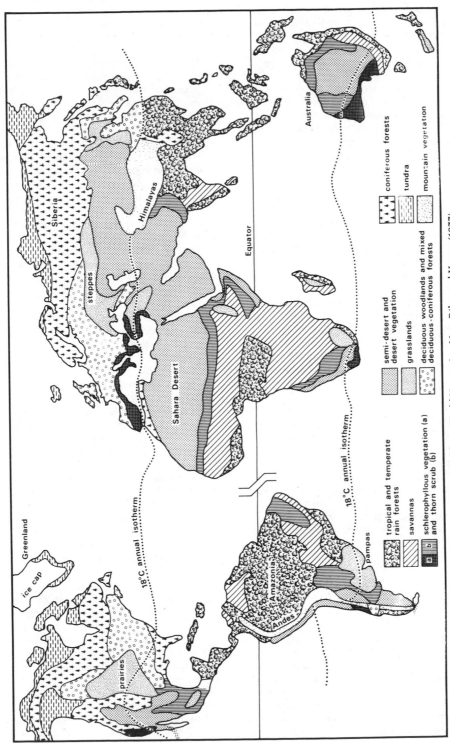

Figure 4.1 *Natural vegetation regions of the world simplified from World Vegetation Map, Riley and Young (1977)*

tropical and temperate rain forests

savannas

schlerophyllous vegetation (a) and thorn scrub (b)

semi-desert and desert vegetation

grasslands

deciduous woodlands and mixed deciduous-coniferous forests

coniferous forests

tundra

mountain vegetation

Greenland
ice cap

prairies

pampas

Amazonia

Andes

Sahara Desert

steppes

Siberia

Himalayas

Australia

Equator

18°C. annual isotherm

18°C. annual isotherm

purposes of cultivation or livestock raising unless water is available for irrigation. The largest expanse of desert is in Africa and extends from the Atlantic coast to the Red Sea, continuing across Southwest Asia. Middle Asia, Australia and South America also have extensive areas of desert.

(5) In many *warm temperate and subtropical* areas with a moderate rainfall extensive areas of grassland are found. These include the prairies of North America, the steppes of the USSR and the pampas of Argentina. Such areas often have deep fertile soils, rich in humus, and are widely cultivated.

(6) *Deciduous woodlands*, in places intermixed with areas of coniferous trees, cover much of Europe and eastern North America. With suitable conditions of soil and slope such areas have been cleared for cultivation.

(7) *Coniferous forests* cover extensive parts of northern North America, Europe and Asia. In any given area often only a few types such as species of the pine, spruce or larch, predominate. Poor soils, bad drainage and low temperatures limit the use of the coniferous forest zone for agriculture.

(8) The *tundra* of North America and Siberia is a large area with a very scant vegetation. Low temperatures preclude cultivation and in many areas the moisture in the subsoil remains permanently frozen.

(9) Only two major mountain areas are shown in *Figure 4.1*, though many others would be included on larger scale maps. As a result of the reduction of temperature with altitude, at 2–3000 metres above sea level in or near the tropics there is a cooler environment with vegetation characteristic of higher latitudes. The Himalayas in Asia and the Andes in South America contain a great diversity of vegetation, from tropical forest at their bases to permanent snow at their summits. They also form barriers to the movement of air masses.

Figure 4.1 also shows that nearly all of Greenland is covered by an ice cap. So also is Antarctica, not shown on account of its distorted representation along the lower edge of the map.

The capacity of a given piece of land to produce plants and animals varies initially according to its inherent natural qualities and secondly on modifications made by man such as ploughing, the application of fertilizers, and irrigation or drainage. To assess precisely the productive value of different kinds of soil even locally in a small area is extremely difficult. For whole countries on a world scale one can only make a rough estimate of quality. In order to assess the bioclimatic or land resources of the countries of the world a subjective estimate was made by the author. The method and its drawbacks are noted below.

The 60 largest countries of the world in the late 1970s were considered:

(1) The area of arable land plus permanent crops (eg tea, vines) was taken for each country. The quality of the land was then estimated and the area 'increased' in some countries to allow for greater inherent productivity above a minimum level. Thus for example the arable area of the USSR, Mexico and Iran were left unchanged. That in the USA, Argentina and some west European countries was increased by 50 percent. Countries such as China and Indonesia, where much of the cultivated land is irrigated and double cropped had their areas doubled. Japan, Egypt and the Netherlands were weighted even more. Each country ended up with a given number of arable land points.

(2) Areas of permanent pasture and of forest for each country were then taken. The value of plant and animal products from these areas is much lower per unit of area than the value of products from arable land. Some of the pasture and forest lands may however be brought into cultivation in the future. A distinction could have been made with regard to the quality of pasture and of forest in different places but for simplicity both types were considered to be of uniform productivity and each was reduced to a tenth of its area to reflect its low productivity compared with that of arable. Each country was then credited with its score of points for pasture and forest.

(3) The three land unit points for arable, pasture and forest were combined to give a global land or bioclimatic resource score for each country. Since a point does not represent an exact area the scores are notional. To facilitate comparison of the bioclimatic resources with other kinds of natural resource it was considered best to calculate each country's share of a grand total for the 60 countries studied. *Table 4.2* shows the bioclimatic resource scores of the 15 largest countries of the world in population and also the 15 countries with the largest bioclimatic scores. From the data it is evident that some countries are better endowed than others with land resources. For example the UK and France have roughly the same number of inhabitants so France has about 2.5 times as much per inhabitant as the UK.

The data in *Table 4.3* show that there has been a remarkable reduction in the area defined as arable (and permanent crops) in relation to population in a mere 15 years. The total area in the world defined as arable (*see* columns 2 and 3) increased by only six percent while population increased by 35 percent. As

Table 4.2 *Bioclimatic resource points in per thousands of the total for sixty countries*

15 largest countries in population		15 countries with largest quantity of bioclimatic resources	
Country	Points	Country	Points
1 China	105	1 USA	155
2 India	126	2 USSR	150
3 USSR	150	3 India	126
4 USA	155	4 China	104
5 Indonesia	19	5 Canada	33
6 Brazil	26	6 Argentina	31
7 Japan	9	7 Brazil	26
8 Bangladesh	9	8 Australia	23
9 Pakistan	15	9 Indonesia	19
10 Nigeria	13	10 Thailand	18
11 Mexico	14	11 Spain	17
12 German FR	7	12 France	15
13 Italy	10	13 Turkey	15
14 UK	6	14 Pakistan	15
15 France	15	15 Mexico	14

a result between 1963 and 1978 the area of arable land per 100 inhabitants in the world as a whole decreased from 42 to 33. Malthusianism is happening right under our noses.

What reservations may be made about the trend? On the whole yields have been higher per unit of area in the late 1970s than in the early 1960s. On the other hand the new arable land added to the world total either tends to be inferior in fertility or in climatic conditions to that already in use or is very costly to reclaim, as for example by irrigation or drainage works or even with terracing.

Of the 20 countries chosen for inclusion in the table the arable area per inhabitant decreased in all but one, Argentina. In many poorer countries the rapid increase in population size is the cause of decline. In some richer countries marginal land has actually been withdrawn from cultivation and, notably in Japan and the Netherlands, urban and industrial buildings and premises eat into the best farmland because there is nowhere else suitable to expand. In the view of the author the possession of ample arable land will be a valuable asset for a country to have in the future.

Table 4.3 *Arable land and population in selected countries 1963–1978*

Country	Total area 1	1961–1965 2	1978 3	% arable 1961–1965 4	%arable 1978 5	1963 6	1978 7	1963 8	1978 9
USA	913	180.3	191.5	20	21	189.2	218.1	95	88
Canada	922	40.7	44.3	5	5	17.0	23.5	239	189
Australia	769	33.7	42.7	4	6	11.0	14.2	306	301
Argentina	278	28.1	35.1	10	13	21.6	26.4	130	133
USSR	2 227	229.5	231.8	10	10	225.1	261.6	102	89
UK	24	7.4	7.0	31	29	53.7	55.8	14	13
France	55	21.1	18.9	38	34	47.8	53.3	44	35
Italy	30	15.5	12.4	52	41	51.2	56.7	30	22
Poland	31	16.0	15.0	52	48	30.7	35.0	52	43
Japan	37	5.9	4.9	16	13	96.8	114.9	6	4
Brazil	846	30.3	40.7	4	5	76.4	115.5	40	35
Mexico	202	23.6	23.2	12	11	38.6	66.9	61	35
Nigeria	92	22.2	24.0	24	26	46.0	72.2	48	33
South Africa	122	12.6	14.6	10	12	18.5	27.7	68	53
Algeria	238	6.9	7.5	3	3	11.2	18.5	62	41
Turkey	78	25.8	28.0	33	36	29.7	43.2	87	65
India	297	162.0	168.5	55	57	462.2	638.4	35	26
Thailand	51	12.6	17.5	25	34	29.2	45.1	43	39
Indonesia	181	14.4	16.4	8	9	99.8	145.1	14	11
China	960	104.5	99.6	11	10	720.0	933.0	15	11
World	13 307	1333.2	1414.2	10	11	3160	4258	42	33

Sources: Areas FAOPY 1979 Rome 1980 Table 1. Populations UNDYB 1979 Historical Supplement, Table 1

1, 2, 3 = millions of hectares, *1* total area, *2* and *3* arable areas plus permanent crops; *4, 5* = arable as % of total area in *4* 1961–65 and *5* 1978; *6, 7* = population in millions; *8, 9* = arable area in hectares per 100 inhabitants.

4.3 Fossil and nuclear fuels

In the early 1980s about 90 percent of all the energy consumed in the world came from the fossil fuels coal, lignite, oil and natural gas. Oil shales and tar sands as yet make only a very small contribution to world fuel supply. Other types of energy include both inanimate sources such as peat, firewood, hydro-electric power and nuclear fuels and animate energy provided by work animals and human work. The prospects are that the relative importance of some 'new' sources of energy, especially 'renewable' ones, such as wind, wave and direct solar power, may increase in the future. Even so, the world energy situation for the next two or three decades is likely still to be dominated by fossil fuels.

Each type of fossil fuel has particular uses. While all types can be used to generate electricity, oil is at the moment commercially far cheaper to use as a source of petrol than any of the others. Fortunately, however, the various types of fossil fuel can be compared in terms of thermal equivalent. The United Nations conventionally measures different fuels in terms of coal equivalent. Not surprisingly British Petroleum uses oil equivalent.

Two major problems make the estimation of world fossil fuel reserves difficult. Firstly, there is no objective way of estimating or allowing for the comparative importance of proved, probable and possible reserves or of as yet completely undiscovered reserves. Secondly, there is much secrecy about the size of reserves and in particular official estimates of the amount of oil that could be extracted seem to be understated.

The *United Nations Statistical Yearbook* publishes estimates annually of the reserves of coal, lignite, oil and natural gas by country. Coal and lignite reserves consist of proved commercially usable reserves, probable ones and notional ones. In the assessment of fossil fuel resources used in the present book greater weight was given to the economic reserves of coal and lignite than to the possible ones. With oil and natural gas only one estimate is given, but possible reserves are probably far greater than the 'official' ones published. In addition it is known that large quantities of oil are trapped in oil shales and tar sands but are very costly to extract and in 1982 two major fuel producing companies abandoned experimental production projects. Even larger reserves of combustible gas probably exist deep in the earth's crust. Hodgson (1978) estimates that geopressure systems exist in some 400 000 square kilometres of porous shale and sandstone, saturated with hot brine at very high pressures and containing enough combustible gas to last 2500 years at present yearly rates of production. Deep gas and shale and tar sand deposits are not included in the estimates used in this section.

Whatever the true situation there is no doubt that the reasonably accessible reserves of coal and lignite are far larger than the official oil and gas reserves in terms of coal (or oil) equivalent. At rates of consumption current around 1980 official oil and gas

Table 4.4 *Fossil fuel reserve points in per thousands of the total for sixty countries*

15 largest countries in population		15 countries with largest fossil fuel reserves	
Country	Points	Country	Points
1 China	78	1 USSR	303
2 India	7	2 USA	183
3 USSR	303	3 China	78
4 USA	183	4 Saudi Arabia[a]	
5 Indonesia	8	5 Kuwait[a]	
6 Brazil	1	6 Iran	70
7 Japan	1	7 German FR	21
8 Bangladesh	0	8 Iraq	21
9 Pakistan	0	9 United Arab Emirates[a]	
10 Nigeria	11	10 Libya[a]	
11 Mexico	5	11 UK	18
12 German FR	21	12 Algeria	16
13 Italy	1	13 Australia	15
14 UK	18	14 Venezuela	14
15 France	1	15 Nigeria	11

[a] Not one of the 60 largest countries of the world.

Table 4.5 *Nuclear fuels: uranium reserves in thousands of tonnes in 1979*

Country	Reserves	Country	Reserves
1 USA	531	7 Brazil	74
2 Australia	290	8 France	40
3 South Africa	247	9 Gabon	37
4 Canada	215	10 India	30
5 Niger	160	11 Algeria	28
6 Namibia	117	12 Argentina	23

Source: UNSYB 79/80

reserves should only last a few decades while coal and lignite could last several centuries. Since the 1950s, however, oil and natural gas together have made a larger contribution to total world fuel consumption than coal and lignite.

World energy reserves are of relevance to world affairs in two distinct ways. Firstly, fossil fuel reserves are very unevenly distributed among the countries of the world. Two countries with large populations but virtually no fossil fuel reserves at all are Bangladesh and Japan. Some countries, including for example Kuwait, Saudi Arabia and Libya, have small populations but very large oil reserves. Secondly, overall world consumption of oil and gas has risen rapidly in the period between the Second World War and the 1970s. If it continues to rise in a similar fashion in the 1980s and 1990s then the more accessible reserves will be depleted or used up in the not too distant future. If consumption stays at the level of the 1970s the reserves will last considerably longer. Even so Flower (1978) expects a decline in oil production early in the 21st century.

A subjective assessment has been made by the author of the quantity of fossil fuel reserves in each of the 60 largest countries of the world in population.

Table 4.4 shows the share of world fossil fuel reserves in the 15 largest countries of the world in the late 1970s and also in the 15 countries of the world with the largest reserves.

From the data in *Table 4.4* it is clear that some countries have virtually no fossil fuel reserves at all while others are very well endowed. According to the criteria taken by the author in the assessment of fossil fuel reserves the USA and USSR together have nearly half of the world's total. In the USA oil and natural gas reserves do not have long to last. Abundant coal reserves are available in the western part of the country though ecological and locational problems make their extraction a problem. In the USSR, many of the considerable oil reserves and large natural gas reserves are in remote parts of the country where they are difficult to reach and extract, as are the enormous possible coal deposits of Siberia.

Southwest Asia and northern Africa account for more than half of the world's oil reserves. More modest reserves of fossil fuels, in relation to population, ensure that the German FR and the UK have a backup of home reserves. On the other hand, as *Table 4.4* shows, several major world powers are very poorly provided with fossil fuels and either like Japan and France import fuels or like Pakistan and Bangladesh, have very low levels of consumption.

Though not strictly a fossil fuel, uranium is also a non-renewable source of energy and it is likely to contribute increasingly to the generation of electricity in the future. The size of Soviet uranium reserves is not made public but the data in *Table 4.5* show there to be a concentration of reserves elsewhere in the world, with six or seven countries apparently in a very strong position.

In view of the key role of oil in the world economy it is interesting to conclude this section with a review of oil reserves. *Table 4.6* shows all countries credited with more than one percent of the world's oil reserves. Such reserves are constantly under revision

Table 4.6 *Sixteen countries each with over one percent of world oil reserves. Reserves in millions of tonnes in 1978*

Country	Reserves	Country	Reserves
1 Saudi Arabia	15 910	10 China	2 740
2 Kuwait	10 180	11 Venezuela	2 490
3 USSR	7 990	12 Nigeria	1 680
4 Iran	6 150	13 UK	1 390
5 Iraq	4 700	14 Algeria	1 310
6 United Arab Emirates	4 320	15 Indonesia	1 070
7 Mexico	3 880	16 Canada	790
8 USA	3 800		
9 Libya	3 720	World	77 690

Source: UNSYB 79/80 Table 52

as oil is extracted and new reserves located. Some reserves may be upgraded as new methods of extraction are applied to reduce the percentage left in the ground. *Figure 4.2* compares the amount of oil produced in the world between 1859 (the first oil well, Pennsylvania) and 1980 with the oil still in the ground.

According to the British Petroleum estimates in *Figure 4.2* about 60 000 million tonnes of oil have been extracted since 1859. What is striking is that 47 percent of the total extracted, nearly 30 000 million tonnes, were extracted in the 1970s (in spite of oil price rises). The situation can be simplified to say that it took barely one-tenth as long to extract (and consume) the second half of the total extracted since 1859 as the first half.

United Nations estimates of world oil reserves around 1980 are about 75 000 million tonnes while British Petroleum estimates are about 86 000 million. If oil consumption continued at the level of the 1970s and no new oil were discovered, reserves would

hardly last until the year 2010. While some industrial countries appear to be consuming less oil in the early 1980s than a few years previously, many developing countries hope for fast economic growth and have fast growing populations. They could benefit from much higher consumption levels of oil.

Whatever the true situation not everyone was pessimistic in the early 1980s. Mackrell (1980) of 'Shell' Transport and Trading Company noted in a speech to shareholders:

'Oil will remain of vital importance – by the year 2000 it will still provide more than half of the global demand for energy, though supplies are likely to be tight . . . There is no problem about the existence of the main fossil fuels. Only a very small proportion of the world's reserves have been produced so far . . . about 15 percent of oil and gas reserves and about 5 percent of coal.'

Shareholders could return home with some confidence especially in view of Shell's growing involvement in the coal business.

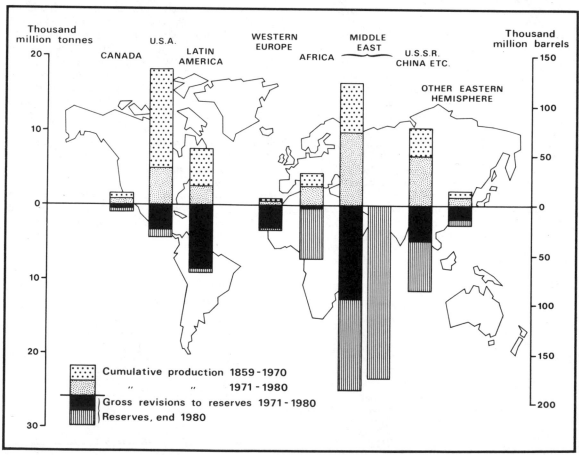

Figure 4.2   *Oil production and reserves since 1859 by regions of the world. Source: The British Petroleum Company Limited, BP Statistical review of the world oil industry 1980*

4.4 Non-fuel minerals

It was evident from the preceding studies of the distribution of bioclimatic resources and of fossil reserves that both these resources are very unevenly distributed over the earth's surface. They are also very unevenly distributed according to population. Oil reserves are highly concentrated in certain developing countries while coal reserves are mostly found in a few developed ones. Most types of non-fuel mineral are similarly concentrated in a small number of countries.

Whereas present day climatic conditions greatly affect the distribution of bioclimatic resources, the distribution of most non-fuel minerals is closely related to very long-term processes in the earth's crust. A simple account of tectonic place theory is given by Dewey (1972) and the processes and their implications will not be described here. The study of tectonic plates has led to the view (eg of Rona (1973)) that the occurrence of certain types of mineral is closely associated with particular structural features on the margins of some tectonic plates. Some parts of the earth's surface are therefore much more likely to have mineral reserves than others.

While the existing distribution of continents, oceans and 'new' mountain ranges is the result of tens of millions of years of change, the process of tectonic plate formation and movement has gone on at least for hundreds of millions of years. Thus older mountain regions, now often much worn down with erosion, also contain many minerals of economic significance. An example is the Ural Range in the USSR. Five main 'shield' areas of old rock have traditionally been regarded as sources of various non-fuel minerals. They are in Canada, Siberia, Brazil, southern Africa and Australia.

If the distribution of reserves of various non-fuel minerals in the earth's crust is very complex, so also is the problem of arriving at an estimate of total world non-fuel mineral resources. As with fossil fuels it is not possible to make more than a subjective guess as to the extent of undiscovered deposits. In contrast to fossil fuels, which can be reduced to a common unit of measurement in terms of energy, non-fuel minerals differ greatly in their composition and also their use, adding to the difficulty of assessing reserves. Given the great difference in value between for example a tonne of phosphates and a tonne of silver the comparison of different minerals by their weight is meaningless.

In order to make a rough assessment of the importance of the reserves of each type of non-fuel mineral it was considered best to relate the importance in the world economy of each major mineral to the total *value* of its production in a given year. The above assumption is however complicated by the fact that the relative importance of different non-fuel minerals changes over time. New minerals, like aluminium in the 20th century, come into widespread use and others, like nitrates or asbestos, decline relatively through the gradual exhaustion of accessible deposits or some technical or ecological obstacle to their use.

From data published by the US Bureau of Mines (1970) it could be calculated that the relative value of production of key minerals was very roughly as follows:

10 copper, aluminium, iron.
 4 lead, gold, nickel, tin, zinc, sulphur.
 2 silver, industrial diamonds, phosphates, potash.
 1 tungsten, molybdenum, mercury, chromium, asbestos, antimony and cobalt.

With the help of information provided by Kessler (1976) and Crowson (1980) it was possible to arrive at a global estimate of non-fuel mineral 'points' for the 60 largest countries of the world. The share of world reserves in each major source country of each of the non-fuel minerals listed above was weighted by the ratios indicated above. The proportion of the world total of the non-fuel minerals possessed by each country could then be calculated. *Table 4.7* shows the share in each of the 15 largest countries of the world in population and in the 15 countries with the largest shares of non-fuel minerals in relation to the total for the 60 largest countries in population.

Table 4.7 *Non-fuel mineral reserve points in per thousands of the total for the sixty largest countries of the world in population*

15 largest countries in population		15 countries with largest resources	
Country	Points	Country	Points
1 China	41	1 USSR	144
2 India	23	2 USA	115
3 USSR	144	3 Canada	103
4 USA	115	4 Australia	90
5 Indonesia	22	5 South Africa	77
6 Brazil	51	6 Brazil	51
7 Japan	3	7 China	41
8 Bangladesh	0	8 Chile	38
9 Pakistan	0	9 Zaire	37
10 Nigeria	0	10 Mexico	24
11 Mexico	24	11 India	23
12 German FR	6	12 Indonesia	22
13 Italy	2	13 Morocco	22
14 UK	2	14 Peru	21
15 France	0	15 Poland	20

Source: Based on data in Crowson (1980)

Table 4.8 *The mineral reserves of five major sources of minerals in percentages of the totals for the sixty largest countries of the world in population*

	USA	Canada	Australia	South Africa	USSR	Total
Coal						
known recoverable	36.9	1.2	3.3	2.5	19.3	63.2
total resources	28.1	1.2	1.4	0.5	49.1	80.3
Lignite						
known recoverable	14.4	0.3	6.3	–	33.4	54.5
total resources	24.3	0.5	3.3	–	65.4	93.5
Oil	4.9	1.1	0.4	–	9.6	16.0
Natural gas	7.2	3.3	1.3	–	34.8	46.6
Uranium	29.6	13.3	22.5	17.2	?	(82.6)
Iron ore	3.9	11.7	11.5	1.2	30.2	58.5
Copper	19.5	6.4	x	x	7.3	33.2
Bauxite	1.0	–	24.3	–	5.0	30.3
Lead	20.5	9.4	13.4	3.9	12.6	59.8
Gold	9.2	3.8	2.0	48.5	20.0	83.5
Nickel	x	14.4	9.3	2.8	9.6	36.1
Tin	–	–	3.2	–	6.1	9.4
Zinc	18.1	14.2	12.3	–	7.0	51.6
Sulphur	37.0	6.0	x	–	13.0	56.0
Manganese	–		8.7	42.8	37.9	89.4
Silver	24.7	11.6	3.2	–	20.0	59.5
Industrial diamonds	–	–	new	7.4	3.7	11.1
Phosphate	8.1	–	–	11.1	5.2	24.4
Potash	1.5	75.7	–	–	15.1	92.3
Tungsten	6.2	10.7	3.8	–	8.0	28.7
Molybdenum	43.3	7.5	–	–	8.6	59.4
Mercury	6.8	–	–	–	9.7	16.5
Chromium	x	x	–	67.6	9.0	76.6
Asbestos	4.6	42.5	–	5.7	30.0	82.8
Antimony	3.0	–	2.5	7.2	6.2	18.9
Cobalt	–	8.0	3.3	–	8.0	19.3
Platinum group	1.0	4.0	–	47.0	47.0	99.0
Population	5.1	0.5	0.3	0.6	6.0	12.5

x small reserves;   – no reserves of note

The estimates of non-fuel minerals are only very approximate due to the lack of straightforward comparability between different types of mineral and to the rough nature of estimates of reserves. Moreover some countries, such as the UK and Japan, have been much more intensively explored than others, such as Brazil and Zaire. Whatever exact criteria are taken for assessing non-fuel minerals, however, some countries of the world are far better endowed than others in relation either to area size or to population size or both.

From *Table 4.7* (column 2) it is evident that about half of the non-fuel mineral resources are concentrated in five countries. These countries do admittedly cover about 40 percent of the world's land area but together they have only about 12 percent of the population of the world. A more detailed breakdown of their estimated shares of both fossil fuels and non-fuel minerals is given in *Table 4.8*. On the other hand several countries that are large in population but small in area have few or virtually no proved reserves of non-fuel minerals. These include two types of country, industrialized countries like Japan and France, which can import non-fuel minerals from other countries, and countries like Bangladesh and Nigeria with very limited industrial capacity and little prospect either of producing their own non-fuel mineals or of importing them.

When the composition of the non-fuel mineral reserves of individual countries is examined in detail it is evident that some countries score mainly from one or two types of mineral while others have a broader profile. Although, for example, Chile has iron ore, nitrates and other commercial minerals its

Table 4.9 *Five countries having largest shares of world total of reserves of selected minerals*

Country	Copper A	B	Country	Alumina/bauxite A	B
1 USA	19.5	19.5	1 Guinea (West Africa)	31.1	31.1
2 Chile	19.5	39.0	2 Australia	24.3	55.4
3 USSR	7.3	46.3	3 Brazil	9.5	64.9
4 Zambia	6.7	53.0	4 Jamaica	7.6	72.5
5 Peru	6.4	59.4	5 India	6.1	78.6

Country	Iron A	B	Country	Lead A	B
1 USSR	30.2	30.2	1 USA	20.5	20.5
2 Brazil	17.5	47.7	2 Australia	13.4	33.9
3 Canada	11.7	59.4	3 USSR	12.6	46.5
4 Australia	11.5	70.9	4 Canada	9.4	55.9
5 India	6.0	76.9	5 Mexico	3.9	59.8
			(South Africa	3.9	63.7)

Country	Nickel A	B	Country	Tin A	B
1 New Caledonia	25.2	25.2	1 Indonesia	24.8	24.8
2 Canada	14.4	39.6	2 China	14.9	39.7
3 USSR	9.6	49.2	3 Thailand	11.9	51.6
4 Australia	9.3	58.5	4 Bolivia	9.8	61.4
5 Indonesia	8.3	66.8	5 Malaysia	8.2	69.4

Country	Zinc A	B	Country	Phosphates A	B
1 Canada	18.1	18.1	1 Morocco	66.7	66.7
2 USA	14.2	32.3	2 South Africa	11.1	77.8
3 Australia	12.3	44.6	3 USA	8.1	85.9
4 USSR	12.0	56.6	4 USSR	5.2	91.1
5 Peru	5.2	61.8	5 Western Sahara[a]	1.5	92.6

Source: based on data in Crowson (1980)

A = individual percentage; B = accumulated percentage

[a] now *de facto* part of Morocco

high score in the world total of non-fuel mineral reserves comes overwhelmingly from its large reserves of copper ore. Canada and Australia, on the other hand, each hold reserves of international importance of many different minerals.

From the data for selected non-fuel minerals in *Table 4.9* it is evident that there is a high degree of concentration of reserves in a small number of countries. A glance at the countries represented shows that some are in the top five for several minerals while others, as New Caledonia with nickel and Morocco with phosphates, have only one particular mineral. Further details of non-fuel mineral resources can be found in Crowson (1980).

Theoretically it would be possible for a few countries to join together, as the OPEC countries have for oil, to influence world prices and even control supplies of various minerals. There has been talk of an organization of producers of tin ore. It is thought that the USSR and South Africa, ideologically opposed though they are, reach agreement on the production and price of gold and platinum. Between them they are credited with about 70 percent of the world's gold reserves and over 90 percent of the world's platinum reserves.

A country possessing large deposits of a particular non-fuel mineral is not necessarily a large producer. In the case of bauxite/alumina, for example, Guinea is credited with about 30 percent of the world's reserves, but accounts for about 13.5 percent of the world's output of this material. Jamaica, on the other hand, has 7.6 percent of the reserves but accounts for 13.7 percent of world output. At such rates of production the reserves of Guinea would last several times as long a those of Jamaica.

The contrast between reserves, extraction and the eventual processing and use in industry of various minerals is even more noticeable in many cases. Countries with the largest reserves of bauxite/alumina are shown in *Table 4.9*. The six largest producers of aluminium are the USA, USSR, Japan, Canada, the German FR and Norway, none of which has major deposits of bauxite. Of the countries with the largest reserves, India and Brazil produce small amounts of aluminium.

The data given so far in this section do not indicate the ultimate size of the world's reserves of non-fuel minerals. Some, however, have much shorter lives than others. While there is enough iron ore, bauxite and phosphates to last many centuries at current rates of production, the known reserves of tin, mercury and industrial diamonds would last only a matter of decades. It is possible to some extent to replace scarce minerals by other materials and to recycle most metals. It must also be appreciated that fossil fuels can themselves be used as raw materials rather than as sources of energy and thereby replace some kinds of non-fuel mineral. Even so, the search for new reserves of some minerals is not bringing to light enough to keep up with current production.

4.5 Natural resources and population

The difficulties of assessing the size of various kinds of natural resource were discussed in section 1 of this chapter. It was noted, however, that whatever method is used to estimate natural resources, some countries have more natural resources than others. Such disparities are to a considerable extent a result of the great difference in size between the largest and the smallest countries in the 60 considered. In this section it is necessary to move from a review of the absolute size of natural resources in different countries to the quantity of natural resources per inhabitant. From tables in preceding sections it is evident that the USSR, the USA, China and India each have a considerable share of the natural resources of the world. They are also very large in population size. Canada and Australia are also well endowed but they are much smaller in population. In terms of natural resources per inhabitant, Canada and Australia are far better endowed than India or China.

To calculate natural resources per inhabitant it is necessary to use the following procedure, the results of which are shown in *Tables 4.10* and *4.11*. Only the 60 largest countries of the world in population in the late 1970s were taken into account:

(1) The percentage of the total bioclimatic, fossil fuel and non-fuel mineral resources for the 60 countries, possessed by each country, was tabulated (*see* columns 2–4 in *Table 4.10* for the 15 largest in population).

Table 4.10 *Natural resurce scores for the fifteen largest countries*

Country	Area 1	Bioclimatic resources 2	Energy 3	Non-fuel minerals 4	All resources 5	Popn 6	Total resources per inhabitant 7
1 China	7.0	10.4	7.8	4.1	7.3	23.9	31
2 India	2.2	12.6	0.7	2.3	4.5	16.5	27
3 USSR	16.5	15.0	30.3	14.4	19.1	6.5	294
4 USA	6.9	15.5	18.3	11.5	13.1	5.4	243
5 Indonesia	1.1	1.9	0.8	2.2	1.5	3.5	43
6 Brazil	6.3	2.6	0.1	5.1	3.5	3.0	117
7 Japan	0.3	0.9	0.1	0.3	0.4	2.9	14
8 Bangladesh	0.1	0.9	0	0	0.3	2.2	14
9 Pakistan	0.2	1.5	0	0	0.4	2.1	19
10 Nigeria	0.7	1.3	1.1	0	0.8	1.9	42
11 Mexico	1.5	1.4	0.5	2.4	1.5	1.7	90
12 Germany FR	0.2	0.7	2.1	0.6	0.9	1.5	60
13 Italy	0.2	1.0	0.1	0.2	0.4	1.4	29
14 UK	0.2	0.6	1.8	0.2	0.7	1.4	51
15 France	0.4	1.5	0.1	0	0.5	1.3	38

1–6 = percentages of total for all 60 countries; 7 = values in column 5 as percentages of values in column 6.

(2) In order to allow for the possibility of other sources being discovered or developed, the area of each country was expressed as a percentage of the total area of the 60 countries (column *1* in *Table 4.10*). The justification for adding an 'area resource' to the calculation was that all other things being equal the larger a country the more 'other' resources it might be expected to have. Potential solar and wind power, water and land that could one day be brought into use are roughly related to area.

(3) It was arbitrarily decided that equal weight should be given to each of the four natural resource categories. It could be argued that bioclimatic resources are more important than the others since food is more vital than inanimate sources of energy and many manufactured goods. A weighted score for bioclimatic resources or any of the others would not however greatly affect the outcome of the exercise.

(4) The four resource scores for each country were summed and then divided by four (column *5* in *Table 4.10*). A single 'all resources' score was thus made for each country, being the percentage that country possessed of the total resources of the 60 countries, regardless of the profile of resources.

(5) The percentage share of natural resources of each country (column *5*) was then divided by its percentage share of population (column *6*) and multiplied by 100 (column *7*). The resulting score is one that expresses the total natural resources per inhabitant in each country in relation to an average of 100 resource units per inhabitant for all the 60 countries.

Table 4.10 shows the profile for the four categories of natural resources for the 15 largest countries. It can be seen that the USSR and the USA both have high per inhabitant scores through the possession of a wide range of natural resources. Brazil, in contrast, scores very badly on energy reserves, while Japan, Bangladesh and Pakistan have low scores on everything, ending up with dismally low overall scores. The West European countries do better than might be expected, the German FR and the UK thanks to their energy reserves and France with agricultural land.

The scores for total natural resources per inhabitant for each of the 60 countries considered in this chapter are given in *Table 4.11* rounded to make them easier to read and to take away the spurious precision of an exact percentage. In the view of the author, whatever reasonable assessment was made of natural resources, the countries would be ranked roughly in the order they appear on the scores used here. Whether each Australian apparently had about 85 times as many natural resources to his name as

Table 4.11 *Natural resource scores per inhabitant of sixty large countries (World average = 100)*

Rating more than double the world average

1	Australia	1280	30	Colombia	60
2	Canada	920	31	German FR	60
3	Chile	430	32	Netherlands	60
4	South Africa	360	33	Romania	60
5	USSR	290	34	Yugoslavia	60
6	Argentina	260	35	Turkey	50
7	Iraq	250	36	Czechoslovakia	50
8	USA	240	37	UK	50
9	Iran	240	38	Ethiopia	50
10	Algeria	220	39	Kenya	50
11	Peru	210			
12	Zaire	210			
13	Venezuela	210			

Rating above world average   *Rating less than half world average*

14	Morocco	160	40	German DR	50
15	Sudan	130	41	Burma	50
16	Poland	130	42	Philippines	40
17	Malaysia	120	43	Indonesia	40
18	Brazil	120	44	Portugal	40
			45	Nigeria	40
			46	Egypt	40
			47	France	40
			48	Ghana	30

Rating below world average

19	Madagascar	90	49	China	30
20	Bulgaria	90	50	Italy	30
21	Mexico	90	51	Uganda	30
22	Greece	90	52	Nepal	30
23	Cuba	80	53	Sri Lanka	30
24	Mozambique	80	54	India	30
25	Hungary	80	55	Belgium	20
26	Afghanistan	80	56	Pakistan	20
27	Spain	80	57	Viet Nam	15
28	Thailand	70	58	Bangladesh	15
29	Tanzania	70	59	Japan	15
			60	Korean R	10

each Japanese, or 50 or even 100 times as many does not affect the general picture of very great differences in the resource endowments per inhabitant of different countries.

In the next chapter, the production of goods and services will be discussed. It will become clear that some countries are using products of bioclimatic and mineral resources at a much faster rate than other countries. Thus the developed or industrial countries in *Table 4.11* are depleting their own non-renewable natural resources and also those of the developing countries. If the control of natural resources has been one of the causes of colonization and conflict throughout history, then increasing pressure on natural resources in the future would seem to bring the prospect of further conflict.

CHAPTER FIVE
The Production of Goods and Services

In chapter 4 the subject of natural resources was discussed in some detail. The distribution of natural resources in the world has received less attention than the distribution of production and of consumption since natural resources were largely considered for practical purposes to be 'unlimited'. Those countries that industrialized early could obtain the food, fuel and raw materials they could not produce at home elsewhere in the world. Although the geography of production and consumption is better known and more widely publicized than that of natural resources the subject will also be dealt with at some length but only selected goods and services will be discussed.

5.1 The production of food and beverages

Much of the productive capacity of the bioclimatic resources of the world goes into producing food and beverages. In addition some products of the soil and the seas are used as raw materials for industry, or as fuel. There is not therefore a straightforward relationship between agriculture and food production.

Until the great improvement in communications in the 19th century it was normal for the food needs of most communities in the world to be supplied locally or regionally. Since the development of the steamship and the railway large quantities of food have been transported about the world. Certain countries, such as the UK, the German FR and Switzerland, now have no possibility of being self-sufficient in food and beverages under present agricultural conditions.

In this section the distribution by country of the production of major foods and beverages will be described briefly. Inevitably the larger countries tend to figure prominently among the top producers. The following points regarding food production should be borne in mind:

(1) Some small countries have virtually no bioclimatic resources at all. Hong Kong and the Netherlands Antilles are examples of places with a large number of people on a very small area. In Greenland physical conditions are so adverse that even the land not covered by the ice cap cannot be used commercially to produce plants or to support livestock.

(2) Countries such as the UK, the German FR and Japan have some very good agricultural land but even though this is farmed intensively and high yields are obtained, none of these countries feeds itself.

(3) Mainly for climatic reasons many countries, especially those in the cooler latitudes, are not able to produce certain foods and beverages that they choose to consume, even if the products are not essential. Coffee, tea and bananas, for example, have to be grown in tropical or subtropical conditions.

(4) Many countries have large areas of agricultural land compared with population but obtain only low yields. Such is true in much of Africa and South America. With growing populations and the frequent occurrence of droughts in many areas, food supply is increasingly unsatisfactory.

(5) Until the 1940s Europe was the only major net importing region of food and beverages. Many countries in the rest of the world had considerable surpluses for export. In the 1940s some Asian countries became net importers of food. These were joined in the 1960s and 1970s by increasing numbers of African and Latin American countries. Even the USSR has become a large importer of grain. As a result, only a few countries still have a substantial surplus of agricultural products. These countries include the USA, Canada, Australia, New Zealand, Argentina, Uruguay and France. Therefore the situation has changed from that a century ago when a few industrial countries were able and indeed happy to exchange their manufactured goods for food and raw materials with their own colonies or with other countries. A problem in world affairs in the

future could be increasng competition to obtain food supplies from the few countries still able to export.

Tables 5.1 and *5.2* show food and beverage production. The largest ten or five producers are listed and the absolute amount of their production is given. Alongside is their percentage of total world production and the concentration of total world production found in the countries listed. Though perhaps the concentration of production of certain minerals in only a few countries is more likely to lead to the creation of monopolies and price agreements between producers there is also considerable concentration of production of some foods and beverages.

It can be seen in *Table 5.1* that, very roughly, the same *quantities* of wheat, rice and maize are produced in the world. They are the principal cereal crops. Barley, oats, rye, millets, sorghum and some crops associated with cereals are less widely grown.

Wheat is cultivated mainly in temperate latitudes, though it is also grown locally in highlands and some other places in the tropics. About 17 percent of all the wheat produced in 1979 entered international trade. Wheat yields vary greatly at the moment, being particularly high in West Europe but fairly low in North America, the USSR and Australia.

Rice is cultivated mainly in the tropics but also in warm temperate areas such as north Italy. It is the staple cereal food of many Asian countries but is much less widely grown in Africa and Latin America. In contrast to wheat there is very little international trade in rice, the proportion of production being only about three percent in 1979. As with wheat, great differences occur in yield, with Japan and Australia achieving very high yields but with India, China and other Asian countries having much lower yields.

Maize is grown in both tropical and temperate countries. It is a basic human food in many tropical and subtropical countries but in North America, Europe and the USSR it is mainly cultivated as feed for livestock. Though a New World crop, and still therefore very widely grown in the Americas, maize has also become a basic food crop in both Africa and Europe, but except in parts of China it is of comparatively limited importance in Asia. In 1979 almost 20 percent of production entered world trade, over three-quarters of the exports coming from the USA alone.

The remaining cereals of the world may either serve as supporting crops to a main cereal, as often barley and oats do with wheat, providing also animal feed, or they may form the staple diet of the population in many developing countries. As a carbohydrate crop, the *potato*, like cereals, is of basic importance in some regions, including parts of Latin America where it actually originated, and in northern and eastern Europe where it became very popular especially in the 19th century. Of all the carbohydrate plants mentioned so far, only wheat at the moment is a product of great significance in international relations. In the end, however, carbohydrate crops are to some extent interchangeable and in the event of serious world food shortages it would be the total carbohydrate supply that would have to be assessed, not just the preferred cereal of most developed countries.

Sugar has been included as a major world crop in view of its special character as a food product and its prominent place in world trade. There is also controversy as to whether it should be left largely to developing tropical countries to supply it from sugar-cane or whether developed countries in temperate latitudes should provide their own needs from beet sugar. Of the ten largest producers listed in *Table 5.1* the USSR and France produce sugar from beet, the USA from both beet and cane, and the rest mainly or entirely from cane. Sugar is or has been a major item of export from Caribbean countries such as Cuba, the Dominican Republic and Jamaica, as well as from several African and Asian countries.

The six products in *Table 5.2* illustrate several features and problems relevant to world affairs.

Soybeans have not been widely grown until the 1960s except in northern China and Korea. Production has recently increased greatly in the USA and in a particular part of southeast Brazil. Since soybeans contain a larger proportion of protein than most plant products (as well as oil), they are a possible substitute for meat and other livestock products, and could provide much of the protein consumed at the moment in developed countries.

Groundnuts have a somewhat similar position to soybeans in world agriculture but their cultivation is more restricted by temperature conditions than that of soybeans.

Palm oil has been included to provide a contrast with groundnuts and soybeans. The oil palm tree can only be cultivated in equatorial regions and indeed one country alone accounts for nearly half of the world production. A very large part of the production enters world trade.

Three of the principal beverages in the world are included in *Table 5.2*. While all three are grown for local consumption, for some countries they are also a principal item for export, as with *coffee* for Colombia, *cocoa beans* for Ghana and *tea* for Sri Lanka. The cultivation of all three crops is limited to certain areas by special physical conditions but they do not occupy large areas and it has been possible to increase production fairly easily. Wages for agricultural workers engaged in cultivating such commercial crops in many of the producing countries are exceedingly low. There has tended to be competition between exporting countries as well as agreements on quotas and prices. The industrial countries in

Table 5.1 *The ten largest producers of six basic food items in millions of tonnes in 1979*

Country	Wheat A	B	Country	Maize A	B	Country	Rice, paddy A	B
1 USSR	90.1	21.2	1 USA	197.2	50.1	1 China	143.4	36.0
2 China	60.0	14.1	2 China	40.6	10.3	2 India	69.0	17.3
3 USA	58.3	13.7	3 Brazil	16.3	4.1	3 Indonesia	26.4	6.6
4 India	35.0	8.2	4 Romania	12.4	3.1	4 Bangladesh	19.4	4.9
5 France	19.4	4.6	5 France	10.3	2.6	5 Thailand	15.6	3.9
6 Canada	17.7	4.2	6 Yugoslavia	10.1	2.6	6 Japan	15.6	3.9
7 Turkey	17.6	4.1	7 Mexico	9.3	2.4	7 Viet Nam	10.5	2.6
8 Australia	16.1	3.8	8 Argentina	8.7	2.2	8 Burma	10.0	2.5
9 Pakistan	9.9	2.3	9 USSR	8.4	2.1	9 Korean R	8.1	2.0
10 Italy	9.1	2.1	10 South Africa	8.2	2.1	10 Brazil	7.6	1.9
World	425(100%)	78%	World	394(100%)	82%	World	398(100%)	82%

Country	Barley A	B	Country	Potatoes A	B	Country	Raw Sugar (centrifugal) A	B
1 USSR	46.0	26.7	1 USSR	90.3	31.7	1 Cuba	8.0	9.0
2 China	19.5	11.3	2 Poland	49.6	17.4	2 USSR	7.6	8.5
3 France	11.2	6.5	3 USA	15.8	5.5	3 Brazil	7.0	7.9
4 UK	9.6	5.6	4 China	14.0	4.9	4 India	6.4	7.2
5 Canada	8.5	4.9	5 German DR	12.5	4.4	5 USA	5.1	5.7
6 USA	8.2	4.8	6 India	10.0	3.5	6 France	4.2	4.7
7 German FR	8.2	4.8	7 German FR	8.7	3.1	7 China	3.6	4.0
8 Denmark	6.7	3.9	8 France	7.1	2.5	8 German FR	3.3	3.7
9 Spain	6.2	3.6	9 UK	6.5	2.3	9 Mexico	3.1	3.5
10 Turkey	5.2	3.0	10 Netherlands	6.3	2.2	10 Australia	3.0	3.4
World	172(100%)	75%	World	285(100%)	78%	World	88.9(100%)	58%

Source: FAOPY 79 Tables 10, 13, 11, 12, 19, 60

A = total production in millions of tonnes; B = percentage of world total

Table 5.2 *The five largest producers of non-cereal crops and of three beverages in thousands of tonnes in 1979*

Country	Soybeans A	B	Country	Groundnuts A	B	Country	Palm oil A	B
USA	61 700	65.5	India	5 800	30.2	Malaysia	2 180	48.1
China	13 100	13.9	China	2 900	15.1	Nigeria	680	15.0
Brazil	10 000	10.6	USA	1 800	9.4	Indonesia	610	13.5
Argentina	3 700	3.9	Sudan	1 100	5.7	China	180	4.0
Mexico	700	0.7	Senegal	1 000	5.2	Zaire	170	3.8
World	94 200(100%)	95%	World	19 200(100%)	66%	World	4 530(100%)	84%

Country	Coffee A	B	Country	Cocoa beans A	B	Country	Tea A	B
Brazil	1300	26.2	Ivory Coast	350	22.1	India	550	30.2
Colombia	760	15.3	Brazil	309	19.5	China	303	16.6
Ivory Coast	280	5.6	Ghana	270	17.0	Sri Lanka	210	11.5
Indonesia	270	5.4	Nigeria	180	11.4	Turkey	117	6.4
Mexico	230	4.6	Cameroon	115	7.3	Japan	106	5.8
World	4970(100%)	57%	World	1585(100%)	77%	World	1821(100)%	71%

Sources: FAOPY 79 Tables 28, 29, 39, 69, 70, 187

A = total production; B = percentage world total

Table 5.3 *The largest producers of meat (1979) and fish (1978) in the world in millions of tonnes*

Country	Meat		Country	Fish	
	A	B		A	B
USA	25.9	18.8	Japan	10.8	15.0
China	22.1	16.0	USSR	8.9	12.4
USSR	15.5	11.2	China	4.7	6.5
France	5.3	3.8	USA	3.5	4.9
German FR	4.6	3.3	Peru	3.4	4.7
World	138.1(100%)	53%	World	72.0(100%)	44%

Source: Meat: FAOPY 79 Table 88
        Fish: UNSYB 79/80 Table 46

A = absolute amount; B = percentage of world total

temperate latitudes have thus been able to obtain coffee, cocoa beans and tea easily and comparatively cheaply.

Since meat and fish (*see Table 5.3*) are both produced very widely in the world it is to be expected that most of the large producers are themselves among the larger countries of the world. Meat production takes place under greatly different conditions in different countries. In the USSR it is closely related to arable farming, and livestock are fed from cereals and other feed crops grown in association with plants for human consumption. In parts of Brazil and Australia, in contrast, beef cattle are raised in lands used almost exclusively for grazing purposes and unsuitable for arable farming. In China livestock raised for meat consist largely of pigs and poultry, raised in close association with human settlements. In India cattle are used as work animals but are rarely raised for their meat, and very little meat of any kind is consumed.

It may be noted that agriculture is a fast changing sector of the economy of many countries and changes could occur in the following ways:

(1) The data tabulated and described in this chapter show that the production of some agricultural items is highly concentrated in certain countries or continents, while other crops are widely grown. Further improvements in world food supply might be achieved by a change in crop composition in some regions with the adoption of crops that have proved successful in other regions. Such a process has gone on for a long time but only gradually until the 16th century with the discovery of the Americas. Then Old World (eg wheat, coffee, sugar cane) and New World (eg maize, the potato) crops and livestock could be exchanged. The success of the soybean experiment has been mentioned but the campaign of Soviet leaders in the 1950s to grow grain maize widely in the USSR. in hopeless climatic conditions, was a dismal failure.

(2) Yields could theoretically be raised greatly in many existing areas of cultivation if they could be raised to the highest yields actually obtained in the world. Japanese rice yields and French wheat yields are several times as high as those in many developing countries. Large quantities of fertilizer would, however, have to be applied, a water supply ensured in drier areas and the quality of seeds and plants improved.

(3) New areas could theoretically be brought into cultivation. Data given in *Table 4.3*, chapter 4, show that in the last two decades the cultivated area of the world has not been growing as fast as world population.

(4) The livestock population of the world could be reduced in order to release land for the cultivation of food to be supplied directly to humans. From the same area of land, several times as much food can theoretically be obtained by taking a short cut through soybeans and other plants instead of growing feed crops to raise and fatten livestock. The reason is that before they are slaughtered, livestock have to be fed for a considerable period. However, meat and dairy products are regarded as highly desirable items of diet in many countries, both rich and poor, and traditional diets are difficult to relinquish.

(5) The application of modern means of production to agriculture can cause problems. Fertilizers can pollute the water supply in areas where they are applied. Pesticides may harm the health of humans. The introduction of mechanized methods of ploughing, while tilling the land more consistently and thoroughly, may increase the

Table 5.4 *The ten largest producers of six sources of energy in 1977*

Country	Coal A	Coal B	Country	Lignite A	Lignite B	Country	Oil 1977 A	Oil 1977 B	Oil 1980 A
1 USA	604	24.4	1 German DR	254	28.2	1 USSR	546	18.3	603
2 USSR	500	20.2	2 USSR	164	18.2	2 Saudi Arabia	458	15.3	493
3 China	490	19.8	3 German FR	123	13.6	3 USA	403	13.5	428
4 Poland	186	7.5	4 Czechoslovakia	93	10.3	4 Iran	283	9.5	74
5 UK	122	4.9	5 Poland	41	4.5	5 Iraq	122	4.1	130
6 India	100	4.0	6 Yugoslavia	37	4.1	6 Venezuela	117	3.9	116
7 German FR	91	3.7	7 Australia	29	3.2	7 Nigeria	103	3.4	102
8 South Africa	86	3.5	8 USA	27	3.0	8 China	100	3.3	106
9 Australia	71	2.9	9 Bulgaria	25	2.8	9 Libya	100	3.3	86
10 Korean DPR	45	1.8	10 Greece	24	2.7	10 Kuwait	99	3.3	71
World	2476(100%)	93%	World	902(100%)	91%	World	2986(100%)	78%	3074

Country	Natural gas A	Natural gas B	Country	Uranium A	Uranium B	Country	Hydro-electricity A
1 USA	4 932	41.7	1 USA	11.2	31.4	1 USA	69
2 USSR	2 889	24.4	2 USSR approx.	7.0	19.7	2 USSR	45
3 Netherlands	733	6.2	3 South Africa	6.7	18.8	3 Canada	40
4 Canada	686	5.8	4 Canada	6.1	17.1	4 Japan	26
5 UK	377	3.2	5 France	2.2	6.2	5 China	20 (?)
6 Romania	321	2.7	6 Niger	1.6	4.5	6 Brazil	19
7 Iran	202	1.7	7 Australia	0.4	1.1	7 France	18
8 German FR	152	1.3	8 Spain	0.2	0.6	8 Norway	17
9 Mexico	128	1.1	9 Argentina	0.1	0.3	9 Italy	15
10 Italy	126	1.1	10 Portugal	0.1	0.3	10 Spain	13
World	11 821(100%)	89%	World	35.6(100%)	100%		

Sources: UNSYB 78 oil Table 52; gas Table 53; coal Table 50; lignite Table 51; uranium Table 55; Hydro-electricity Table 143

A = total production – coal, lignite, oil in millions of tonnes; natural gas in thousands of tonnes; uranium in thousands of tera-calories; hydro-electricity in millions of KW capacity.
B = percentage of world fuel

risk of soil erosion and will certainly release large numbers of workers from the agricultural land. Mechanization uses large quantities of inanimate energy.

(6) The Portuguese introduced the cultivation of sugar cane into their colony of Northeast Brazil in the 16th century. Ever since that time there has been a choice in tropical countries between growing 'subsistence' food crops such as manioc, maize or millets for local consumption or more valuable exportable 'commercial' crops such as coffee, cotton, sisal or rubber.

5.2 The production of energy

The distribution and possible reserves of the main sources of energy were described in the last chapter. It was noted that many parts of the world have hardly been explored yet for energy resources and that new sources such as oil in shales and deep natural gas deposits await new technology before they can be extracted on a large scale. It was clear, however, that currently known reserves of fossil fuels are concentrated in certain areas of the world, as for example oil in the Middle East, coal in the USSR, USA and China, and lignite in Europe.

Some uses of energy require a particular type of fuel. The making of steel has traditionally required charcoal from wood or coking coal from special types of coal. Fuel for motor vehicles is most cheaply derived from oil though it can be derived from coal or natural gas. For some purposes, such as heating and the generation of electricity, most kinds of fuel are suitable, even including minor sources such as peat. Animal dung and firewood are widely used for cooking purposes, heating and local industry in many developing countries.

In order to have a broad idea of the total amount of energy used in different places it is customary to express all sources in some conventional equivalent unit. One of the most commonly used units is the quantity of one tonne of coal of a given quality. A tonne of oil contains more energy than a tonne of coal and it is equivalent roughly to 1.5 tonnes of coal. Natural gas and electric power are measured in different units but can also be converted to coal equivalent. The distribution of production of individual types of energy will be discussed first and total production of all types after. *Table 5.4* shows the ten largest producers of each of six sources of energy in 1977.

In 1980 coal and lignite provided nearly 30 percent of the world's energy consumption, while nearly 44 percent came from oil and nearly 19 percent from natural gas. Other sources include hydro-electric (6 percent), nuclear power (2.4 percent), geothermal

power, peat and wood. Although the non-fossil-fuel sources, together with animate sources (draught animals, humans), are important locally, the four fossil fuels account for well over nine-tenths of the total.

Some countries are using up their non-renewable sources of energy far more quickly than others. If the current rate of gas extraction in 1980 were to continue, the USA would use up its reserves of natural gas known in that year in about a decade, the USSR in about a century, but Iran in a much longer period still. In all three the rate of extraction could change and new reserves might also be found. Even so the contrasts are striking.

Table 5.5 shows the positions of 26 selected countries of the world with regard to their production and consumption of energy. The values in columns *1* to *3* are in millions of tonnes of *coal* equivalent. The amounts given in the table cannot therefore be directly compared with amounts in *Table 5.4*. The countries in *Table 5.5* are ranked according to the ratio of production to consumption, the relation expressed in column *4* of the table, the result of dividing millions of tonnes produced (column *1*) by millions of tonnes consumed (column *2*). Of the countries represented in the table, the most favourably placed are the world's leading exporters of oil, while the countries depending very heavily on imported oil come at the lower end. At one extreme, Saudia Arabia produced in 1980 over 60 times as much energy as it consumed, and the United Arab Emirates over 30 times as much. At the other extreme, Japan, Italy and France all need to import over three-quarters of their energy requirements in spite of having hydro-electric sources and a small output of fossil fuels. The following points may be noted with regard to *Table 5.5*.

(1) The volume of oil moved in international trade grew roughly ten times between 1950 and 1973. Between 1973 and 1980 the volume of oil moved internationally has not increased but has fluctuated from year to year. Oil is by far the largest single source of energy in world trade.

(2) A large part of the flow of oil in the late 1970s was crude oil from the Middle East, North Africa and West Africa destined for the USA, Japan and West Europe. Other major net exporters of oil include Venezuela, the USSR and Indonesia. Over 50 percent of the oil produced in the world enters international trade.

(3) Although they are large producers of oil, and in some cases of natural gas as well, and in spite of having between them about 70 percent of the world's oil reserves, several of the top ten countries in *Table 5.5* (column *5*) had much lower levels of energy consumption per inhabitant in 1980 than energy-deficient industrial

Table 5.5 *The energy balance in selected countries in 1980*

Country	Production 1	Consumption 2	Absolute difference 3	Relative difference 4	Consumption in kg per inhabitant 5
Large relative surplus					
1 Saudi Arabia	730	12	+718	6083	1 480
2 United Arab Emirates	127	4	+123	3175	5 450
3 Libya	138	6	+132	2300	2 180
4 Kuwait	126	9	+115	1400	6 720
5 Nigeria	153	11	+142	1391	140
6 Algeria	116	15	+101	773	810
7 Venezuela	192	47	+145	409	3 380
8 Indonesia	134	33	+101	406	220
9 Iran	138	47	+91	294	1 250
Small relative surplus					
10 Mexico	205	127	+78	161	1 770
11 Australia	115	88	+27	131	6 030
12 USSR	1939	1486	+453	130	5 600
13 Canada	280	245	+35	114	10 240
14 Netherlands	98	88	+10	111	6 210
15 China	614	565	+49	109	600
16 Poland	212	199	+13	107	5 590
17 South Africa	92	86	+6	107	2 600
18 Argentina	51	49	+4	104	1 820
Deficit					
19 UK	275	276	−1	100	4 940
20 USA	2090	2370	−280	88	10 410
21 India	101	126	−25	80	190
22 German FR	160	353	−193	45	5 730
23 Brazil	34	94	−60	36	760
24 France	50	234	−184	21	4 350
25 Italy	26	189	−163	14	3 320
26 Japan	43	408	−365	11	3 490

Source: UNSYB 79/80: data derived from Table 189

1–3 = millions of tonnes of coal equivalent

countries such as Italy and Japan. The oil producing countries with high energy consumption levels per inhabitant are limited to those with only a very small number of inhabitants, and to Venezuela, which has been a major exporter since about 1930 and has already used about two-thirds of its total oil reserves.

(4) Oil has had the advantage of being both very cheap to extract compared with most coal and lignite and, unlike natural gas, is easy and cheap to transport by sea. Many industrial countries changed emphasis from coal to oil in the 1950s and 1960s.

(5) The five-fold increases in crude oil prices during 1973–74 can be seen as an attempt firstly to prolong the life of oil reserves in exporting countries and secondly to enable those countries to import more goods per quantity of oil exported. Such goods could be either capital or consumer goods. At the same time some of the oil-producing countries find it attractive to invest some of their surplus in other countries in view of their own limited natural resources other than oil and gas.

(6) The big oil producers of the developing world listed in *Table 5.5* differ greatly among themselves. For simplicity they may be seen as three types: Indonesia and Nigeria are very poor and have big populations; Iran, Iraq, Algeria and Venezuela have other natural resources and not such big populations; Saudi Arabia, Kuwait, the United Arab Emirates and Libya are largely desert lands with small populations and little but oil and gas.

5.3 The production of raw materials

Raw materials are produced from bioclimatic resources, from fossil fuels and from non-fuel minerals. The plant and animal raw materials produced from the land and sea are competing with food production. Fossil fuels would ideally be used as raw materials rather than for energy and their contribution is already considerable (synthetic fibres, rubber, chemicals) but they will be used primarily for energy production until enough non-exhaustible energy is assured.

Table 5.6 shows the five leading producing countries of nine non-fuel minerals. For the seven metallic ores the actual metal content is shown. Manganese and nickel are two of several metals used as alloys in the manufacture of steel. The leading producers of the two main ingredients of steel, iron ore and coke oven coke, are given in *Table 5.7*. Phosphate rock and potash are two of the principal raw materials for fertilizers. Several countries or groups of countries have a very strong position in the world with regard to both reserves and present production of non-fuel minerals. Some parts of the world have, however, been explored for minerals much more thoroughly than others and some also have already been exploited far longer than others. The following *regions* seem to be of outstanding importance for non-fuel minerals at present: Asiatic USSR, Canada, western USA, southern Africa and Australia. Their reserves are shown in *Table 4.8* in chapter 4.

Table 5.8 gives the principal producers of a selection of plant and animal raw materials. All products are susceptible to competition from artificial and synthetic materials. This may be one reason

Table 5.6 *The world's five largest producers of nine major non-fuel minerals in 1977*

Country	Copper A	B	Country	Lead A	B	Country	Zinc A	B
USA	1364	17.1	USA	538	16.5	Canada	1055	18.3
USSR	1100	13.8	USSR	510	15.6	USSR	720	12.5
Chile	1058	13.2	Australia	418	12.8	Peru	478	8.3
Zambia	819	10.2	Canada	284	8.7	Australia	475	8.3
Canada	781	9.8	Peru	182	5.6	USA	416	7.2
World	8000(100%)	64%	World	3270(100%)	59%	World	5750(100%)	55%

Country	Manganese A	B	Country	Nickel A	B	Country	Tin A	B
USSR	2904	30.5	Canada	235	29.3	Malaysia	59	31.9
South Africa	2338	24.6	USSR	168	21.0	Bolivia	31	16.8
Gabon	941	9.9	New Caledonia	116	14.5	Indonesia	25	13.5
Brazil	900	9.5	Australia	81	10.1	Thailand	24	13.0
Australia	811	8.5	Cuba	37	4.6	Australia	11	5.9
World	9510(100%)	83%	World	801(100%)	80%	World	185(100%)	81%

Country	Bauxite A	B	Country	Phosphate rock A	B	Country	Potash A	B
Australia	22.8	28.7	USA	47	37.3	USSR	8.5	31.8
Jamaica	11.4	14.3	USSR	24	19.0	Canada	5.9	22.1
Guinea	10.8	13.6	Morocco	18	14.3	German DR	3.2	12.0
Surinam	4.9	6.1	China	4	3.2	German FR	2.8	10.5
USSR	4.6	5.8	Tunisia	4	3.2	USA	2.2	8.2
World	79.6(100%)	69%	World	126(100%)	77%	World	26.7(100%)	85%

Sources: UNSYB 78 Tables 60, 62, 72, 64, 67, 69, 58, 74, 75

A = Quantity – copper, lead, zinc, manganese and nickel are in thousands of tonnes; bauxite, phosphate rock and potash are in millions of tonnes
B = percentage of world total.

Table 5.7 *Production of iron ore and of coke oven coke in millions of tonnes in 1977*

Country	Iron ore (iron content) A	B	Country	Coke oven coke A	B
1 USSR	131	27.1	1 USSR	87	24.0
2 Australia	60	12.4	2 USA	52	14.4
3 Brazil	57	11.8	3 Japan	43	11.9
4 USA	35	7.2	4 China	29	8.0
5 China	33	6.8	5 German FR	28	7.7
6 Canada	32	6.6	6 Poland	19	3.9
7 India	27	5.6	7 UK	14	3.9
8 South Africa	17	3.5	8 Czechoslovakia	11	3.0
9 Sweden	16	3.3	9 France	11	3.0
10 Liberia	12	2.5	10 India	10	2.8
11 France	11	2.3			
12 Venezuela	8	1.7			
World	483(100%)	91%	World	362(100%)	84%

Source: UNSYB 78 Tables 56, 123.

A = total production in millions of tonnes; B = percentage of world total.

Note: Major users of iron ore with a small or negligible production include: Spain (4.2), UK (0.9), German FR (0.8), Japan (0.5), Poland (0.2), Italy (0.2), Belgium, Czechoslovakia.

Table 5.8 *The five leading producers in 1979 of six raw materials obtained from plants*

Country	Jute A	B	Country	Cotton lint A	B	Country	Wool scoured A	B
India	1170	29.3	USA	3 200	22.7	Australia	482	29.4
China	1089	27.2	USSR	2 800	19.9	USSR	283	17.3
Bangladesh	996	25.0	China	2 200	15.6	New Zealand	229	14.0
Thailand	370	9.3	India	1 200	8.5	Argentina	93	5.0
Brazil	88	2.2	Pakistan	700	5.0	South Africa	57	3.0
World	4000(100%)	93%	World	14 000(100%)	72%	World	1638(100%)	70%

Country	Natural rubber A	B	Country	Coniferous roundwood A	B	Country	Deciduous roundwood A	B
Malaysia	1617	44.0	USSR	319	28.4	Indonesia	141	10.0
Indonesia	851	23.1	USA	259	23.1	Brazil	131	9.3
Thailand	498	13.5	Canada	134	11.9	India	129	9.1
Sri Lanka	155	4.2	China	90	8.0	China	105	7.4
India	140	3.8	Sweden	42	3.7	USA	83	5.9
World	3679(100%)	89%	World	1123(100%)	75%	World	1415(100%)	42%

Source: FAOPY 79 Tables 76, 78, 95; UNSYB 78 Table 114

A = jute, cotton lint, wool scoured, natural rubber in thousands of tonnes; roundwood in millions of cubic metres: B = percentage of world total.

why production of all the items has tended to grow unevenly since the Second World War. Fluctuations in the requirements for jute and rubber in particular affect developing countries. Most of the world's wool exports come from the temperate countries of the southern hemisphere, while flax and hemp production are concentrated in the USSR and East Europe.

Much of the wood pulp and construction timber used in the industrial countries comes from the coniferous forest of cool latitudes, especially from Canada and northern USA, Scandinavia and the more accessible parts of the Soviet forest lands. Although much of the broadleaf forest of the world is in the tropical rain forest of South America, Africa and Southeast Asia, these areas have so far been cut on a large scale only in certain areas. One problem in their exploitation is the difficulty of locating and selecting certain commercial species from among many hundreds of species altogether. Increasingly, however, it is possible that the whole mass of vegetation in a tropical forest area can be cleared, processed and used as a raw material. Another possibility is to clear the forest and plant one or a few commercial species for cutting, a procedure already practised for the cultivation of trees such as the rubber tree, grown for a special product, not for the whole tree. Experiments have been carried out in the Brazilian Amazon to plant a few species of fast growing tree but reservations about such a drastic modification of the fragile environment are expressed, for example, by McIntyre (1980).

5.4 The production of manufactured goods

Before raw materials are eventually turned into specific manufactured goods they may pass through various stages of transformation. It is customary to refer to the initial transformation or refining of materials as processing. Often processing is carried out at or near the places where materials are produced in order to reduce their weight and/or bulk (metallic ores, sugar cane). Later stages of transformation may be referred to as manufacturing, while the term industry may broadly cover all stages referred to above, even mining (as in the USSR), and sometimes all branches of production. The purpose of this section is to examine the distribution of selected branches of manufacturing and to trace the way some of the present distributions came about.

The processing and manufacturing of raw materials has been carried on in traditional societies alongside agriculture for thousands of years. The production of some manufactured goods became highly concentrated and specialized in some civilizations. Since the 16th century technological developments in parts of Europe gradually led to the emergence of more sophisticated manufacturing processes. Through their various empires European powers were able to obtain raw materials from their colonies and to export manufactured goods there.

With great improvements in the means of transporting goods over long distances in the 19th century, Europe's dominance of world trade and industry was firmly established. In the second half of that century China and Japan as well as most of Africa were forced into the system. By the beginning of the 20th century a large proportion of the world's modern, sophisticated industry was concentrated in western Europe and eastern USA. Techniques that were developed in Britain in particular and also in France, Belgium and Germany were subsequently adopted in other countries such as Italy, Russia and Japan. The export of producer goods in the form of equipment and machines enabled other parts of the world to set up processing and manufacturing on modern lines. Such countries as Australia, Brazil, India and South Africa began to industrialize. The reduction of trade with Europe during the First World War was an incentive to some non-industrial countries to develop industry.

The end product of the above process may be illustrated by the distribution of world steel production in selected years over the last few decades for representative countries, both developed and developing. The modern iron and steel industry requires heavy investment in means of transport and production and uses large quantities of coal, iron ore and other materials. Economies of large scale have made the use of large plants desirable.

It is clear from the data in *Table 5.9* that the 'industrialized' countries of the world produced nearly all the steel in the 1930s, with only two to three percent coming from a few 'non-industrial' countries. By 1978 the so-called developing countries produced about 80 million tonnes of steel or about ten percent of the world total. Relatively, then, the share of steel production in developing countries has grown. On the other hand the absolute gap is much greater: from about 132 against 3 million tonnes in 1937 to about 620 million against 80 million in 1978.

Table 5.10 shows the leading producers of pig iron and steel in 1978. Though China, India and Brazil appear among the largest producers they are all large in total population. Thus India produces about as much pig iron as Czechoslovakia yet it has about 40 times as many inhabitants.

A large share of international trade consists of the movement of manufactured goods. The military strength and strategic aspirations of world powers are related to their scientific, technological and industrial capabilities. A large element of the study of world affairs, then, is related to the distribution of industry. The concentration of the iron and steel industry in certain countries was evident from the data in *Table 5.10*. Some other industries will now be discussed briefly.

Table 5.9 *Steel production in selected countries, 1937–1978 in millions of tonnes*

Country	1937	1940	1945	1950	1955	1960	1965	1970	1975	1978
Industrialized				.						
USSR	18	18	12	27	45	65	91	116	141	151
USA	51	61	72	88	106	90	119	119	106	124
Japan	6	7	2	5	9	22	41	93	102	102
German FR	20[a]	22[a]	–	12	21	34	37	45	40	41
France	8	4	2	9	13	17	20	24	22	23
Italy	2	2	–	2	5	8	13	17	22	24
UK	13	13	12	17	20	25	27	28	20	20
Poland	1	–	–	3	4	7	9	12	15	19
Czechoslovakia	2	2	1	3	4	7	9	11	14	15
Canada	1	2	3	3	4	5	9	11	13	15
Non-industrialized										
China	–	1	–	–	3	18	15	18	29	32
Brazil	–	–	–	1	1	2	3	5	8	12
India	1	1	1	1	2	3	6	6	8	10
Mexico	–	–	–	–	1	2	2	4	5	7
Turkey	–	–	–	–	–	–	1	1	1	2
Korean DPR	–	–	–	–	–	1	1	2	3	3
Venezuela	–	–	–	–	–	–	1	1	1	1
World	135	142	113	189	266	328	443	594	643	704

Source: United Nations Statistical Yearbook 1957, 1966, 1978, 1979/80

[a] The whole of Germany
– under half million tonnes or data not available (but production small)

Table 5.10 *Production of pig iron, steel and cement in 1978 in millions of tonnes*

Country	Pig iron 1	Steel 2	Country	Cement 3
1 USSR	111	151	1 USSR	127
2 Japan	88	102	2 Japan	85
3 USA	80	124	3 USA	78
4 China	35	32	4 China	63
5 German FR	30	41	5 Italy	38
6 France	18	23	6 German FR	34
7 UK	12	20	7 Spain	29
8 Italy	12	24	8 France	28
9 Canada	11	15	9 Brazil	23
10 Poland	11	19	10 Poland	22
11 Brazil	11	13	11 India	20
12 India	10	10	12 UK	16
13 Czechoslovakia	10	15	13 Turkey	16
14 Belgium	10	13	14 Korean R	15
15 Romania	8	12	15 Mexico	14
16 Australia	7	8	16 Romania	14
17 Spain	7	12	17 German FR	13
18 South Africa	7	8	18 Greece	11
World	520	704	World	827

Sources: UNSYB 79/80, Tables 112, 113, 144

Textiles and clothing

Some of the earliest innovations in the Industrial Revolution came with the mechanization of spinning, weaving and associated processes. For a time Britain and other countries that industrialized early were able not only to provide their own home markets with factory made goods but also to sell these widely in other countries. Local domestic production was disrupted in many parts of the world. Already by the middle of the 19th century, however, textile manufacturing and associated equipment could be set up elsewhere in the world. Russia, Latin America, Japan and India began to make their own cotton goods.

The textile and clothing industry is now widely dispersed in the world and most countries aim to be largely self-sufficient in production. Two ironical results of the history of modern textile and clothing manufacturing may be noted here. Firstly, thanks to their low labour costs, some of the later adopters of the new technology have been able to compete for markets in the very countries from which the innovations came, as, for example, textiles from India and Pakistan compete in Britain. Secondly, modern factories by definition produce more goods per worker than craft industries where much of the work is done by hand. Some years ago a comparison was made in Tanzania between two textile mills, one set up by the British and one by the Chinese. The latter was praised because it provided *more* jobs for a similar amount of production.

Cement

The cement industry is not one that often makes the headlines. It is, however, a useful branch to consider here. Cement is clearly a product with a wide range of uses in any kind of country. It is a comparatively simple product and one for which the main raw material, limestone, is very widely found in the world. Cement is a bulky product in relation to its value and is also perishable. One heard stories in the 1950s of bags of Soviet cement being caught on the wharves of Rangoon in Burma by the monsoon. The equipment for making cement is simple and there is no great economic disadvantage in making cement in a small factory.

Column *3* in *Table 5.10* shows the 18 largest producers of cement in the world in 1978. By that year the *United Nations Statistical Yearbook* listed 115 producing countries altogether, over half of them developing ones. As with the production of steel, there is a contradictory trend in the distribution of the industry in the last few decades. In 1937 about 100 million tonnes of cement were produced in the world, all but about two percent of which was accounted for by the industrialized countries. Non-industrial countries either used other materials for construction or *imported* cement when essential. In 1978 about 830 million tonnes of cement were produced in the world (for a total population about double that in 1937). Over 25 percent of the total was produced in developing countries, which, however, had about 70 percent of the total population of the world. Thus the *relative* importance of the developing countries in the cement industry has grown greatly in the last few decades but the *absolute* gap was much greater in 1978 at 600:230 million tonnes compared with 100:2 four decades previously. The cement industry is an example of an industry using modern equipment and now dispersed throughout the world broadly in relation to the consumption of each country.

Motor vehicles

A major branch of the engineering industry, the manufacture of motor vehicles (*see Table 5.11*), will be taken to illustrate a different picture from that of the steel, textile or cement industries. More than most modern industries, the engineering industry requires skilled labour. Most branches require a large market (home plus foreign if possible) for a particular product or range of products. A large country therefore tends to have advantages over a small one, as can be seen with the manufacture of aircraft in the USA compared with that in the smaller West European countries or the manufacture of motor vehicles in Brazil compared with that in smaller Latin American countries.

The motor vehicles industry requires the assembly of many different products at one place. It is an industry in which many technological changes have taken place since it started about a century ago. Mechanization, automation and resultant mass production have characterized the industry since the 1920s in the USA and have followed elsewhere.

In the 1930s the USA accounted for a very large proportion of the passenger cars produced in the world and in 1937 it had about 75 percent of the passenger cars in use. West European countries produced and had in use most of the rest, but the expanding production of passenger cars there was by this time being stifled by military preparations in Germany and Italy.

By 1978 Japan had joined North America and West Europe as a major producer of motor vehicles while the USSR and East Europe roughly satisfy their needs. The manufacture of passenger cars and to a somewhat lesser extent of commercial vehicles is still concentrated in the developed countries. A considerable number of motor vehicles were being assembled in developing countries but only a few countries, notably Brazil, Argentina, India, the Korean Republic and Mexico were making complete vehicles.

Table 5.11 *Production of motor vehicles in 1978*

Country	Passenger cars		Country	Commercial vehicles	
	A	B		A	B
1 USA	9.2	29.2	1 USA	3.3	31
2 Japan	6.0	19.0	2 Japan	3.3	31
3 German FR	3.9	12.4	3 USSR	0.8	8
4 France	3.6	11.4	4 Canada	0.6	6
5 Italy	1.5	4.8	5 France	0.5	5
6 USSR	1.3	4.1	6 Brazil	0.5	5
7 UK	1.2	3.8	7 UK	0.4	4
8 Canada	1.1	3.5	8 German FR	0.3	3
9 Spain	1.0	3.2	9 Spain	0.2	2
10 Brazil	0.5	1.6	10 Italy	0.1	1
World	31.5(100%)	93%	World	10.5(100%)	96%

Other producers of motor vehicles:
  Argentina, Australia, Austria, Bulgaria, Czechoslovakia, China, German DR, Hungary, India, Mexico, Netherlands, Poland, Romania, Sweden, Yugoslavia

Countries assembling but not producing:
  Belgium, Chile, Colombia, Egypt, Indonesia, Iran, Ireland, Malaysia, New Zealand, Philippines, Portugal, South Africa, Thailand, Turkey, Venezuela, Zaire

Source: UNSYB 79/80, Table 125

A = millions; B = percentage of world total

Fertilizer production

This is a further example of a major sector of industry concentrated in the developed countries. *Table 5.12* shows that China and India are producers but that their output of fertilizers is limited in relation both to their area under cultivation and to the number of persons engaged in agriculture. Even where the raw material is abundant in a developing country (as phosphates in Morocco) that country tends to export its raw material, to be processed in a developed country. Where the raw material is abundant in developed countries, as is potash, then the industry is dominated by developed countries.

It is hoped that enough evidence has been given in this section to show the following features and trends. On the whole the same picture would emerge from a study of other branches of industry such as non-ferrous metals, the manufacture of synthetic fibres or electrical and electronic goods.

(1) Most branches of industry are dominated by the developed countries.
(2) Nevertheless some industries, for which the means of production could be easily obtained, have expanded greatly in recent decades in developing countries.

(3) Industrial production in the world has increased several times in the last four decades while population has only doubled. Great pressure has, therefore, been put on sources of some materials, especially oil reserves, fisheries and forests (and woodlands).
(4) In the next decade or two there could be competition from an increasing number of industrial countries to obtain fuel and raw materials from a diminishing number of countries with a surplus of primary products.
(5) Industry is not a large employer of labour in most countries and it has a declining share of total employment in the most highly industrial countries.

5.5 The production of services

The production of fuel, raw materials and manufactured goods can be measured in units that are reasonably obvious and straightforward. A tonne of cement or a railway locomotive are straightforward material items even if a cubic metre of natural gas or a kilogram of uranium ore may not be familiar to most people. The products of the service sector of any economy are more difficult to visualize. Even so,

Table 5.12 *Ten largest producers of three principal types of fertilizer in millions of tonnes[a] in 1979–1980*

Country	Nitrogenous fertilizers A	Country	Phosphate fertilizers A	Country	Potash fertilizers A
1 USA	11.2	1 USA	9.1	1 Canada	7.1
2 China	9.2	2 USSR	5.9	2 USSR	6.6
3 USSR	9.1	3 China	1.9	3 German DR	3.4
4 India	2.2	4 France	1.4	4 German FR	2.7
5 France	1.8	5 Brazil	1.3	5 USA	2.1
6 Romania	1.7	6 Poland	0.9	6 France	1.9
7 Canada	1.7	7 Australia	0.9	7 Israel	0.8
8 Netherlands	1.6	8 India	0.8	8 Spain	0.7
9 Italy	1.5	9 German FR	0.7	9 UK	0.3
10 Japan	1.5	10 Japan	0.7	10 Italy	0.1
World	59.8	World	33.5	World	25.7

Source: UNSYB 79/80, Tables 105, 106, 107

A = millions of tonnes

[a] Production is measured in terms of active content, Nitrogen (N), Phosphoric acid (P_2O_5) and Potassium oxide (K_2O)

Western economists are accustomed to counting the production of services as part of total gross national product a practice not so readily accepted by Soviet economists.

In this section an indirect approach has been used to examine the distribution of the products of certain services. Schools and hospitals are regarded as means of production, places that process people, and the products are educated and healthy people. How can the success of educational health and other service sectors be assessed?

Education

Education has increasingly become the concern of the state in almost every country of the world over the last century or so. Formal education at various levels has been available for the few in Europe since the Middle Ages, in China, and in some other parts of the world. Compulsory universal education up to a given official school-leaving age is a much more recent feature. Scandinavian countries were among the first to adopt compulsory formal education for all children. In Britain the school system in the 19th century partly owes its development to the need to keep children from working in factories.

Virtually every national government in the world now has some educational goal for its citizens even if lack of facilities prevents the goal from actually being achieved. Many reasons are given for having compulsory education to a certain age. In Western industrial countries the complex economic and administrative set-up is said to require citizens with a minimum level of ability in such skills as literacy, numeracy and graphicacy as well as an 'elite' able to carry out research, manage the economy and carry out specialized professional work. In the USSR more perhaps than in most countries the educational system is required to give extra social and ideological instruction to citizens.

Two aspects of education have been taken for consideration here. Countries representative of different regions of the world have been selected. One aspect measures the 'means of production', the availability of teachers, given the number of potential pupils, the other the 'product', the number of literate people.

Availability of teachers. Table 5.13 shows the number of school-age children per teacher in selected countries. Potential school-age children are between 5 and 19, a notional age group in excess everywhere of the age group theoretically obliged to attend school (eg 5–16 in UK). School enrolment is far from complete in many developing countries and each teacher in Upper Volta is not actually confronted with a class of 534 children. Even so, classes can be very large in many developing countries in areas where attendance is high. The actual quality or qualifications of teachers and the quantity and quality of buildings and facilities are not evaluated in the data.

If the teacher situation in North America (24), the USSR (28) and Europe (30) is satisfactory, then the less developed countries, with 63 children per teacher on average would immediately need more than twice as many teachers as they have now to reach a

Table 5.13 *Number of school-age children per teacher in 1975 in selected countries (a low score is 'good')*

Country	Best 15 Pupils/teacher	Country	15 largest in total popn Pupils/teacher	Country	Worst 15 Pupils/teacher
1 Sweden	17	1 China	46	1 Tanzania	168
2 Israel	18	2 India	80	2 Rwanda	168
3 Norway	20	3 USSR	28	3 Gambia	173
4 Denmark	20	4 USA	24	4 Haiti	186
5 Iceland	22	5 Indonesia	91	5 Senegal	196
6 Canada	23	6 Brazil	43	6 Mali	203
7 UK	23	7 Japan	38	7 Somalia	244
8 Belgium	23	8 Bangladesh	120	8 Afghanistan	258
9 Switzerland	23	9 Pakistan	115	9 Burundi	265
10 USA	24	10 Nigeria	140	10 Mozambique	272
11 France	24	11 Mexico	58	11 Yemen	303
12 Finland	24	12 German FR	30	12 Ethiopia	342
13 Lebanon	24	13 Italy	28	13 Niger	420
14 Libya	24	14 UK	23	14 Chad	468
15 Australia[a]	25	15 France	24	15 Upper Volta	534

Averages for selected regions

Region	Pupils/teacher	Region	Pupils/teacher
North America	24	World	50
Europe	30	More developed	27
Latin America	53	Less developed	63
Asia	61		
Africa	111		

Source: Population Reference Bureau: World's Children Data Sheet, 1979

[a] Also at 25 German DR, Luxembourg, Kuwait

European level and after a possible 50 percent increase in the number of children by the year 2000, three times as many teachers as now.

Whatever the defects of the educational system in the present developed countries literacy is now nearly universal except among older age groups. The level of literacy in a country may be taken as a rough measure of success of a conventional education system. Up-to-date literacy data are not readily available for most countries but the data for selected countries in *Table 5.14* give a fair idea of the situation in the last two decades.

Most developed countries record near complete literacy but Italy still has members of older generations unable to read, especially in the South, while Portugal still had a poor record in 1970. Literacy levels differ markedly among developing countries and the data in *Table 5.14* show Argentina to have a level comparable with that in developed countries while only two decades ago hardly anyone at all was literate in Saudi Arabia. Literacy levels in developing countries are highest in Latin America

and lowest in much of Africa and in parts of Southwest Asia.

From the data provided it emerges clearly that in developing countries some sectors of the population are 'deprived' with regard to literacy and presumably to education in general. Urban literacy levels are much higher than rural ones. The more compact urban populations are for one thing easier to reach and organize than dispersed populations in many rural areas. Life is difficult in an urban environment if you cannot read. Another disparity is the fact that a larger proportion of men are literate than women in most developing countries and even in Portugal, though not among the Coloured or Bantu population of South Africa. The disparity reflects presumably the priority given in most societies to educating males, when facilities are limited, rather than greater mental ability.

The data for India show the effect of the double disparity urban/rural and male/female in the literacy rate of 72 percent for urban males against only 12 percent for rural females. The revealing figures

Table 5.14 *Literacy levels in selected countries*

Country	Date	Age		Percentage literate Total	Male	Female
USA	1969	14+		99	99	99
Japan	1960	15+		98	99	97
Italy	1971	15+		94	95	93
Argentina	1970	15+		93	94	92
Sri Lanka	1971	15+		78	86	68
Mexico	1970	15+	Total	74	78	70
			Urban	83	87	80
			Rural	60	66	55
Portugal	1970	15+		71	78	65
South Africa	1960	15+	Total	57	57	57
			White	98	98	98
			Asiatic	74	87	60
			Coloured	69	67	70
			Bantu	41	42	41
India	1971	15+	Total	33	47	19
			Urban	60	72	45
			Rural	26	40	12
Afghanistan	1975	15+	Total	12	19	4
			Urban	27	36	14
			Rural	11	16	2
Ethiopia	1965	15+		6	8	4
Ivory Coast	1965			5	8	2
Saudi Arabia	1962	15+		2	5	0

Source: UNESCO Statistical Yearbook 1980, Table 1.3

obligingly produced by South Africa for 1960 show also great contrasts between ethnic groups in that country, with virtually universal literacy among all Whites compared with little over 40 percent among Bantu, a level however far higher than that among Africans in some other parts of the continent. One therefore finds contrasts in educational levels due to age, sex, type of settlement, cultural and religious background and income or social class.

Education is one of the most difficult aspects of development to assess. The relevance of education is often questioned, especially by those who expect it to be exclusively vocational. Compulsory education for older children and 'necessary' higher education for younger adults is a valuable way of keeping job seekers from the labour market. On the other hand the time 'lost' acquiring secondary and higher education may sharpen the expectations of those participating and cause frustration when the expectations cannot be fulfilled. All kinds of formal school education have been criticized by some, as for example I. D. Ilich (1973), and in some African countries it has been noted that quite irrelevant subjects have been taught in schools while the traditional 'education' whereby children simply grow up learning from their elders in the fields and homes has been lost or disrupted.

Health services

Two widely used measures of the availability of health services are the number of physicians and of hospital beds in a given country or region. Unfortunately the definition of physician varies from country to country while the usefulness of a hospital bed (or place) depends on the standard of equipment and nursing. The availability of hospital beds to population is used here since it is more straightforward to measure and compare than health personnel.

If 100 persons or fewer per *hospital bed* is taken as a desirable (or acceptable) level of hospital facilities then the data in *Table 5.15* show that much of West Europe together with the USSR, Japan, Australia and New Zealand are well provided. The USA (155), Canada (108) and the UK (113) are not so well provided as most other developed countries. The hospital bed 'gap' in the world is, however, clearly shown by the presence of 15 countries with more than 1000 people per hospital bed. Haiti is the only Latin American country here while Argentina (176) is not far above the USA. With regard to educational facilities, southern and southeastern Asia are rather better provided than Africa, but the reverse is true for the provision of hospital facilities. If every part of the world were to have no more than 100 persons per

Table 5.15 *Population per hospital bed in the mid-1970s in selected countries (a low score is 'good')*

Best 15		15 largest in total popn		Worst 15	
Country	Popn/bed	Country	Popn/bed	Country	Popn/bed
1 Finland	65	1 China	na	1 Haiti	1037
2 Sweden	66	2 India	1645	2 Burma	1128
3 Norway	68	3 USSR	83	3 Nigeria	1168
4 Australia	81	4 USA	155	4 Upper Volta	1174
5 USSR	83	5 Indonesia	83	5 Niger	1200
6 German FR	84	6 Brazil	264	6 Chad	1248
7 Scotland	86	7 Japan	95	7 Korean R	1406
8 Switzerland	87	8 Bangladesh	4868	8 Mali	1426
9 Austria	88	9 Pakistan	1903	9 India	1465
10 German DR	93	10 Nigeria	1168	10 Indonesia	1625
11 Italy	94	11 Mexico	863	11 Pakistan	1903
12 Mongolia	94	12 German FR	84	12 Ethiopia	3277
13 Ireland	94	13 Italy	94	13 Bangladesh	4868
14 Japan	95	14 UK	113	14 Afghanistan	5879
15 New Zealand	97	15 France	97	15 Nepal	6626

Source: UNSYB 78 Table 207

na · not available

hospital bed then something like a tenfold increase in the number of hospital beds would be needed in the developing countries, and for the Scandinavian level to be achieved throughout the world roughly a twentyfold increase would be needed.

How successful are health personnel and hospital establishments in reducing, curing or preventing disease and ill health? For simplicity one measure of the success of health services is taken here, the rate of deaths of infants in their first year of life. This measure correlates highly with mortality levels in general. Census data in the later 18th century show that two hundred years ago infant mortality rates in Scandinavia were roughly comparable with those now found in the countries with the highest levels in Africa and Asia. In two hundred years, infant mortality levels in Sweden dropped by the early 1980s to about one-twentieth of what they had been.

Improvements in health conditions are a high priority of virtually every government in the world. In developing countries they have led to a rapid reduction in mortality rates, which has not usually been offset by a corresponding reduction in fertility rates. In developed countries health care has brought about the virtual eradication of many diseases that afflicted and wiped out large numbers of people, regardless of age, and has left old age itself as the ultimate problem, with vast amounts of research effort and money being spent to reduce mortality only marginally.

Table 5.16 shows the infant mortality indices for selected countries. Rates of up to several hundred have been estimated in some parts of the world in the past, as for example in parts of China at times. A rate of zero is technically unlikely and even the very low rates achieved in Japan and some European countries are maintained through elaborate health systems. The countries with the worst rates are mainly in Africa in spite of the apparent superiority of that continent compared with southern Asia with regard to availability of hospital beds. Bolivia (168) has the worst rate in Latin America and contrasts with very low rates in parts of the Caribbean and Central America (Puerto Rico 18, Cuba 19, Costa Rica 22). The Soviet level of 36 is surprisingly high but may be to some extent attributed to comparatively poor health facilities in some Asian Republics of that country. To bring infant mortality levels throughout the world to 20 per 1000 (the average level for all developed countries) or better would require enormous improvements in the quantity and quality of food intake and in basic health facilities in the developing world.

5.6 Gross national product

The usefulness of an abstract unit of measurement such as money to assess production and consumption is that it allows the comparison and aggregation of unlike goods and services. Gross national product is one measure of the total production of goods and services in a year.

Table 5.16 *Annual number of deaths of infants under one year of age per 1000 live births mostly in 1978 or 1979 (a low score is good)*

Best 15		15 largest in total popn		Worst 15	
Country	Infant mort	Country	Infant mort	Country	Infant mort
1 Sweden	7	1 China	56	1 Bolivia	168
2 Japan	8	2 India	134	2 Zaire	171
3 Finland	8	3 USSR	36	3 Laos	175
4 Netherlands	8	4 USA	13	4 Somalia	177
5 Norway	9	5 Indonesia	91	5 Ethiopia	178
6 Denmark	9	6 Brazil	84	6 Congo	180
7 Switzerland	9	7 Japan	8	7 Upper Volta	182
8 France	10	8 Bangladesh	139	8 Afghanistan	185
9 Canada	12	9 Pakistan	142	9 Mauritania	187
10 Australia	12	10 Nigeria	157	10 Chad	190
11 Belgium	12	11 Mexico	70	11 Central African R	190
12 USA	13	12 German FR	15	12 Angola	192
13 UK	13	13 Italy	15	13 Niger	200
14 Hong Kong	13	14 UK	13	14 Mali	210
15 Singapore	13	15 France	10	15 Guinea	220

Source: Population Reference Bureau 1981 World Population Data Sheet

Global measures such as gross national product have been criticized for various reasons. Different countries measure it differently. In many there is a considerable sector of production that does not officially enter the money economy, subsistence production. Gross national product is not necessarily correlated with happiness. It is in reality a measure of the speed at which non-renewable natural resources are being used up. The real value of a currency changes through time (usually declining). Exchange rates are highly changeable and are often fixed arbitrarily so when, as is common in the Western industrial countries, other currencies are expressed in US dollars, these drawbacks must be appreciated.

Whatever the problems in using gross national product it is a concise way of comparing the size of different economies and the amount of goods and services produced per inhabitant. It is sometimes used as a criterion for determining which countries most need development assistance. Gross national product may be used as a measure of the size and possible influence of a country in world affairs. *Table 5.17* shows the 20 countries with the largest gross national products in 1979.

Table 5.17 *The twenty countries with the largest gross national products in 1979 in thousands of millions of US dollars*

Country	Dollars	Country	Dollars	Country	Dollars
1 USA	2384	8 Canada	229	15 India	126
2 USSR	1085	9 China	219	16 Mexico	108
3 Japan	1020	10 Brazil	201	17 German FR	107
4 German FR	718	11 Spain	163	18 Belgium	107
5 France	531	12 Netherlands	143	19 Sweden	99
6 UK	354	13 Poland	136	20 Switzerland	90
7 Italy	298	14 Australia	131		

CHAPTER SIX

Development and Living Standards

6.1 Differences within countries

Before the development gap is examined in the world as a whole it is of interest to examine contrasts *within* countries because such regional inequalities make a useful comparison with international inequalities. Four countries will be briefly examined here, the largest in the world in population size. Given the presence in the world of some 4500 million people in the early 1980s it is necessary to look at world problems in terms of aggregates of people, grouped either by territory in countries or provinces or by class, defined according to income level, type of employment or in some similar way. The territorial subdivisions used for the four countries that follow mostly have populations counted in millions. *Within* each unit, therefore, one can expect to find smaller groupings or individual families far richer or poorer than the average.

USA

The USA is characterized by a high absolute income per inhabitant compared with that in most countries. In 1979 the personal income was about 8900 dollars per inhabitant. *Figure 6.1* shows the 'richer' and 'poorer' states of the USA. Much of the Northeast, the Middle West and the extreme West had above average incomes while the states with an income per inhabitant more than 10 percent below the US average were mostly in the southeast (the old 'Deep South'), in the western half excluding the Pacific states and in some New England states. The poorest state in 1979 was Mississippi, with 71 percent of the US average. Not counting the 'freak' cases of Alaska (127 percent, subsidized?), District of Columbia (125 percent, the capital Washington) and Nevada (177 percent, gambling) the richest states were Connecticut (114 percent) in the northeast and California (114 percent) on the Pacific coast.

The contrast between rich and poor states in the USA goes back to times before the Civil War (1861–65) when the 'South' had tended to stagnate on a slave-based agricultural economy while the 'North' was already becoming industrialized, had an extensive rail system and was bringing the fertile agricultural lands of the Middle West into the economy. In the 1860s the United States was still basically a rural country but with its large area of potential agricultural land it did not experience the accumulation of a very large agricultural population on a limited extent of cultivated land, as happened over thousands of years in India, China and some other parts of the world, leaving there enormous numbers of rural poor.

By comparison with many countries, regional contrasts in the USA are not now very marked. As recently as 1960 the gap between incomes per inhabitant of the poorest and richest states was much greater than now, being 54 percent for Mississippi against 129 percent for Connecticut (the US average still 100). In real terms personal income in the USA has increased by more than 70 percent between 1960 and 1979 but on the whole the increase has been faster in the states that were poorer. In the interwar period the gap between the poorest and richest states was even higher than in 1960.

USSR

The USSR publishes sets of data that show the degree of variation among major regions in purchasing power and in the availability of health and educational facilities. The Communist Party of the USSR has inherited from the Russian Empire the largest country in the world in area and one in which marked economic and cultural contrasts existed in the decades before the Revolution of 1917. One of the priorities of Soviet leaders since 1917 has been to reduce differences in living standards between different parts of the country.

The data in *Figure 6.2* are for retail sales. The Soviet average per inhabitant was 968 roubles in 1979 (about 1500 US dollars) but the measure is more restricted than personal income in the USA and the

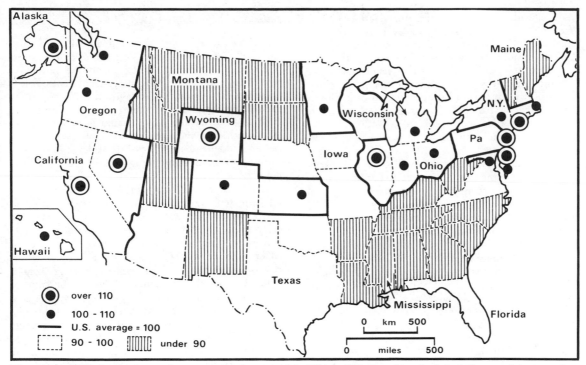

Figure 6.1 *Variations in average income among the states of the USA in 1979. Data from Statistical Abstract of the United States 1980, p 447*

Figure 6.2 *Retail sales per inhabitant in republics and regions of the USSR. Data from Narodnoye khozyaystvo SSSR v 1979 godu, Moscow 1980 p 455*

state provides its citizens with many services either 'free' or at a nominal cost. The 24 regions on the map consist of 14 individual Soviet Socialist Republics and 10 subdivisions of the largest, the RSFSR (Russian Republic). The divisions with retail sales per inhabitant in excess of the national average are either in the northwest part of European USSR or in Siberia (including the Far East). The divisions with the lowest retail sales are mostly in non-Russian areas colonized in the 19th century, particularly in the Middle Asia group.

Purchasing power in the Soviet Union varies considerably at the level of aggregation of the 24 units used here and even more at a more detailed regional level. Tadjikistan (568 roubles) and Azerbaijan (571) are barely 40 percent of the level of Latvia (1386), Estonia (1422) or Moscow and Leningrad.

Two factors make a major contribution to present contrasts in living standards in the USSR. Firstly, the 'colonial' contrasts between comparatively prosperous regions of European USSR and the backward regions of the South have not been eliminated. Secondly, natural resources per inhabitant are more abundant in Siberia and Kazakhstan than in much of European USSR or the south. Productivity per worker in both agriculture and industry is much higher in some regions than in others and the question no doubt arises in Soviet planning circles as to whether people in different regions should be remunerated for what they actually produce or whether the less productive should be subsidized at the expense of the more productive.

In the USSR at present industrial workers tend to be better remunerated than agricultural workers. Thus the highly industrialized, mainly urban, Centre region (1257 roubles) is much more prosperous than the largely agricultural and deeply rural adjoining Blackearth Centre region (689 roubles). The policy of reducing regional inequalities in the USSR is further impeded by the need to pay workers more for working in harsh environmental conditions.

The Soviet case has been discussed here at some length because in a way the USSR is a world in miniature. It can be argued that if marked regional inequalities still exist in the USSR after more than six decades of central planning committed to building communism, there is little immediate prospect of substantially reducing the development gap in the world as a whole. While however living standards in Moscow, Latvia and Estonia are roughly three times as high as they are in Soviet Middle Asia, those in Sweden (close to Estonia and Latvia) are perhaps 30 times as high as those in Afghanistan, which adjoins Soviet Middle Asia. The comparison must be made with many reservations and without necessarily supporting the methods used in the USSR whereby colonial areas have been ruthlessly brought into the Soviet political and economic system.

India

India (*see Figure 6.3(a)*) was gradually absorbed into the British Empire in the 18th and 19th centuries. Some parts of the region subsequently became more developed than others as commercial agriculture was favoured in certain regions. In the 20th century modern factories were established in existing towns such as Bombay and Calcutta and in new towns in areas with coal and non-fuel minerals. When India became independent in 1947 considerable regional differences in development already existed in the country.

Attempts by successive Indian governments have not affected regional inequalities greatly. Johnson (1979) quotes data (*Figure 6.3(a)*) showing that in the early 1970s, against an average for India of about 700 rupees of income per inhabitant, the state of Maharashtra had 1075. Maharashtra includes Bombay as well as some of India's most fertile agricultural land. At the other extreme the level in the state of Bihar was 403 in spite of the establishment there of much new industry. The presence of Delhi, the national capital, contributed to the above average level in Haryana, but West Bengal (568) remains poor even by Indian standards in spite of containing the large industrial city of Calcutta.

Regional inequalities in income levels in India may to some extent be due to a lack of comparability in the way income is assessed in different states and also to cultural attitudes towards development. They do not appear to be related to any great extent to differences in bioclimatic resources since the areas with superior conditions for agriculture (soil fertility, rainfall, irrigation) have accumulated a greater density of population per unit of cultivated area over the long time they have been settled.

China

China remained largely isolated from the world economic system until the middle of the 19th century. It came under pressure from European powers including Russia and from the USA to trade with the industrial countries of the time. For some decades outside influence came to China largely through ports along the coast. The British colony of Hong Kong, on the coast of Guangdong province, influenced the south of China, while treaty ports further north along the coast of China brought innovations and pressure to several provinces.

After 1911 China itself started to modernize, a process continued by the Japanese after 1930 as they occupied the northeast of the country and then attempted to take the rest. When the Communist Party under Mao finally gained control of 'mainland' China in 1949 much of China's industrial capacity had been destroyed.

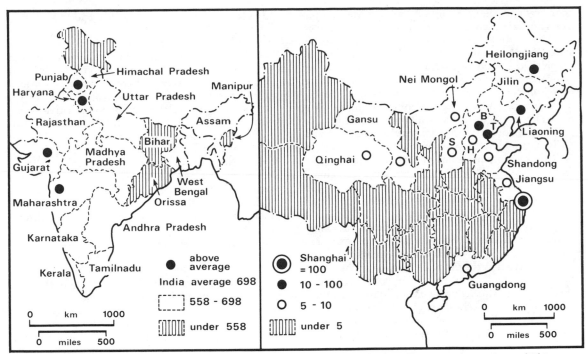

Figure 6.3(a) *Income per inhabitant in the states of India. Source: Johnson (1979)*

Figure 6.3(b) *Level of indusrialization in the provinces of China. Source: Chung-Tong Wu (1981)*

Unfortunately, data are not readily available to indicate the extent of regional inequalities in China. Industrial and mining workers, especially those in the larger, more sophisticated factories and mines, may be expected to have a considerably higher standard of living than most agricultural workers. The data in *Figure 6.3(b)*, given by Chung-Tong Wu (1981) do reflect, reasonably well, contrasts in living standards among 29 administrative divisions. Three are special urban provinces and some are special ethnic divisions.

If the gross value of industrial output per inhabitant in Shanghai, the most highly industrialized province of China, is expressed as 100 units, then that in the other two urban provinces Beijing (44) the national capital and in Tianjin (49), its port, is about half the Shanghai level. The three provinces of northeast China, industrialized by the Japanese in the 1930s, have comparatively high scores. About half of the population of China, however, lives in provinces with less than one-twentieth of the amount of industrial output per inhabitant of Shanghai, while the score for the lowest division, Xizang (Tibet), is only 1.

China has frequently been praised in the West for the way starvation and poverty have been eliminated since 1949 or at least the way in which poverty is shared. Symbolically, everyone has a bicycle and

no-one has a car, actually an impossibility when there are less than 100 million bicycles yet 1000 million people. During the 1950s, when Soviet influence was strong, the policy in China was to decentralize industry, establishing and expanding it in hitherto basically agricultural provinces. The estimates of Chung-Tong Wu show that between 1952 and 1974 very little reduction actually occurred in the concentration of Chinese industry inherited from the decades before 1949, although total capacity has grown. It seems reasonable to conclude that great contrasts in living conditions still exist throughout China between town and country, and also between the industrial provinces of the north and the coastal ports and the rest of the country. Even brief visits by the author to some major towns of China and a deeply rural area (Guilin) shows clearly what contrasts could be found (Cole 1980).

Marked regional differences in living standards exist in many countries of the world. According to a model proposed by Williamson (1968) they are least marked when a country is not developed at all and is uniformly poor, though even here certain social classes may enjoy high living standards. Regional contrasts grow as a country develops. They are particularly marked in Latin American countries where often one or a few urban areas dominate the economy and attract wealth. Brazil and Mexico, for example, have income differences per inhabitant of as much as tenfold between the richest and poorest states. In poor countries like Nigeria and Indonesia urban centres and oilfield areas have higher living standards than most other areas. Spain and Italy are two European countries in which industrialization came first to certain regions. Both countries have marked regional differences in living standards.

In some countries that are highly developed and fairly small in total population, a further stage in Williamson's model seems to have been reached. In Sweden, Australia and the German FR regional inequality is still found but it appears to be diminishing and is not so marked as in the countries considered so far.

This section may be concluded with two questions. If marked regional inequalities exist and in some cases are even increasing within individual countries, what hope is there that they may be reduced in the world as a whole? And, given that many countries devote effort and funds to reducing inequalities *within* their own limits is it likely the richer countries will be deflected to divert effort and funds to poorer countries *outside* their own limits? After all, charity begins at home.

6.2 The consumption of goods and services

The gap between rich and poor countries is a subject that has been given much publicity since the Second World War. Politicians, international bodies and philanthropic institutions have thought up various ways to narrow the development gap. In chapter 4 it was shown that in relation to their population, some countries have far more natural resources than others. Some resource-poor countries such as Japan and Switzerland have high standards of living but some resource-rich regions like the interior of Brazil and much of Zaire are poor. In chapter 5 it was shown that the countries with high material standards are those in which the economically active population in agriculture forms only a small proportion of total economically active population. High productivity from workers in agriculture, industry and services is a feature of the richer or developed countries.

In chapter 5 gross national product was used to measure the size of economies according to the total size of production of goods and services in different countries. A large proportion of the total production of goods is consumed, while some is used to maintain or expand the productive capacity of the country and some is exported. National income is a closer

Table 6.1 *Gross national product per inhabitant in US dollars in 1979 in selected countries*

Over 4 times world average (9 320)		15 largest countries in popn		Under 10% of world average (234)	
Country	*GNP/ inhabitant*	*Country*	*GNP/ inhabitant*	*Country*	*GNP/ inhabitant*
Kuwait	17 270	1 China[a]	230	Sri Lanka	230
Qatar	16 590	2 India	190	Rwanda	210
United Arab Emirates	15 590	3 USSR	4 110	Malawi	200
Switzerland	14 240	4 USA	10 820	India	190
Luxembourg	12 820	5 Indonesia	380	Burundi	180
Sweden	11 920	6 Brazil	1 690	Upper Volta	180
Denmark	11 900	7 Japan	8 800	Afghanistan	170
German FR	11 730	8 Bangladesh	100	Viet Nam	170
Belgium	10 890	9 Pakistan	270	Burma	160
USA	10 820	10 Nigeria	670	Mali	140
Norway	10 710	11 Mexico	1 590	Ethiopia	130
Brunei	10 680	12 German FR	11 730	Somalia[b]	130
Iceland	10 490	13 Italy	5 240	Nepal	130
Netherlands	10 240	14 UK	6 340	Chad	110
France	9 940	15 France	9 940	Bangladesh	100
Canada	9 650			Kampuchea[b]	90
		World	2 340	Laos	90
				Bhutan	80

Source: Population Reference Bureau, 1981 World Population Data Sheet

[a] contrast with 460 on 1980 sheet!
[b] estimated by the author

measure of the consumption of goods and services than gross national product but the two indices show very similar contrasts in the world in values per inhabitant.

Table 6.1 shows the gross national product *per inhabitant* in selected countries of the world in 1979. Among those with over four times the world average are a number that are very small in population but produce large quantities of oil, mainly for export. The rest are in North America or West Europe (Australia nearly makes it with 9100). All the poorest are in Asia or Africa and they include India, with its very large population. The lowest scoring country in Latin America is Haiti, at 260, almost in the list of countries with under one-tenth of the world average per inhabitant.

In most developing countries the leaders are committed to trying to raise material standards and many of their fellow citizens realize that they are missing something at present by being poor, even if they only aspire to modest improvements in the availability of water and food supply, housing and health care. The development issue is a major one in world affairs. In order to appreciate specific features and problems of the gap between rich and poor countries in the world it is necessary to examine different individual sectors of consumption of goods and services.

The average citizen of the USA has a gross national product valued in US dollars as about 100 times as great as that of a citizen of Bangladesh. It is not proposed here to debate whether he (or she) is 100 times happier. Only when the money value is expressed in real goods and services can the problem be properly appreciated. It needs little thought to appreciate that the gap between richest and poorest is not always at the ratio of 100 to 1 for gross national product. The world food intake gap is very serious yet it is roughly a gap of 2 to 1 whereas the private car gap may be largely unnoticed although most families in the USA have at least one car whereas in China no family possesses a car exclusively for its own use.

The six items of consumption to be examined in this section for 60 countries are tabulated in *Table 6.7* together with other items for the 60 largest countries of the world in population in the late 1970s. For ease

Table 6.2 *Consumption of energy and steel and availability of cars*

Energy consumption in kilograms of coal equivalent per inhabitant		Steel consumption in kilograms per inhabitant		Passenger cars per 1000 inhabitants	
Country	Cars	Country	Cars	Country	Cars
1 USA	11 550	1 Czechoslovakia	700	1 USA	510
2 Canada	9 950	2 USA	618	2 Australia	411
3 Czechoslovakia	7 400	3 German DR	591	3 Canada	395
4 German DR	6 790	4 USSR	567	4 German FR	324
5 Australia	6 660	5 Canada	550	5 France	316
6 Netherlands	6 220	6 Poland	540	6 Belgium	295
7 Belgium	6 050	7 German FR	538	7 Italy	293
8 German FR	5 950	8 Japan	512	8 Netherlands	281
9 UK	5 270	9 Romania	464	9 UK	254
10 USSR	5 260	10 Belgium	388	10 Japan	179
World	2 070	World	177	World	80
51 Nigeria	90	51 Tanzania[a]	5	51 Uganda	2
52 Tanzania	70	52 Zaire	3	52 Sudan	2
53 Madagascar	70	53 Mozambique	2	53 Ethiopia	2
54 Zaire	60	54 Ethiopia	2	54 Viet Nam	2
55 Burma	50	55 Sri Lanka	2	55 Afghanistan	2
56 Uganda	50	56 Bangladesh	2	56 Burma	1
57 Afghanistan	40	57 Burma	1	57 India	1
58 Ethiopia	30	58 Afghanistan	1	58 Bangladesh[b]	–
59 Bangladesh	30	59 Nepal	1	59 Nepal[b]	–
60 Nepal	10	60 Uganda	1	60 China[b]	–

Sources: UNSYB 1978, New York 1977, Tables 142, 174, 158

[a] Madagascar also 5
[b] means under 0.5 per 1000

of reference only scores for the highest and lowest scoring ten countries are given in *Tables 6.2, 6.3* and *6.4.* In anticipation it will be seen in *Table 6.7* that there is a fairly continuous distribution of scores from the highest scoring country to the lowest scoring one, a feature not apparent when the extreme values only are taken.

Energy consumption (see Table 6.2)

This covers a wide variety of uses including fuel and power for industry and transport as well as for domestic heating and cooking. The ingredients range from oil and electricity to firewood and animal dung. The world energy consumption gap is of the order of 1000 to 1, a somewhat nebulous statistic. Given however that modern industrial growth has been based largely upon the increasing use of inanimate sources of energy to supplement or replace human and animal muscle, then the ten countries consuming 100 kilograms or less per inhabitant have a long way to go even to reach the present world average of around 2000 kilograms. It is true that they are mostly in the tropics and do not need domestic heating but this use makes up only a small part of the total energy consumed in the top ten consumers.

A carefully applied cut by the big users of 10 or even 20 percent of their energy consumption would not affect greatly their own life styles yet if transferred to the very low consumers could help immensely by reducing the use of firewood as a cooking fuel and by leaving animal dung to be returned directly to the soil or to be transformed into fuel gas and fertilizer. It would not however be easy to persuade the profligate users of energy, whether the owners of large cars in North America or the wasteful burners of coal and lignite in the Soviet bloc, to cut their consumption and to give some of their fuel away.

Steel consumption

Steel consumption, like energy consumption, covers a variety of uses. Much steel goes into construction or into the engineering industry. Eventually a considerable amount ends up in consumer goods.

Table 6.3 *Availability of telephones, secondary schools and food*

Telephones in use per 1000 inhabitants		Secondary educational attendance %		Calorie supply per inhabitant as % of requirements	
Country		Country		Country	
1 USA	744	1 Romania	97	1 Portugal	141
2 Canada	618	2 USA	96	2 Czechoslovakia	141
3 Japan	424	3 Japan	95	3 Italy	140
4 Netherlands	418	4 German FR	92	4 Bulgaria	139
5 UK	415	5 Canada	90	5 Belgium	138
6 Australia	404	6 Netherlands	89	6 Greece	138
7 German FR	374	7 Belgium	86	7 USSR	136
8 France	329	8 UK	85	8 France	135
9 Belgium	315	9 Chile	85	9 Hungary	134
10 Italy	285	10 German DR[b]	84	10 USA	134
World	110	World	51	World	107
51 India[a]	3	51 Burma	26	51 Sudan	88
52 Pakistan	3	52 Uganda	22	52 Ethiopia	88
53 Indonesia	3	53 Nigeria	19	53 Nigeria	88
54 Afghanistan	2	54 Bangladesh	19	54 Algeria	86
55 Nigeria	2	55 Sudan	18	55 Philippines	86
56 Zaire	2	56 Ethiopia	15	56 Mozambique	85
57 Bangladesh	1	57 Afghanistan	13	57 Tanzania	84
58 Nepal	1	58 Pakistan	12	58 Bangladesh	84
59 Burma	1	59 Mozambique	12	59 Zaire	83
60 China	1	60 Nepal	10	60 Afghanistan	82

Sources and definitions of variables see Table 6.8

[a] also at 3: Viet Nam, Ethiopia, Sudan
[b] also at 84: Yugoslavia

The steel consumption gap is several hundred to one, a gap comparable to that for energy. Data for the 1970s show that the consumption per inhabitant has actually tended to decline in recent years. Imports are stagnating and population is increasing. The ten countries with the lowest consumption of steel do not actually produce any steel themselves. They have few or no proved coking coal or iron ore reserves and their home markets are too small even to need a modern iron and steel works of a minimum viable size.

Passenger cars

Passenger cars in circulation indicate yet another enormous disparity between the richest and poorest countries. One man's necessity is another man's luxury. If all the world were as motorized as the USA then there would be about 2000 million cars in use and running them would cause the world's oil reserves to disappear in a few years. While a car of some sort is within the reach of most families in North America, Australia and New Zealand, West Europe and Japan, it is still a privilege of the few in the USSR and East Europe. In developing countries car ownership is fairly high among the better off urban dwellers in Latin America and even among the rich in much of Africa and Asia. In Bangladesh the owner of a car is an 'aristocrat' and in China passenger cars are only used as taxis or by institutions for non-private use.

Telephones in use (see Table 6.3)

Telephones in use reveal an ever more marked gap between the extremes of North America and parts of Asia. As with cars, telephones are available to most families in the western industrial countries whereas in developing countries subscribers are generally the better off urban dwellers. Internal communications between different regions are mainly for special users such as government officials or the military.

Secondary education

This is an example of a service far more comprehensively provided in developed countries than in developing areas. The exact percentage of participation should not be taken too seriously. Even so it is clear that an education gap exists, one more marked in reality than the data indicate since teaching facilities and standards on the whole are much higher in the rich countries than in the poor ones. Like many development variables education is both a result of development and a cause of it.

It has frequently been argued that if basic and technical educational levels were raised in developing countries and more places were available then young people would strive to achieve high standards of living and would have the skills to handle the means of production needed to achieve them. On the other hand educational qualifications may raise the hopes and expectations of the young yet not be required at least locally. The author recalls a class of girls in a remote village in the Andes learning office skills. Those skills were not in demand locally at all. Some might have left to try their luck in a distant town.

Food intake

Food intake can be measured both in terms of total calorie intake and more specifically with regard to the quantity and proportion in the diet accounted for by such items as proteins and fats. Given the great importance of and preoccupation with food consumption two measures have been used here to show the gap between the richest and poorest countries. They both measure total calorie intake but in different ways. The data in *Table 6.4* show the average daily calorie intake per inhabitant during 1975–77. In contrast the data in column 3 of *Table 6.3* measure actual intake in terms of whether it falls above or below a theoretical average calorie intake per inhabitant regarded as necessary to maintain moderate activity. The estimate, made by the Food and Agriculture Organisation, takes into account among other things the age–sex structure of the population, climate and average body weight.

The data in *Table 6.4* show that the calorie intake gap between extreme countries in the world was 2 to 1 between Poland and Chad. All but two of the top 15 consumers are in Europe, but Canada and Australia, as well as the UK and the German FR, do not fall far behind. Japan, at 2850, is however considerably below the level in other industrial countries. On the other hand Argentina and Uruguay have high levels. The world average was 2590 and China with an index of 2440 did not fall far below

Table 6.4 *Calorie intake in selected countries in 1975–1977 in calories per inhabitant per day*

Country	Calories per inhabitant per day	Country	Calories per inhabitant per day
Poland	3650	Zambia	2020
German DR	3610	Ghana	2010
Bulgaria	3590	Upper Volta	2000
Belgium	3570	Laos	1980
Austria	3550	Afghanistan	1970
USA	3540	India	1950
Ireland	3520	Bangladesh	1950
Hungary	3490	Mozambique	1930
Yugoslavia	3470	Guinea	1920
Italy	3460	Yemen DR	1900
France	3460	Mauritania	1890
Greece	3440	Kampuchea	1860
New Zealand	3440	Ethiopia	1840
USSR	3440	Chad	1790

this level. The countries with the lowest average food consumption are mostly in Africa or southern Asia in areas that have been afflicted frequently by droughts, floods and in several cases long conflicts.

Column *3* in *Table 6.3* contains the mildly optimistic index of 107 for average food consumption in the world implying that more food is actually consumed than is needed according to the criteria taken by FAO. While the citizens of the rich countries might not like to be accused of over-eating, they apparently consume much more food than they really need, even Japan with an index of 121 going to excesses. Food intake in developing countries is on average four percent below the desirable level and what the data do reveal is that while China and Latin America are coping with food supply except locally, hundreds of millions of people in large areas of Africa and southern Asia are badly undernourished. Food statistics are, however, very suspect because they are usually only very rough estimates. Politically it may be useful for some countries to understate what they think their consumption levels to be in order to attract aid, and for others to overstate consumption, as perhaps China has done since 1949, in order to 'prove' the merits of a particular political system.

One cannot help feeling that the FAO data indicate more an imbalance in food availability in the world as a whole than an overall shortage. It is hardly likely, however, that the rich countries would voluntarily forgo consumption of food in order to help out the deprived countries. Even if they did, then the means of redistributing more than just some crumbs from the rich man's table seem totally inadequate at present.

6.3 Agricultural employment and urbanization

One variable that reflects development more perhaps than any others is employment in agriculture. In the late 1970s slightly less than half of the world's economically active population was engaged in agriculture. The extremes were Bhutan and Nepal (in the Himalayas) with 94 and 93 percent engaged in agriculture compared with a mere two percent in the UK and USA.

The total population of any country can be divided into four subsets of people: agricultural and non-agricultural, each of which includes economically active and dependent people. The situation is illustrated by the example of Turkey which has a population of about 44 million and about 10 million persons working in agriculture.

Total population (1979)	44 200 000
Agricultural	24 700 000
Economically active	10 300 000
Dependent	14 400 000

Non-agricultural	
Economically active	8 200 000
Dependent	11 300 000

Turkey is roughly self-supporting in food so each person working in agriculture supports 4–5 people altogether (including themselves). In deeply rural Bangladesh, about 85 percent of the economically active population is in agriculture. Out of a total population of 86 million, 25 million are economically active in agriculture, while under 5 million are in all other branches of activity. Thus each person working in agriculture supports only 3–4 people altogether.

Like Turkey and Bangladesh the USA produces most of its own agricultural needs and even exports some products. In 1979 a mere 2 250 000 persons were engaged in agriculture out of a total population of 220 million. Each person working in agriculture, therefore, supports about 100 fellow citizens. In the UK, also, only one person per hundred is engaged in agriculture but the country only produces about two-thirds of its food needs and very few agricultural raw materials or non-alcoholic beverages.

The data in *Table 6.5* show the countries with the highest and lowest proportions of economically active population in activities other than agriculture and the indices for the 15 largest countries in population. In the world as a whole some 820 million persons working in agriculture supported a total population of 4340 million, rather more than 1 to 5. In Europe the ratio was about 1 to 14 and in the USSR about 1 to 12.

Development is characterized by an increase in the productivity of each person engaged in agriculture, which releases workers to move into other activities. Whether the process started through pressure of population in rural areas compelling people to leave the land or whether increased yields in existing areas of cultivation released people will not be debated here. Certainly the UK was 'non-agricultural' long before 1841, by which date only 22 percent of the economically active population was engaged in agriculture. Not until the 1860s was less than half the economically active population in agriculture in the USA and France. Italy and Japan only crossed the half way mark in the 1920s and the world as a whole in the early 1970s.

Such predominently agricultural countries as those in column *3* of *Table 6.5* with over four-fifths of their population in agriculture would need *at least* several decades before they could expect to have less than half of their economically active population in agriculture. On this evidence development characterized by widespread industrialization and high energy consumption is not just round the corner.

As people move out of the agricultural sector they tend to move from comparatively small rural villages and hamlets into urban settlements. The definition of

Table 6.5 *Economically active population not engaged in agriculture as a percentage of total economically active population in 1979*

15 with 'highest' score		15 with largest popn		15 with 'lowest' score	
Country	%	Country	%	Country	%
1 UK	98	1 China	39	1 Tanzania	18
2 USA	98	2 India	36	2 Upper Volta	18
3 Singapore	98	3 USSR	83	3 Papua-New Guinea	17
4 Hong Kong	98	4 USA	98	4 Mauritania	17
5 Belgium-Luxembourg	97	5 Indonesia	40	5 Bangladesh	16
6 German FR	96	6 Brazil	61	6 Malawi	16
7 Puerto Rico	96	7 Japan	88	7 Lesotho	16
8 Canada	95	8 Bangladesh	16	8 Chad	16
9 Sweden	95	9 Pakistan	46	9 Burundi	16
10 Switzerland	95	10 Nigeria	46	10 Madagascar	16
11 Australia	94	11 Mexico	63	11 Mali	12
12 Netherlands	94	12 German FR	96	12 Niger	11
13 Israel	93	13 Italy	88	13 Rwanda	10
14 Denmark	93	14 UK	98	14 Nepal	7
15 Norway	92	15 Viet Nam	29	15 Bhutan	6
		World	54		

Source: FAOPY 79, Table 3

urban varies considerably among the countries of the world. Usually it takes into account the size of a settlement (eg over 2000 people urban) or its status (eg district capital) or the function (eg whether it is predominantly non-agricultural). *Table 6.6* has been included largely for reference. The reader will find a fairly close correlation between percentage of economically active population not in agriculture and percentage of population defined as living in urban settlements. Again it is not easy to say that spontaneous urbanization has been a cause or a result of development, but many non-agricultural sectors of the economy, whether 'goods' like mining and manufacturing or 'non-goods' like administration, commerce, finance and education, benefit from concentration, and in many countries the creation of urban centres has been an essential part of the drive towards development.

Table 6.6 *Urban population as a percentage of total population*

Country	%	Country	%	Country	%
Singapore	100	China	25	Ethiopia	13
Belgium	95	India	22	Tanzania	12
German FR	92	USSR	65	Uganda	12
Hong Kong	90	USA	74	Bangladesh	11
Israel	89	Indonesia	20	Yemen	10
Netherlands	88	Brazil	61	Malawi	10
Iceland	88	Japan	76	Swaziland	9
Australia	86	Bangladesh	11	Mozambique	9
New Zealand	85	Pakistan	28	Upper Volta	8
Uruguay	84	Nigeria	20	Oman	7
Denmark	84	Mexico	67	Lesotho	5
Sweden	83	German FR	92	Nepal	5
Argentina	82	Italy	69	Rwanda	4
Chile	81	UK	76	Bhutan	4
France	78	Viet Nam	19	Burundi	2
		World	41		

Source: Population Reference Bureau 1981 World Population Data Sheet

Table 6.7 *Data matrix for sixty countries by twenty-two variables*

Country	1 Area	2 Popn	3 Area/ popn	4 Land res	5 Energy res	6 Mineral res	7 Non agric	8 Urban	9 Popn change	10 Cement prod
1 China	9 561	975.0	10	45	37	19	39	26	12	50
2 India	3 046	676.2	5	75	5	15	36	21	19	30
3 USSR	22 402	266.0	84	230	473	240	83	62	8	500
4 USA	9 363	222.5	42	285	340	230	98	74	7	340
5 Indonesia	1 492	144.3	10	55	22	67	40	18	20	20
6 Brazil	8 512	122.0	70	85	3	180	61	61	28	190
7 Japan	370	116.8	3	30	3	12	89	76	9	660
8 Bangladesh	143	90.6	2	40	0	0	16	9	26	0
9 Pakistan	212	86.5	2	70	1	0	46	26	28	40
10 Nigeria	924	77.1	12	70	69	0	46	20	32	20
11 Mexico	1 973	68.2	29	85	30	150	63	65	31	220
12 German FR	248	61.1	4	45	137	43	96	92	−2	520
13 Italy	301	57.2	5	75	6	15	88	67	3	680
14 UK	244	55.8	4	40	126	15	98	78	1	280
15 France	547	53.6	10	115	7	0	91	73	3	550
16 Viet Nam	333	53.3	6	55	0	0	29	19	23	20
17 Philippines	300	47.7	6	70	0	73	53	32	24	100
18 Thailand	514	47.3	11	155	0	73	24	13	23	120
19 Turkey	781	45.5	17	135	0	10	44	45	25	340
20 Egypt	1 000	42.1	23	55	14	0	49	44	27	90
21 Iran	1 648	38.5	43	95	837	0	61	47	30	180
22 Korean R	98	38.2	3	35	6	0	60	48	16	410
23 Spain	505	37.8	13	180	6	78	82	70	9	790
24 Poland	312	35.5	9	135	103	250	69	57	10	630
25 Burma	678	34.4	20	100	0	38	47	24	24	10
26 Ethiopia	1 184	32.6	36	105	0	0	20	13	25	0
27 Zaire	2 345	29.3	80	75	8	530	25	30	28	20
28 South Africa	1 223	28.4	43	130	86	1280	71	48	28	260
29 Argentina	2 777	27.1	102	470	33	0	87	80	16	240
30 Colombia	1 138	26.7	43	75	12	0	72	60	21	140
31 Canada	9 976	24.0	416	550	161	1910	95	76	8	440
32 Yugoslavia	256	22.4	11	125	9	39	61	39	9	390
33 Romania	238	22.3	11	150	23	0	52	48	10	620
34 Morocco	444	21.0	21	95	0	460	48	42	30	150
35 Algeria	2 382	19.0	125	90	371	23	49	55	34	110
36 Sudan	2 506	18.7	134	120	0	0	23	20	31	10
37 Tanzania	937	18.6	50	90	0	0	18	13	31	10
38 Peru	1 285	17.6	73	110	13	530	62	62	28	130
39 German DR	108	16.7	6	85	79	0	90	76	0	720
40 Afghanistan	658	15.9	41	140	0	0	22	11	27	10
41 Kenya	583	15.9	37	55	0	0	22	10	39	90
42 Czechoslovakia	128	15.4	8	110	68	0	89	67	7	660
43 Sri Lanka	66	14.8	4	60	0	0	46	22	22	30
44 Australia	7 695	14.6	527	650	441	2730	94	86	8	370
45 Netherlands	34	14.1	2	40	170	0	94	88	4	280
46 Nepal	141	14.0	10	40	0	0	7	4	24	0
47 Malaysia	333	14.0	24	120	60	190	51	27	25	150
48 Venezuela	912	13.9	66	120	467	65	81	75	30	270
49 Uganda	236	13.7	17	80	0	0	19	7	30	10
50 Iraq	435	13.2	33	130	736	67	59	66	34	230
51 Ghana	239	11.7	20	50	0	37	48	36	31	70
52 Chile	757	11.3	67	100	8	1460	81	80	14	110
53 Hungary	93	10.8	9	160	35	0	83	52	3	440
54 Mozambique	783	10.3	76	135	0	0	35	8	26	20
55 Cuba	115	10.0	12	115	0	170	76	64	12	280
56 Portugal	92	9.9	9	115	0	0	73	29	7	480
57 Belgium	31	9.9	3	60	0	0	97	95	1	790
58 Greece	132	9.6	14	90	13	230	62	65	7	1170
59 Bulgaria	111	8.9	12	155	45	50	66	60	5	540
60 Madagascar	587	8.7	67	135	0	0	16	16	26	10
Unweighted mean	1 774	68.0					58	47	19	270

Sources and definitions of variables given in Table 6.8

11 Energy prod	12 Steel prod	13 Energy cons	14 Steel cons	15 Pass cars	16 Tele-phone	17 Popn/doctors	18 Infant mort	19 Pupils/teacher	20 Secondary education	21 Food reqt	22 GNP
730	30	710	38	0	1	3 000	56	46	30	97	460
200	20	220	16	1	3	3 960	134	80	27	89	180
6 580	580	5 260	567	20	75	300	31	28	80	136	3700
9 600	530	11 550	618	510	744	620	13	24	96	134	9700
840	0	220	8	4	3	16 390	91	91	37	94	360
250	100	730	107	55	41	1 650	109	43	53	106	1570
340	920	3 680	512	179	424	850	8	38	95	121	7330
10	0	30	1	0	1	14 180	153	120	19	84	90
120	0	180	9	3	3	3 530	142	115	12	92	230
2 440	0	90	27	3	2	14 810	157	140	19	88	560
1 520	90	1 230	100	45	59	1 900	70	58	55	116	1290
2 680	630	5 920	538	324	374	500	15	30	92	129	9600
510	420	3 280	368	293	285	490	18	28	73	140	3840
3 540	360	5 270	349	254	415	760	14	23	85	133	5030
850	420	4 380	368	316	329	680	11	24	82	135	8270
110	0	120	7	2	3	17 180	115	54	40	106	170
20	0	330	31	9	13	3 150	80	52	60	86	510
20	10	310	31	6	8	8 370	68	64	34	104	490
310	40	740	112	12	28	1 760	119	68	45	112	1210
730	10	470	26	8	14	1 170	90	80	38	105	400
14 160	0	1 490	137	28	23	2 750	112	56	53	97	1440
480	80	1 020	186	4	54	1 950	38	71	63	117	1160
540	310	2 400	249	167	261	550	16	35	64	130	3520
5 890	510	5 250	540	46	84	620	22	36	76	133	3660
60	0	50	1	1	1	5 410	140	111	26	99	150
0	0	30	2	2	3	93 970	162	342	15	88	120
100	0	60	3	4	2	28 800	160	88	44	83	210
2 990	280	2 990	161	85	83	2 020	97	68	74	117	1480
1 650	110	1 800	140	80	90	520	45	27	64	124	1910
800	10	690	31	19	56	1 820	77	62	55	93	870
11 350	600	9 950	550	395	618	570	12	23	90	128	9170
1 380	110	2 020	239	90	71	790	34	38	84	133	2390
3 960	540	4 040	464	5	56	750	31	33	97	123	1750
50	0	270	44	20	12	11 140	133	115	28	107	670
5 450	10	730	110	17	16	4 940	142	67	34	86	1260
0	0	140	10	2	3	9 490	141	155	18	88	320
0	0	70	5	3	4	18 480	125	182	28	84	230
450	20	640	37	19	26	1 560	92	54	73	99	740
4 690	410	6 790	591	132	171	520	13	25	84	132	5660
190	0	40	1	2	2	26 840	226	258	13	82	240
10	0	150	15	9	10	10 790	83	52	49	92	320
5 660	1020	7 400	700	124	190	400	19	32	69	141	4720
10	0	110	2	7	5	4 010	42	46	54	94	190
8 880	550	6 660	365	411	404	720	12	25	81	126	7920
8 620	360	6 220	322	281	418	600	10	35	89	124	8390
0	0	10	1	0	1	37 060	133	157	10	92	120
1 010	0	640	50	49	30	6 580	44	52	50	115	1090
16 580	70	2 840	285	80	60	870	45	59	56	97	2910
10	0	50	1	2	4	24 700	120	133	22	92	250
15 110	0	730	60	14	28	2 560	104	49	46	99	1860
50	0	160	11	7	7	10 310	115	63	47	100	390
560	50	990	60	29	48	2 180	40	42	85	112	1410
2 120	350	3 550	343	71	103	440	24	29	74	134	3450
60	0	130	2	11	6	16 390	148	272	12	85	140
30	40	1 230	53	8	33	1 120	25	28	65	118	810
90	40	1 050	156	118	120	820	39	34	55	141	2020
900	1150	6 050	388	293	315	490	12	23	86	138	9070
850	50	2 250	176	69	250	480	19	43	70	130	3270
1 620	300	4 710	276	15	107	450	22	34	80	139	3200
0	0	70	5	7	4	10 340	102	121	33	104	250
2 460	190	2 170	177	80	110	7 330	75	73	55	110	2400

Table 6.8 *Sources of data in Table 6.7*

S1 *United Nations Statistical Yearbook 1978*, New York 1977.
S2 *1980 World Population Data Sheet*, of the Population Reference Bureau, Inc., Washington DC.
S3 *Food and Agriculture Organization Production Yearbook, 1979*.
S4 Crowson, P., *Non-fuel minerals data base*, Royal Institute of International Affairs, Jan. 1980.
S5 *World's Children Data Sheet*, of the Population Reference Bureau Inc., Washington DC.

Definitions of variables and sources

Variable

1 *Area* in thousands of square kilometres (S1).
2 *Population* in millions to nearest hundred thousand (S2).
3 *Square kilometres* per 1000 inhabitants (S1 and S2).
4 *Land resource* units per inhabitant, the world average being 100 (based on data in S3, Table 1).
5 *Fossil fuel* reserve units per inhabitant, the world average being 100 (based on data in S1, Tables 50–53).
6 *Non-fuel mineral* reserve units per inhabitant, the world average being 100 (S4, various tables).
7 Economically active population *not* engaged in *agriculture* as a percentage of total economically active population (S3, Table 3).
8 *Urban* population as a percentage of total population (S2).
9 Annual rate of *Change of population* in per thousands (S2).
10 *Cement* production in kilograms per inhabitant (S1, Table 124).
11 *Energy production* in kilograms of coal equivalent per inhabitant (S1, Table 142).
12 *Steel production* in kilograms per inhabitant (S1, Table 126).
13 *Energy consumption* in kilograms of coal equivalent per inhabitant (S1, Table 142).
14 *Steel consumption* in kilograms per inhabitant (S1, Table 174).
15 *Passenger cars* in circulation per 1000 inhabitants (S1, Table 158).
16 *Telephones* in use per 1000 inhabitants (S1, Table 167).
17 Persons per *doctor* (S1, Table 207).
18 *Deaths of infants* (age under one year) per 1000 live births (S2).
19 *Number of School-age Children* per teacher (S5).
20 Percentage of eligible pupils aged 12–17 actually attending *secondary education* (S5).
21 *Calorie supply* per inhabitant as a percentage of requirements (S5).
22 Gross national product per inhabitant in US dollars (S2).

6.4 A set of development data

One of the most commonly used words in world affairs since the Second World War has been 'development'. It would be impossible to arrive at a clear definition of the term that would be to everyone's satisfaction. For the purposes of the present book it seems more appropriate to adopt the definition of development used by the United Nations (*UNSYB 78*, Table 4 and Footnotes). A division of the countries of the world is used for grouping them according to type of economy.

(1) Market Economies
   (*a*) Developed
   (*b*) Developing
(2) Centrally Planned Economies

Category 2 is not explicitly subdivided into developed and developing. The complete line-up is as follows:

(1) Market economies
   (*a*) Developed
      Countries in North America and West Europe together with Australia and New Zealand, Israel, Japan and South Africa.
   (*b*) Developing
      All the countries of Latin America (including Cuba, in reality centrally planned), of Africa except South Africa, of Asia except Israel, Japan and other Asian countries listed below in 2a and 2b, and of Oceania except Australia and New Zealand.
(2) Centrally planned economies
   (*a*) The USSR, East Europe (German DR, Poland, Czechoslovakia, Hungary, Romania, Bulgaria, Yugoslavia and Albania), Mongolia.
   (*b*) China, Korean DPR, Viet Nam.

Variations exist in the definition of 'development' and also in the grouping of countries according to perceived levels. These aspects need only a brief reference. At one time countries were referred to as *developed* and *underdeveloped* (or undeveloped). Hope (false perhaps) was given to the underdeveloped countries by redefining them as *developing*. Recently the FAO has used the terms *more developed* and *less developed*. With regard to the grouping of countries, Soviet statistical publications recognize broadly the UN line-up but use the term 'socialist' for centrally planned and have a somewhat larger list of such countries (*see* section 8.7).

The purpose of the rest of this chapter is to present for reference and comment briefly on the set of data for 60 countries, given in *Table 6.7*. The data include two variables measuring absolute size and the rest expressed in terms of per inhabitant or a percentage. Natural resources, production, consumption and demographic indices are all included. The data are then processed through a series of steps to show how closely each pair of variables is correlated and how similar various countries are to one another.

One of the outstanding features of development admirably illustrated by many of the variables in *Table 6.7* is that there are not two distinct types of country, developing and developed. On all variables there is a spread throughout the values though on some there are concentrations near each extreme of the scale. Another feature shown by the correlation in *Table 6.7* is that development is indivisible. There are broad correlations between almost all the variables that express level of production and consumption of goods and services as well as demographic characteristics and employment structure. A country that has good health care facilities can be expected to have good educational services. A country with a low consumption per inhabitant of energy will have few motor vehicles and will consume very little steel.

The use of principal components analysis to handle large amounts of data for a set of countries now has a considerable tradition. Berry (1961) was the first geographer to apply the method to world data. He was followed by Russett (1965) and others. Studies of sets of world data can be found in previous editions of *Geography of World Affairs*. As in many multivariate studies, the choice of data affects the results to some extent. It must also be appreciated that much 'crushing' of the information takes place, the price for obtaining some very concise results. The data have been chosen within the constraints of their availability to give a picture of various indicators considered to be of relevance to development. *Table 6.8* gives a full definition of each of the variables used in the data set and of their sources. The way the natural resource scores were calculated was described in chapter 4.

The data matrix contains data for the 60 largest countries of the world in population excluding the Korean DPR and Taiwan, for which adequate data were not readily available. The 60 countries included contain more than 92 percent of the total population of the world though unfortunately they vary greatly in size among themselves. In all but the first two variables the direct influence of the size of the country has been eliminated by the use of relative rather than absolute values (eg percentages, quantities per inhabitant).

Before the results of the analysis of the data set in *Table 6.7* are discussed some further technical points must be noted briefly:

(1) In most of the data the distributions of values are not *normally* distributed. Some are very skewed while some tend to be bi-modal.

(2) In most variables (eg consumption per inhabitant of energy, steel) a high value is 'good' and a low value 'bad'. With some variables, namely *9* (population change), *17* (persons per doctor), *18* (infant deaths) and *19* (pupils per teacher) a high value is 'bad' while a low value is 'good'. Strong *negative* correlation coefficients therefore occur when variables *9, 17, 18* and *19* are associated strongly with any of the other variables except *1, 2* and *11*.

(3) There were gaps in the data even for some of the countries included. Estimates were made for the following variables: China *15, 17, 20*; USSR *15*; Viet Nam *16* and *20*; Iran *22*; Romania *15*; Afghanistan *15*; Uganda *22*; Mozambique *15*; Belgium *14*; Syria *5*.

Table 6.9 shows the correlation between all possible pairs of variables. A value near +1.00 indicates a strong positive correlation while a value near −1.00 indicates a strong negative correlation. Values near 0.00 indicate little or no correlation. A careful look at the coefficients in *Table 6.9* will show that there are quite strong correlations between all pairs of variables *7–10* and *12–22*. Natural resources per inhabitant in variables *3–6* also correlate quite highly, forming a second association of variables.

Table 6.10 shows the 60 countries ordered according to their scores on the main family of variables, the development variables. The scores of the countries are a consensus of the various measures of development as combined to form the first component. Approximately half of the total information expressed by 22 variables in the original data is reduced to one supervariable. The scores in *Table 6.10* have no absolute value but they show the positions of countries in relation to an average of zero for all 60.

Several interesting features of development emerge from *Table 6.10*. There are clearly not just two distinct groups of country, developed and developing. There is some confusion and overlap among the countries ranked 20 to 25. According to the United Nations definition, Argentina, Venezuela and Chile are 'developing' while Yugoslavia, Portugal and South Africa are 'developed'. The less developed parts of Europe are similar in some respects, therefore, to the more developed parts of Latin America.

Among the developing countries, Latin American countries come near the top of the list while seven of the ten with the lowest scores are in Africa. Canada, the USA and Australia stand well above the rest of the developed countries. Again it must be stressed that the result shown in *Table 6.10* comes from the application of a particular technique to a subjectively chosen set of development variables.

Table 6.9 *Matrix of Pearson product moment correlation coefficients (r lies between +1.00 and −1.00)*

	1	2	3	4	5	6	7	8	9	10	11	12	13	14	15	16	17	18	19	20	21	22
1 Area	1.00																					
2 Population	0.46	1.00																				
3 Area/population	0.45	−0.09	1.00																			
4 Land resources	0.47	−0.09	0.84	1.00																		
5 Fossil fuels	0.36	0.00	0.32	0.30	1.00																	
6 Non-fuel minerals	0.34	−0.09	0.83	0.69	0.20	1.00																
7 Not agriculture	0.19	−0.08	0.19	0.34	0.29	0.29	1.00															
8 Urban	0.17	−0.12	0.23	0.32	0.33	0.31	0.92	1.00														
9 Population change	−0.13	−0.10	−0.02	−0.22	0.06	−0.11	−0.73	−0.62	1.00													
10 Cement production	0.02	−0.14	−0.02	0.16	0.07	0.06	0.68	0.67	−0.74	1.00												
11 Energy production	0.29	−0.05	0.37	0.39	0.89	0.28	0.44	0.47	−0.09	0.20	1.00											
12 Steel production	0.21	−0.03	0.15	0.25	0.11	0.19	0.70	0.66	−0.71	0.70	0.27	1.00										
13 Energy consumption	0.34	−0.02	0.32	0.47	0.30	0.34	0.77	0.77	−0.73	0.63	0.53	0.82	1.00									
14 Steel consumption	0.29	−0.03	0.16	0.32	0.27	0.17	0.73	0.70	−0.76	0.76	0.47	0.89	0.92	1.00								
15 Passenger cars	0.22	−0.05	0.39	0.46	0.22	0.39	0.73	0.67	−0.65	0.50	0.38	0.68	0.83	0.70	1.00							
16 Telephones	0.24	−0.03	0.33	0.42	0.19	0.34	0.73	0.68	−0.68	0.56	0.38	0.70	0.87	0.74	0.95	1.00						
17 Doctors	−0.11	−0.07	−0.05	−0.14	−0.12	−0.19	−0.59	−0.55	0.35	−0.44	−0.25	−0.32	−0.37	−0.40	−0.30	−0.31	1.00					
18 Infant deaths	−0.12	0.02	−0.08	−0.26	−0.10	−0.19	−0.80	−0.73	0.80	−0.68	−0.27	−0.65	−0.71	−0.74	−0.61	−0.65	0.57	1.00				
19 Pupils/teachers	−0.16	−0.08	−0.05	−0.17	−0.22	−0.18	−0.68	−0.66	0.51	−0.54	−0.29	−0.45	−0.50	−0.52	−0.42	−0.43	0.81	0.77	1.00			
20 Secondary pupils	0.16	−0.13	0.14	0.28	0.19	0.30	0.85	0.83	−0.74	0.72	0.35	0.71	0.77	0.79	0.65	0.69	−0.56	−0.87	−0.75	1.00		
21 Calorie supply	0.14	−0.11	0.04	0.30	0.05	0.16	0.79	0.70	−0.84	0.83	0.20	0.74	0.71	0.79	0.63	0.64	−0.49	−0.85	−0.65	0.81	1.00	
22 GNP	0.22	−0.06	0.29	0.37	0.26	0.30	0.80	0.77	−0.75	0.65	0.44	0.82	0.90	0.84	0.92	0.92	−0.36	−0.71	−0.49	0.76	0.71	1.00

Table 6.10 *Unstandardized scores of countries on first component (or eigenvalue, with 'length' of 11.37 units). This component summarises 'development' according to the variables 7 to 22, except 11*

Country	Score	Country	Score
1 Canada	26.33	31 Brazil	−2.40
2 USA	25.12	32 Peru	−3.27
3 Australia	23.92	33 Colombia	−3.68
4 Belgium	18.84	34 Malaysia	−4.03
5 German FR	18.01	35 Turkey	−4.67
6 Czechoslovakia	15.30	36 Algeria	−6.04
7 Netherlands	14.94	37 China	−6.11
8 Japan	14.46	38 Philippines	−6.62
9 France	14.18	39 Egypt	−6.77
10 German DR	14.15	40 Sri Lanka	−6.94
11 USSR	13.59	41 Thailand	−8.03
12 UK	13.12	42 Morocco	−8.14
13 Italy	11.58	43 Ghana	−8.33
14 Poland	9.73	44 Indonesia	−9.19
15 Spain	8.81	45 Viet Nam	−9.49
16 Hungary	7.62	46 India	−9.90
17 Bulgaria	7.34	47 Burma	−9.94
18 Greece	7.09	48 Kenya	−10.23
19 Romania	6.75	49 Madagascar	−10.28
20 Argentina	5.07	50 Pakistan	−11.06
21 Venezuela	3.83	51 Zaire	−11.17
22 Yugoslavia	3.54	52 Nigeria	−11.87
23 Portugal	2.49	53 Sudan	−12.13
24 Chile	2.25	54 Uganda	−13.19
25 South Africa	1.66	55 Tanzania	−13.39
26 Cuba	0.54	56 Bangladesh	−13.74
27 Korean R	−0.91	57 Mozambique	−14.03
28 Iran	−1.49	58 Nepal	−14.98
29 Iraq	−1.54	59 Afghanistan	−16.28
30 Mexico	−1.79	60 Ethiopia	−18.60

The way the 60 countries are related to the natural resource variables is not tabulated here but composite scores have been worked out and are given in *Table 4.11*, chapter 4. A further association of variables is the correlation between territorial size and population size. All other things being equal, the larger a country is in area, the more people it will have. The actual correlation for the 60 countries is +0.46.

Figure 6.4 shows the result of grouping similar countries according to the evidence provided by the 22 variables in *Table 6.7*. The idea behind the procedure used can be seen with the help of the data in *Table 6.11*, a small selection from the original data matrix given in *Table 6.7*. It will be seen in the linkage diagram that Hungry and Bulgaria, and also Egypt and the Philippines, are very similar. A comparison of the four rows of data in *Table 6.7* shows that Hungary and Bulgaria, while not identical, do indeed have roughly similar indices, as do also Egypt and the Philippines. In contrast, the developed pair differ markedly in most respects from the developing pair.

The dendrogram or linkage tree in *Figure 6.4* takes in all the sixty countries studied in this section and groups them according to their positions on four components, that is on the four dimensions that take into account some 80 percent of all the information contained on the original 22 variables.

The linkage procedure finds the most similar pair of countries according to their scores and joins them into a group. The next most similar pair is then found and the two countries are joined. The joining continues, with new pairs forming, or new entries joining existing groups. Eventually large groups are formed and finally the 60 countries are all linked.

Interestingly in *Figure 6.4* the two groups of countries that finally emerge are split between Venezuela and the USSR, with 19 developed countries and 41 developing countries coinciding with the United Nations definition with only one exception, the inclusion of South Africa among the developing countries. Its first partner is Chile.

The data used in this study were not deliberately chosen to show any partiular world situation. They are interesting because, quite impartially (if not objectively), they show that the conventional view of development is oversimplified. When Hungary, Bulgaria, Romania and Greece turn out to be quite similar to one another the reaction might be to think that the grouping is obvious because the countries are close together and ideological variables are not included in the study to distinguish Greece from the three socialist countries. On the other hand it is thought-provoking to find that the first link of the UK is with the German DR. More surprising are the various early links between developing countries in quite different parts of the world, of which the Philippines–Egypt example is only the first. Geographical proximity may be used to explain the similarity between Iraq and Iran or between Uganda and Tanzania but who would spontaneously associate Colombia and Malaysia or Burma and Kenya?

The findings of this study show that the distribution of features of development does not necessarily respect the tidy arrangement of the continents and culture regions to which we are accustomed. Countries that are far apart on the earth's surface may share common features and problems. In spite of the complex spatial situation hinted at in the present study, it is convenient for the purposes of this book to take groups of contiguous countries as a basis for the study of world affairs in chapters 9–12.

Some concluding points may now be made regarding the data used here and the findings. The developing countries tend to form groups sooner than the developed ones. Why is this so? It seems that because they all have low scores on development variables they are less differentiated among themselves than the developed countries are among themselves. For example gross national product per

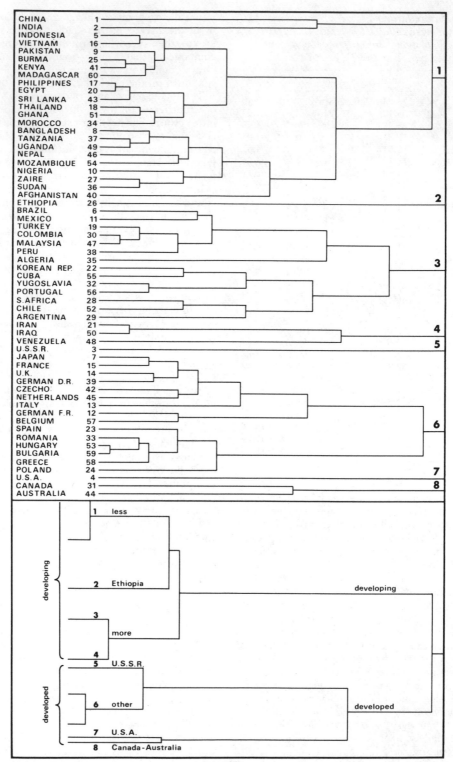

Figure 6.4 *Linkage diagram showing the joining of countries into groups according to their similarity*

Table 6.11a *Variables selected from Table 6.7*

Country	Popn 2	Land resources 4	Energy resources 5	Not agricultural 7	Urban 8	Steel cons 14	Infant mortality 18	Secondary education 20	Food reqt 21
Hungary	10.8	160	35	83	52	343	24	74	134
Bulgaria	8.9	155	45	66	60	276	22	80	139
Egypt	42.1	55	14	49	44	26	90	38	105
Philippines	47.7	70	0	53	32	31	80	60	86

Table 6.11b *Similar countries linking early in linkage tree*

This Country	with	This country		This Country	with	This country
53 Hungary		59 Bulgaria		37 Tanzania		49 Uganda
17 Philippines		20 Egypt		21 Iran		50 Iraq
30 Colombia		47 Malaysia		53(59) Hungary + Bulgaria		33 Romania
25 Burma		41 Kenya		18 Thailand		51 Ghana
17(20) Philippines + Egypt		43 Sri Lanka		30(47) Colombia + Malaysia		19 Turkey

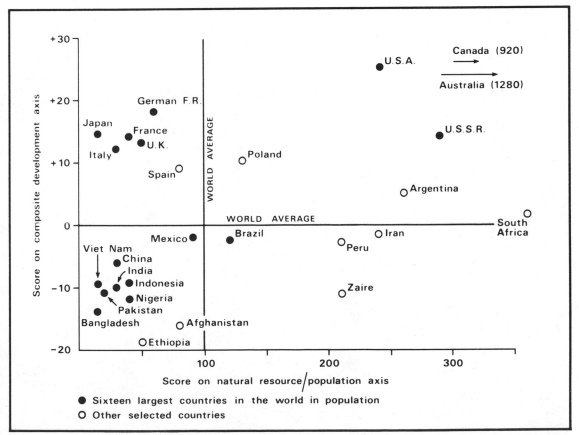

Figure 6.5 *A two dimensional view of development relating conventional development levels to levels of natural resources per inhabitant*

inhabitant ranges between about 100 dollars and 2000 dollars (Argentina) among developing countries but between about 2000 dollars (Portugal) and 10 000 (USA) among developed ones.

How easy is it for some of the developing countries to move into the group of developed countries? The Shah of Iran predicted that his country would be the fourth industrial power in the world before the year 2000. Brazil aspires to be a great world power in the next century. On the whole the prospects are not bright.

How are natural resources related to development? If we use the composite score in *Table 6.10* to represent the level of development and the composite score in *Table 4.11* to represent all natural resources per inhabitant then the result of plotting one dimension against the other on a graph shows four basic types of country in *Figure 6.5*:

(1) Resource poor, high development: Japan, West and East Europe
(2) Resource rich, high development: USA, Canada, Australia, USSR
(3) Resource poor, low development: most of southern and eastern Asia, including China, parts of Africa
(4) Resource rich, low development: Zaire, the Amazon region of Brazil.

The fact that Mexico and Brazil are intermediate on both axes does not invalidate the broad distinction between the four categories. It is thought-provoking to speculate as to how each country might change its position on the graph in the future as production and population increase or decrease and as natural resources are used up or new natural resources are discovered.

International Trade

7.1 General aspects of trade

International or foreign trade accounts for only part of the exchange of goods between different places in the world but as a result of the system of sovereign states in the world, it is sharply distinguished for statistical purposes from internal trade. The internal movement of goods in the USA, USSR and other large countries reaches distances of thousands of kilometres (eg California to New York, or East Siberia to Moscow) whereas much international trade flows between pairs of places only a few hundred kilometres apart or even less (eg France–Belgium–Netherlands–German FR). If the countries in such trading blocs as the EEC and CMEA are counted as a single trading unit, with trade between members defined as internal, then about a quarter of all international trade in the world would cease to be defined as such.

Three basic assumptions may be made about foreign trade:

(1) A country prefers to produce as many of its needs internally as it can and it imports goods (and services) either because it cannot produce them at home at all or because it can obtain them much more favourably (especially cheaply) elsewhere. Exports are therefore produced *in order to obtain* imports.
(2) Foreign trade is closely connected with other aspects of international relations.
(3) A small country like Iceland, the Netherlands or Switzerland, with a very large trade in proportion to its total production of goods and services is more sensitive about foreign relations than a large country like China, India, the USA or the USSR, for which foreign trade is only equivalent to a small part of its total production.

With regard to the first point made above, most countries are unable in practice to produce anywhere near the full range of goods they require. It is not commercially realistic, for example, to cultivate coffee in Canada or Sweden. The USA apparently has virtually no manganese or chrome ore and France no copper, lead and zinc ore reserves of its own. Many developing countries do not possess the equipment or the technical skills needed to produce sophisticated engineering goods. When, however, a country does have a choice between producing something at a comparatively high cost at home or buying it more cheaply abroad it will often settle for the home product, if necessary protecting it by using tariffs or other means to keep out the potential import. Several Latin American countries for example prefer to produce their own high-cost steel rather than buy cheap steel from a low-cost producer like Japan. The development of high-cost North Sea oil started *before* the 1973 price rise in hitherto low-cost Middle East oil.

7.2 The main trading countries

Table 7.1 shows the leading trading countries of the world. The data in columns *1* to *3* show that the top ten countries of the world in value of imports account for almost 60 percent of all world trade though they have only about 20 percent of the total population of the world. At the other extreme, China and India together have nearly 40 percent of the total population of the world but account for less than 2 percent of all world trade.

Column *4* shows how 'favourable' or otherwise the balance of trade is for each country. 'Invisible' items, including tourism, are evident. The countries with the greatest relative excess of exports over imports are the oil exporters Saudi Arabia, Iran and Indonesia, benefiting presumably from the rapid rise in value of oil exports yet not able to absorb greatly increased imports. None of the major Western industrial countries except the German FR had a favourable trade balance in 1980 and the USA was in a particularly unfavourable position due to its rising

Table 7.1 *Largest trading countries of the world in 1980*

Country	Imports 1	Exports 2	% of world imports 3	Exports/ imports 4	Imports per inhabitant 5	Import coefficient 6
1 USA	255.7	216.7	12.5	85	1 149	10
2 German FR	188.0	192.9	9.2	103	3 077	23
3 Japan	140.5	129.2	6.9	92	1 203	12
4 France	134.9	111.3	6.6	83	2 517	21
5 UK	120.1	116.4	5.9	97	2 152	27
6 Italy	99.5	77.7	4.9	78	1 740	27
7 Netherlands	76.9	73.9	3.8	96	5 454	48
8 Belgium-Lux.	71.2	64.1	3.5	90	6 913	56
9 USSR	68.5	76.5	3.3	112	258	6
10 Canada	58.5	64.3	2.9	110	2 438	24
11 Switzerland	36.4	29.6	1.8	81	5 778	35
12 Spain	34.1	20.7	1.7	61	902	17
13 Sweden	33.4	30.9	1.6	93	4 024	30
14 Saudi Arabia	30.2	109.1	1.5	361	3 683	33
15 Brazil	25.0	20.1	1.2	80	205	10
16 Austria	24.5	17.5	1.2	71	3 267	32
17 Singapore	24.0	19.4	1.2	81	10 000	223
18 Hong Kong	22.4	19.7	1.1	88	4 667	111
19 Korean R	22.3	17.5	1.1	78	584	38
20 Australia	20.3	22.0	1.0	108	1 390	14
21 Denmark	19.4	16.5	0.9	85	3 804	29
22 German DR	19.1	17.3	0.9	91	1 144	16
23 China[a]	19.0	19.5	0.9	103	19	7
24 Poland	18.9	17.0	0.9	90	532	14
25 Norway	17.0	18.5	0.8	109	4 171	33
26 Finland	15.6	14.2	0.8	91	3 250	33
27 Czechoslovakia	15.1	14.9	0.7	99	981	17
28 Yugoslavia	14.0	8.4	0.7	60	625	24
29 Romania	13.2	12.2	0.6	92	592	25
30 India	12.9	6.7	0.6	52	19	8
World	2047.5	1988.0	100.0	–	464	18

Sources: UNSYB 79/80, Table 136 for data in columns *1–4*. Population Reference Bureau World Population Data Sheet, 1980 and 1982 for data on which columns *5* and *6* are based.

1 = imports and *2* = exports in thousands of millions of US dollars in 1980; *3* = values in column *1* as percentages of world total; *4* = value of exports in column *2* as a percentage of value of imports in Column *1*; *5* = imports in dollars per inhabitant; *6* = import coefficient, or value of imports as a percentage of total gross national product.

[a] Author's estimates.

imports of oil in the 1970s. The USSR, Canada and Australia looked comparatively comfortable.

Columns 5 and 6 in *Table 7.1* are two ways of assessing the dependence of a country on imports. Column 5 shows the value of imports per inhabitant, with no account taken of gross national product per inhabitant. The USSR and Brazil trade comparatively little in relation to their population, while several smaller European countries trade very heavily. Much of the trade of such countries as Belgium and Switzerland is however with adjoining or nearby European partners.

The value of imports as a percentage of the total value of gross national product is shown in column 6 of *Table 7.1*. This index puts foreign trade in perspective in relation to the total economy of the country. The USSR stands out clearly as a country that depends comparatively little on foreign trade although on account of its large total size *and* large total production it does have a substantial share of total world trade. Belgium, the Netherlands, Switzerland, Norway and Denmark are among the countries most dependent on foreign trade, all of them considerably less 'self-contained' than the German FR or the UK. Japan is surprisingly self-contained. Even more dependent on imports than the smaller West European countries are the freak countries of Singapore and Hong Kong, really no more than large cities. Many developing countries also have high import coefficients.

7.3 Types of goods traded between major regions

The traditional view of world trade is one of a simple two-way exchange of manufactured goods originating in the developed countries for primary products originating in the developing countries. As shown clearly in chapter 4, several countries defined as developed, including the USSR, Canada and Australia, have large natural resources and consequently export large quantities of primary products. Many developing countries, on the other hand, barely produce enough food, fuel and raw materials for their own needs and an increasing number, including for example Brazil, India and Taiwan, import various primary products and export manufactured goods.

In order to obtain a broad view of the direction of flow of different types of products, world trade is examined under four categories of product and between 11 major world regions. The four categories of products are:

(1) Food, beverages and tobacco.
(2) Mineral fuels and related materials.
(3) Crude materials, excluding fuels and oils.
(4) Manufactures under three headings: chemicals, machinery and transport equipment, other manufactured goods (mainly consumer items).

Total world trade in 1977 (*UNSYB 78* Table 150) was 1124 thousand million US dollars, distributed under the above categories as follows (there is a small disparity as the totals below are 1104 thousand million)

	Dollars $\times 10^9$	%
(1) Food etc.	126.5	11.5
(2) Fuel etc.	221.1	20.0
(3) Crude materials	85.1	7.7
(4) All manufactures	671.1	60.8

Table 7.2 *Trade in food, beverages and tobacco in 1977 in thousands of millions of US dollars to the nearest 500 million*

From	To	1 WEu	2 Jap	3 USA	4 C/Au	5 EEu	6 USR	7 LAm	8 Afr	9 MEa	10 SeA	11 Chi	World	Surplus/deficit
1 West Europe		34.1	0.6	2.4	0.7	0.8	0.5	1.1	2.9	1.6	0.7	–	45.7	−14.4
2 Japan		0.1		0.2	0.1	–	–	–	0.1	0.1	0.2	–	0.9	−7.9
3 USA		5.2	2.5		1.3	0.4	0.9	1.8	1.0	0.9	1.5	–	16.0	+3.0
4 Canada/Australia		1.7	1.7	2.0	0.2	0.2	0.5	0.5	0.4	0.6	0.9	0.7	9.8	+6.5
5 E. Europe		1.7	–	0.2	–	1.1	2.1	0.1	0.3	0.3	–	0.1	7.3	+2.3
6 USSR		0.1	0.1	–	–	0.5		0.3	–	–	–	0.1	1.3	−6.2
7 Latin America		8.1	0.8	5.3	0.3	1.3	2.5	1.7	0.7	0.4	0.3	0.4	22.2	+16.6
8 Africa		5.0	0.3	1.2	0.1	0.3	0.3	–	0.5	0.2	0.1	–	8.1	+1.5
9 Middle East		0.6	0.1	0.1	–	0.1	0.1	–	0.1	0.5	–	–	1.6	−4.1
10 Southeast Asia		2.3	2.3	1.4	0.3	0.2	0.3	–	0.5	0.8	2.4	0.2	10.8	+3.6
11 China		0.2	0.3	–	–	0.2	0.2	–	0.1	0.1	1.1		2.3	+0.8
World		60.1	8.8	13.0	3.3	5.0	7.5	5.6	6.6	5.7	7.2	1.5	(126.5)	

Table 7.3 *Trade in mineral fuels in 1977 in thousands of millions of US dollars to the nearest 500 million*

From	To	1 WEu	2 Jap	3 USA	4 C/Au	5 EEu	6 USR	7 LAm	8 Afr	9 MEa	10 SeA	11 Chi	World	Surplus/deficit
1 West Europe		19.4	–	1.3	0.1	0.1	–	0.3	0.9	0.3	0.1	–	24.5	−61.8
2 Japan		–		–	–	–	–	–	–	–	0.1	–	0.2	−28.8
3 USA		1.1	1.1		1.2	0.1	–	0.5	0.1	0.1	0.1	–	4.2	−39.0
4 Canada/Australia		0.3	1.8	4.6	0.1	–	–	–	–	–	0.1	–	7.0	+1.4
5 East Europe		2.0	–	0.1	–	0.9	0.5	0.1	–	0.1	–	0.1	3.8	−5.1
6 USSR		7.6	0.2	0.2	–	6.5		0.7	0.2	–	0.3	0.2	15.8	+14.4
7 Latin America		1.0	–	11.3	1.3	–	–	5.7	0.4	0.1	–	–	20.3	+2.5
8 Africa		13.1	0.2	13.5	0.1	0.2	–	2.4	0.6	0.3	–	–	30.7	+27.2
9 Middle East		41.5	19.3	9.0	2.5	1.1	0.8	7.5	1.2	3.2	10.1	–	100.5	+96.3
10 Southeast Asia		0.3	5.7	3.1	0.3	–	0.1	0.6	0.1	0.1	2.3	–	12.9	−0.4
11 China		–	0.7	–	–	–	–	–	–	–	0.2		0.9	+0.6
World		86.3	29.0	43.2	5.6	8.9	1.4	17.8	3.5	4.2	13.3	0.3	(221.1)	

Table 7.4 *Trade in crude materials in 1977 in thousands of millions of US dollars to the nearest 500 million*

From	To	1 WEu	2 Jap	3 USA	4 C/Au	5 EEu	6 USR	7 LAm	8 Afr	9 MEa	10 SeA	11 Chi	World	Surplus/deficit
1 West Europe		16.4	0.1	0.5	0.2	0.8	0.2	0.2	0.8	0.4	0.4	1.0	20.7	−18.7
2 Japan		0.2		0.1	–	–	–	–	–	–	0.5	0.1	1.1	−10.3
3 USA		5.5	3.0		1.3	0.2	0.2	1.0	0.4	0.2	2.1	0.1	14.2	+6.1
4 Canada/Australia		3.8	1.7	4.7	0.2	0.2	0.2	0.1	0.1	0.1	0.6	0.1	13.5	+11.3
5 West Europe		1.2	–		–	0.4	0.3	–	0.1	0.1	–	0.1	2.3	−3.3
6 USSR		1.7	0.7	–		2.9		0.1	0.1	0.1	0.1	0.1	5.8	+3.6
7 Latin America		2.7	0.9	1.0	0.1	0.2	0.4	0.7	0.1	0.1	0.3	0.2	6.8	+4.2
8 Africa		3.4	0.3	0.3	–	0.3	0.2	0.1	0.2	0.1	0.2	0.1	5.3	+3.3
9 Middle East		0.4	–	–		0.1	0.2	–	–	0.1	0.1	0.1	1.1	−0.4
10 Southeast Asia		2.4	2.7	1.3	0.3	0.2	0.3	0.2	0.2	0.3	2.6	0.3	10.8	−3.4
11 China		0.3	0.3	–	–	0.2	0.2	–	–	–	0.2		1.3	+0.1
World		39.4	11.4	8.1	2.2	5.6	2.2	2.6	2.0	1.5	7.2	1.2	(85.1)	

Table 7.5 *Trade in manufactures in 1977 in thousands of millions of US dollars to the nearest 500 million*

From	To	1 WEu	2 Jap	3 USA	4 C/Au	5 EEu	6 USR	7 LAm	8 Afr	9 MEa	10 SeA	11 Chi	World	Surplus/deficit
1 West Europe		204.6	3.6	22.4	7.0	9.3	7.9	11.9	23.9	22.4	9.3	1.0	338.0	+47.9
2 Japan		12.5		19.4	6.1	0.7	1.8	6.0	5.5	8.0	16.1	2.0	77.7	+65.1
3 USA		20.1	3.8		23.3	–	0.1	14.0	4.5	6.5	6.3	0.1	80.5	+19.0
4 Canada/Australia		4.0	0.5	6.6	1.0	–	–	1.3	0.2	0.4	1.0	0.1	26.9	−12.0
5 East Europe		7.1	0.2	0.5	0.1	11.9	15.2	0.8	1.2	1.8	0.4	0.7	40.7	+5.8
6 USSR		6.6	1.2	0.1	0.1	9.3		1.1	0.5	0.7	0.4	1.0	15.3	−12.9
7 Latin America		3.0	0.4	2.7	0.2	0.1	0.1	4.0	0.3	–	0.1	–	11.1	−30.6
8 Africa		2.0	0.2	0.2	–	0.1	0.2	–	0.8	0.1	–	–	3.7	−35.8
9 Middle East		0.5	0.1	–	–	–	0.1	–	0.1	1.6	–	–	2.8	−43.8
10 Southeast Asia		8.6	3.0	9.4	4.7	0.2	0.3	0.8	1.3	3.0	6.0	0.1	37.6	−4.3
11 China		0.6	0.2	0.1	0.1	0.5	0.3	0.1	0.5	0.3	1.2		4.2	−1.1
World		290.1	12.6	78.6	38.9	34.9	28.2	41.7	39.5	46.6	41.9	5.3	(671.1)	

Eleven groups of countries have been used but not every country is included and small trade flows between some countries are also omitted. Most South African trade data are not apparently divulged to the United Nations, possibly in view of the expectation of sanctions some time in the future. As a result the row, column and 'grand' totals may not be exactly compatible with data inside the tables. Canada, Australia and New Zealand have been made into one region, *4*, while China includes adjoining Asian socialist countries.

Flows of the four sets of products given above are shown in *Tables 7.2* to *7.5* and the larger net flows between regions are shown diagrammatically in *Figures 7.1* to *7.4*. Three features of the data should be noted before brief comments are made about each of the four flows:

(1) There is almost always a flow of a given set of products in *each* direction. These are shown in the tables, but for the diagrams only the net flow is mapped, that is the difference (if any) between the larger and smaller flow except for within region flows, which are shown in their entirety.
(2) Where a region includes more than one country there will be trade between pairs of countries *within* that region, but for Japan, the USA and the USSR there is no within region international trade while none is indicated either for China and its socialist neighbours.
(3) Where a particular international flow exists but is less than 500 million dollars a symbol (–) is used.

All flows are given in thousands of millions of dollars to the nearest 500 million (eg 2.7 in full is 2 700 000 000 and 0.6 is 600 000 000).

Table 7.2 Food, beverages and tobacco, referred to as food for short (see Figure 7.1)

Over one-quarter of all international trade in food is within West Europe. The region has however a large deficit with regard to the rest of the world. Japan, the USSR and the Middle East are the other deficit regions. Of the regions with a surplus, Latin America is the largest contributor to world trade but the USA is both a major exporter and importer. The USA and Canada/Australia account for most of world grain exports but import tropical items. Latin America, Africa and Southeast Asia export items such as coffee, cocoa beans and tea. China and (within Southeast Asia) India take only a small part in world food trade, topping up on their own cereal production in bad years.

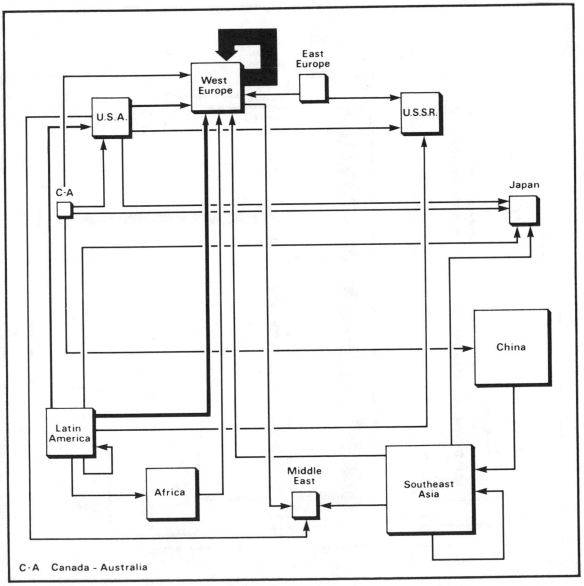

Figure 7.1   *The main flows of food in world trade in the late 1970s*

Table 7.3. Mineral fuels and related minerals (see Figure 7.2)

International trade in fuels consists largely of oil through the quantity of both coal and liquefied natural gas has been increasing in the 1970s. The role of the Middle East as the major supplier of fuels in world trade is seen clearly in *Figure 7.2*. The three main destinations of fuel are West Europe, the USA and Japan. East Europe is also a net importer but is supplied mainly by the USSR. Latin America barely produces a surplus of fuel now though two decades ago Venezuela put the region among the main sources of oil in world trade.

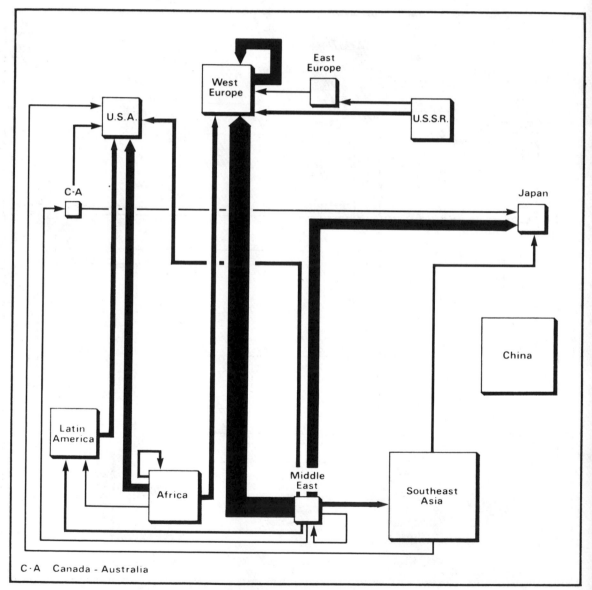

C·A   Canada - Australia

Figure 7.2   *The main flows of fuel in world trade in the late 1970s*

Table 7.4 Crude materials (see Figure 7.3)

This category includes products of plant, animal and mineral origin and, therefore, covers a wide variety of items. West Europe and Japan are the main net importers of crude materials, while Canada and Australia are major exporters. As with fuel, the Soviet Union also supplies most of the crude materials imported by East Europe. The South African contribution, not recorded here, would also be considerable. In contrast to fuel, supplied mainly by developing countries, crude materials are supplied mainly by North America, the USSR and Australia, while Southeast Asia (mainly India) and the smaller Far East industrializing countries are net importers.

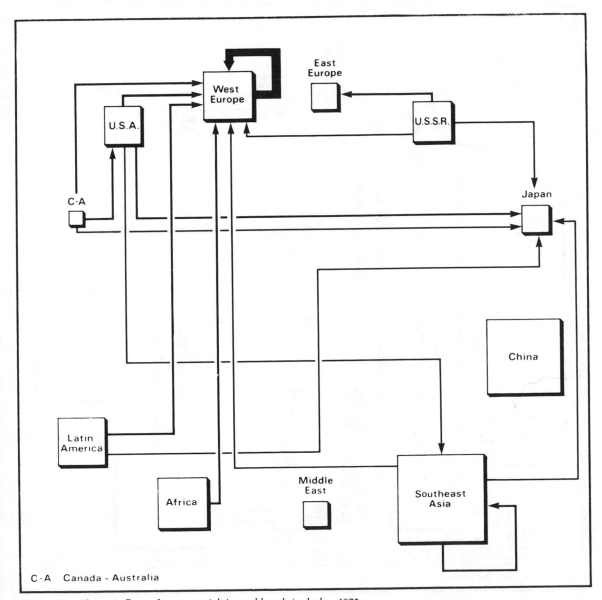

Figure 7.3 *The main flows of raw materials in world trade in the late 1970s*

Table 7.5 Manufactures (see Figure 7.4)

This broad category includes chemicals, which are marginal between crude materials and manufactures, comparatively simple consumer items such as clothing and footwear and also items such as motor vehicles, which are highly sophisticated products. The trade in manufactures shows three main characteristics:

(1) There is a very large trade in manufactured goods between developed countries. Some is 'genuine', as the exchange between the USA and Canada of manufactures for fuel and materials. Some is 'taking in each other's washing', as the sale of Japanese cars in the USA and West Europe.

(2) There is a very large trade in manufactured goods between developed and developing countries, much of it the traditional flow of sophisticated industrial products in exchange for food, beverages, fuel and crude materials. Latin America, Africa and the Middle East are far larger importers of manufactured goods than India or China, especially when the population size of the regions is allowed for.

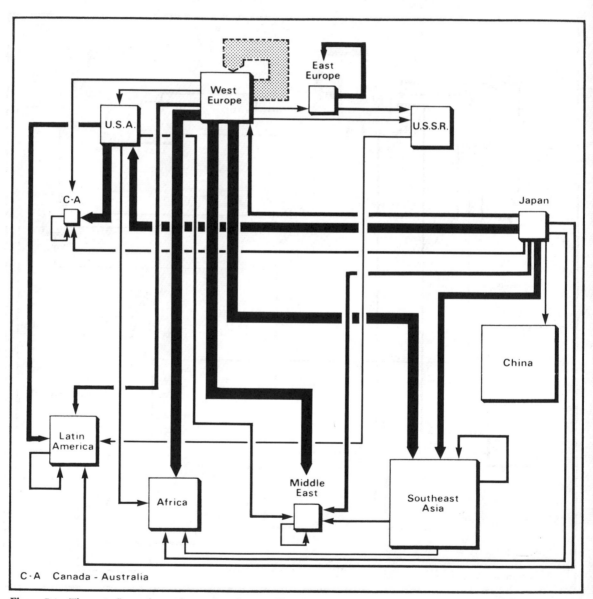

Figure 7.4   *The main flows of manufactured goods in world trade in the late 1970s*

(3) There is a small but increasing trade in manufactured goods between developing countries. Brazilian manufactures are increasingly penetrating markets in other South American countries while India is also building up an export trade, particularly in light manufactures.

For simplicity world trade may be seen to fall into three 'zones' of greatly differing size.

(1) China is a world apart containing over one-fifth of the world's population but accounting for less than one-hundredth of world trade. China really has little to spare that is not desperately needed internally but its export strategy is to produce items that contain only limited amounts of materials and yet have much value added, such as carved ivory, exotic foods, table tennis balls. Oil is however also being exported in increasing quantities. If China were to sink beneath the sea the world trading system would hardly feel a ripple.
(2) The USSR and its CMEA partners form a second zone. Since the Second World War much of Soviet foreign trade has been within the socialist bloc that it dominates. Together with parts of western USSR the East European countries provide a surplus of manufactures while Siberia and the Soviet Far East have a surplus of fuel and raw materials. Within this trade subsystem the USSR tends to export more than it imports, retaining to some degree a control over its socialist partners to which in the 1970s it was exporting oil in particular at prices lower than world prices. In view of the great importance in world affairs of Soviet foreign trade, in spite of its being only a few percent of world trade, it is discussed in a special section further on.
(3) Trade in the rest of the world retains considerable traces of an earlier and simpler pattern based on the West European sea empires. There is no longer a clear distinction between industrial and non-industrial countries and the prospect seems to be that as time passes the most favoured countries will be those like Canada, Australia and the oil producers, which have fuel and/or raw materials with surpluses of food, rather than the increasing number of countries that depend heavily on imported primary products to support predominantly industrial economies.

7.4 Trading partners

To appreciate the influence of international trade on world affairs it is helpful to have a geographical view of the links between individual countries, as well as the links between the groups of countries used in the previous section. If it is accepted that a country imports because it needs certain products that it does not produce at home then by finding which are the main suppliers of the imports of each country a picture of dependence and possible influence can be produced.

Three developed countries, the USA, the German FR and Japan together account for nearly 30 percent of all world trade and fewer than ten developed countries (*see Table 7.1*) together account for half. *Figures 7.5* and *7.6* show all the countries of the world that have one of the seven largest industrial powers as the largest single supplier of imports.

The *USA* is the largest source of imports for almost all the other countries in the Americas. Its share of their imports tends to diminish from Canada and Mexico through the Caribbean into southern South America. Cuba is exceptional in the Americas in receiving about half of its imports from the USSR. During the Second World War the USA was the only major supplier of manufactured goods to the rest of the Americas, but its dominance there has been reduced since the 1940s particularly by the German FR and Japan. The USA also has considerable trade links with a number of countries in Asia and with the increase in its imports of oil since the late 1960s its trade with the Middle East has grown.

The *USSR* conducts about half of its foreign trade with six East European CMEA partners (see inset map in *Figure 7.5*). It is also the largest single source of the imports of its neighbours Finland, Afghanistan and Mongolia as well as of Cuba in the Caribbean.

The *German FR* is the largest single source of imports of most West European countries. It also has a substantial trade with East European countries. Outside Europe the German FR is the second largest source of imports in many African countries but its strong position in Latin America in the 1960s has been somewhat reduced by competition there from Japan.

Japan has worldwide trading connections but the countries for which it is the largest single source of imports are only in Asia. Its largest imports of oil from the Middle East have forced it to attempt to build up exports to that region.

West European countries other than the German FR are the main sources of imports for Africa. Apart from its partners Ireland and Cyprus, the UK is the largest single source of the imports of only a few African countries, all at some stage part of the British Empire. France is the principal source of imports for most of its former African colonies (now the French Community), Italy of its former colonies Libya and Somalia, and Belgium of Zaire, Rwanda and Burundi.

The comparatively simple picture of major trading partners described above is complicated by a number of features. When second and other sources of imports for each country are taken into account the great complexity of world trading links emerges.

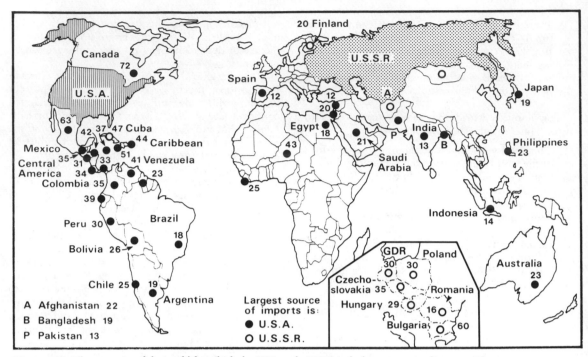

Figure 7.5 *The countries of the world for which the USA or the USSR is the largest source of imports. The percentage provided by the largest source is indicated*

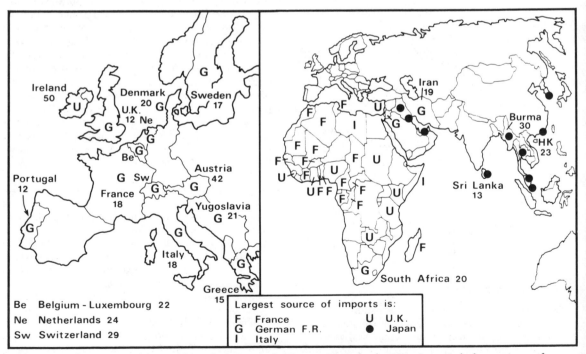

Figure 7.6 *The countries of the world for which France, the German FR, Italy, the UK or Japan is the largest source of imports. The percentage provided by the largest source is indicated for selected countries*

Table 7.6 *Largest sources of exports of countries not indicated in Figures 7.5 and 7.6*

Country	Largest source	%	Country	Largest source	%
USA	Canada	18	Malawi	South Africa	34
USSR	German DR	11	Mozambique	South Africa	20
German FR	Netherlands	12	Burundi	Belgium	–
Norway	Sweden	18	Rwanda	Belgium	21
New Zealand	Australia	20	Zaire	Belgium	15
Uruguay	Argentina	17	Ethiopia	Kuwait	–
Paraguay	Brazil	20	Syria	Iraq	15
Angola	Portugal	12	Turkey	Iraq	12
Trinidad	Saudi Arabia	22	Yemen	Saudi Arabia	12

Source: UNYITS 1979 Volume 1 Tables by countries

When changes of emphasis with regard to partners over time are also considered then the complexity is still greater.

Even when only the first source of imports is considered, as in *Figures 7.5* and *7.6*, the picture needs some further clarification. Which countries are the largest sources of imports of the USA, the USSR and the German FR, and which of the various countries such as Ethiopia and Paraguay, not on the list of any of the big seven in *Figures 7.5* and *7.6*? *Table 7.6* contains the answers.

Changes in world trading partners and patterns are usually gradual but sometimes, mainly for political reasons, they change suddenly. An example of gradual but relentless change has been the declining percentage of world trade accounted for by the UK in recent decades. Another example has been the reorientation of the trade of Australia away from Europe towards eastern Asia and the USA. More sudden changes may result from new political alignments. Around 1960, for example, Soviet trade with China diminished sharply while Soviet trade with Cuba increased rapidly. During 1977–78 the existing substantial trade between Nicaragua and the USA ceased. Uganda's imports in the late 1970s came mainly from (or through) Kenya and, of all places, Brazil.

Since the late 1960s and especially since 1973 the price of oil has increased more rapidly than the price of most other products, whether primary or manufactured. As a result, quite independently of the variations in the actual *quantity* of oil traded, fuels have claimed a larger share of world trade, as assessed by its value. Thus the oil exporting countries have gained a new prominence and influence in the world economy and world affairs.

Further details of international trade between countries can be found in the *United Nations Yearbook of International Trade Statistics*. The United Nations Statistical Yearbook also contains data (eg Table 156 in the 1978 volume) about the leading trading partners of all developed countries.

7.5 Items of trade

In section 7.3 it was shown that the traditional exchange of primary products and manufactured goods between colonial and industrial countries is now only one aspect of world trade. Developed, highly industrialized countries or regions like Australia, Canada and West Siberia have surpluses of food, raw materials and fuel. Developing countries like Brazil, India, China and the Korean Republic are increasingly selling manufactured goods outside as well as inside their own territories. As could be seen in section 7.4, however, a very large part of world trade is still accounted for by the leading western industrial countries. They trade heavily both with developing countries and among themselves.

There follows an examination of the composition of the imports and exports of selected countries. The countries chosen are intended to represent four broad types of country, those noted in *Figure 6.5*. Again the basic source for the data is the *United Nations Yearbook of International Trade Statistics*. Unfortunately the imports are broken down into different categories from the exports. It is therefore not possible to make a precise comparison of exports and imports.

Table 7.7 shows the imports of 17 selected countries distributed according to five types. Column 3 shows great variations in dependence on imported fuel, with Japan and Brazil particularly vulnerable. Column 4 also shows great variations in the importation of machinery and equipment. Japan is very selective and sparing in its imports of products in this category but Canada and Nigeria depend heavily on such products. Columns 6 and 7 show the contribution respectively of primary products (including processed materials) and of manufactures to total imports. On the horizontal scale in the graph in *Figure 7.7* there is no distinction between developed and developing countries. Developed and developing countries are included among both large and small importers (relatively) of manufactures.

Table 7.7 *The imports of selected countries: percentages of total value*

Country	Year of data	GNP 1978	Food 1	Industrial supplies 2	Fuel 3	Mach eqpt 4	Consumer goods 5	1 + 2 + 3 6	4 + 5 7
UK	1979	6 340	13	36	12	27	11	61	38
France	1979	9 940	10	34	21	23	11	65	34
Czechoslovakia	1977	5 290	9	31	15	40	5	55	45
Japan	1979	8 800	13	33	41	7	5	87	12
USA	1979	10 820	9	23	29	25	13	61	38
Canada	1979	9 650	7	23	9	50	10	39	60
Australia	1979	9 100	5	29	11	39	16	45	55
New Zealand	1979	5 940	6	38	16	31	9	60	40
Argentina	1978	2 280	6	38	12	39	5	56	44
Brazil	1979	1 690	10	29	37	22	2	76	24
Peru	1977	730	14	30	18	34	4	62	38
Nigeria	1977	670	13	28	1	47	11	42	58
Saudi Arabia	1978	7 370	10	31	1	40	18	42	58
Tanzania	1976	270	8	34	18	34	6	60	40
India	1977	190	16	36	26	20	2	78	22
Bangladesh	1978	100	20	43	14	19	4	77	23
Indonesia	1979	380	15	39	11	32	3	65	35

Sources: UNYITS 1979 Volume 1 United Nations, New York, 1980. Population Reference Bureau 1981 data sheet for GNP.

1 = Food; *2* = Industrial supplies; *3* = Fuel; *4* = Machinery and equipment; *5* = consumer goods; *6* = *1* + *2* + *3* giving, roughly, primary products; *7* = *4* + *5* giving, roughly, manufactured goods.

Note: Due to rounding and the omission of small categories of products the percentages do not necessarily add up to 100.

Table 7.8 *The exports of selected countries: percentages of total values*

Country	Year of data	Agric 1	Extr 2	Food mf 3	Basic metals 4	Text 5	Wood/paper 6	Chems 7	Non-metal 8	Metal manuf 9	Other manuf 10	Total 1–4	Total 5–10
UK	1979	2	14	6	6	6	2	17	2	42	4	28	72
France	1979	6	1	9	9	7	3	19	2	42	2	25	75
Czechoslovakia	1977	2	4	3	9	10	3	8	3	57	3	17	83
Japan	1979	1	–	1	15	5	1	9	2	65	3	16	86
USA	1979	16	4	6	3	3	4	14	1	47	3	29	71
Canada	1979	9	17	4	7	1	19	9	1	33	2	36	64
Australia	1979	24	20	23	9	2	1	7	–	5	8	76	24
New Zealand	1979	16	1	46	4	14	7	5	–	5	–	67	31
Argentina	1978	41	–	29	4	10	1	5	1	9	–	74	26
Brazil	1979	22	10	25	6	9	4	5	1	18	1	63	37
Peru	1977	17	16	18	36	3	1	7	–	2	–	87	13
Nigeria	1977	5	93	1	–	–	–	–	–	–	1	99	1
Saudi Arabia	1978	0	96	–	–	–	–	4	–	–	–	100	0
Tanzania	1976	74	4	10	1	3	–	6	–	1	–	90	10
India	1977	28	14	8	8	26	1	4	1	1	–	58	32
Bangladesh	1978	34	–	2	–	59	1	2	–	–	1	36	63
Indonesia	1979	24	61	3	3	1	2	5	–	1	–	91	9

Source: UNYITS 1979 Volume 1 United Nations, New York, 1980.

1 = Agricultural; *2* = Extractive industry; *3* = Food manufactures; *4* = Basic metals; *5* = Textiles; *6* = Wood/paper; *7* = Chemicals; *8* = Non-metal; *9* = Metal manufactures; *10* = Other manufactures.

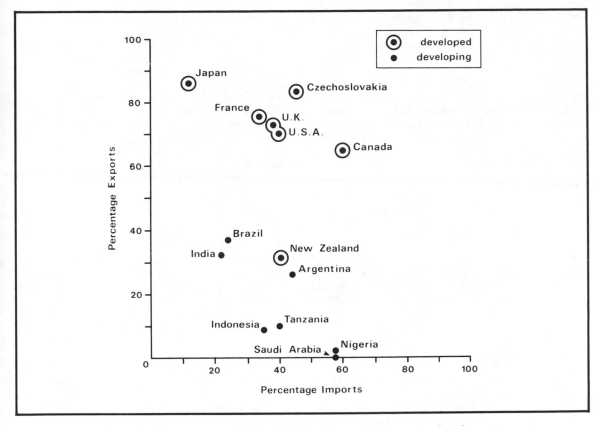

Figure 7.7 *Manufactured goods as a percentage of the total value of imports and exports in selected countries*

Table 7.8 shows the export of the 17 selected countries already discussed. Ten types of exports have been identified. Very broadly the first four are primary products, including processed materials, while the remaining six are manufactures. The two final columns of *Table 7.8* contain the respective contributions of primary and manufactured goods. All the developed countries included apart from New Zealand depend heavily on manufactures for their exports and the percentages would be higher if basic metals were subdivided into those that are merely processed at or near the source of extraction and those like steel that are made mainly in the industrial countries. Most of the developing countries depend heavily if not entirely on primary products for their exports but Bangladesh is a surprising exception. In *Figure 7.7* the distribution of the countries along the vertical axis indicates a broad distinction between the developed and developing countries. Some pairs of countries with very similar patterns of trade emerge: the USA and the UK, Brazil and India, New Zealand and Argentina, Nigeria and Saudi Arabia.

7.6 World trade in major conventional weapons

The world trade in arms and their transfer through aid has been a major and fast growing part of international transactions since the end of the Second world War. According to *SIPRI* (1981) (p xx):

'It is estimated that about 130 wars or armed conflicts have taken place in the world since 1945. Approximately 50 of these took place during the past decade. Furthermore, these armed conflicts were fought almost exclusively in the Third World and, with few exceptions, using weapons supplied by the industrial countries.'

Some 25 million people have been killed in conflicts since 1945.

During the late 1940s and 1950s much of the international flow of arms was in the form of military aid provided by the USA and USSR especially to allies in Europe and, in the case of the USA, also to allies in Asia in a zone seen to 'contain' the possible expansion of Soviet and Chinese influence beyond the socialist bloc of the time. Around 1960 China's relations with the USSR deteriorated, many former colonies were becoming independent, and Soviet trade and aid began to expand in the developing world. Since the late 1950s there has been a dramatic increase in major arms exports while military aid has become relatively much less important than it was.

Between about 1960 and 1980 the annual trade in arms increased about 12 times. In the first half of the 1960s it rose by five percent per year but in the second half by ten percent per year. According to *SIPRI* (1981) (p 105):

'During the 1970s the international trade in conventional arms increased dramatically. New suppliers and new recipients entered the arms market, the weapons became more sophisticated and expensive, and the chance of controlling the arms trade diminished.'

Developing countries have not only been the scene of numerous conflicts, both internal and international, but they have been importing arms in unprecedented quantities instead of importing goods and services that would help in their development.

In the late 1970s the USA and the USSR together provided 70 percent of the total value of major-weapons exports in the world. Industrialized countries and the Middle East each took nearly one-third of all major-weapons imports. A breakdown of exporting and importing countries or regions during 1977–80 is given in *Table 7.9*. A number of countries that are regarded as 'developing' or marginal in this respect have now joined the traditional exporters of weapons. They include Israel, South Africa, Brazil, Argentina, India and China, a reflection of the continual increase in the number of countries able to produce sophisticated weapons.

7.7 Soviet foreign trade

In the late 1970s the Soviet Union took 3.5 percent of all imports in world trade and provided four percent of all exports. Around 1980 exports were only equivalent in value to five to six percent of total Soviet production. Every year between 1949 and 1972 over half of Soviet foreign trade was with its six East European partners in CMEA. In spite of its limited size, Soviet foreign trade is a major influence in world affairs. Less is known about it in the West than about the foreign trade of most other countries and this special section is therefore devoted to it.

Ever since the 1917 Revolution, after which the Bolshevik or Communist Party seized power in Russia, Soviet politicians and planners have controlled foreign trade rigidly. In view of the hostile attitudes and actions of Western countries towards and against the Soviet Union its policy in the 1920s and 1930s was to be as self-sufficient economically as possible. With central planning of the economy and Five-Year Plans to be taken into account the production of goods specifically for export has not usually been easy in view of unpredictable changes in the world market. Throughout the Soviet period, however, it has been necessary for the USSR to obtain certain types of product outside its boundaries. Soviet leaders have no doubt appreciated also that foreign trade is a way in which a country anxious to spread its influence can sometimes put pressure on other countries.

The USSR has large and varied resources in relation to its population size. Although it is not able to produce tropical crops commercially it can exist without for example bananas, tropical oils and coffee and it can make synthetic rubber. It appears to have

Table 7.9 *Exporters and importers of major-weapons[a], 1977–80*

Exporters	Percent of world total	Importers	Percent of world total
USA	43.3	Middle East	32.0
USSR	27.4	Industrialized	31.0
France	10.8	Far East	10.4
Italy	4.0	North Africa	7.4
UK	3.7	Africa South of the Sahara	7.0
German FR	3.0	Latin America	7.0
Other industrial	5.0	South Asia	4.9
China	0.6		
Other Third World	2.2		

Source: SIPRI (1981) p. xxi.

[a] Defined as aircraft, armoured vehicles, missiles and ships.

reserves of nearly every major economic mineral though at one time was thought to lack tin ore. When the USSR began to expand its productive capacity during and after its First Five-Year Plan (1928–32), producer goods were the main type of import and the country had to export food and raw materials in exchange. After the Second World War Soviet industrial capacity increased greatly and manufactured goods began to figure prominently among exports, together with fuel (mainly oil), while raw materials remained a major export type.

In spite of efforts to increase agricultural production the USSR has increasingly become a net importer of agricultural products, especially grain.

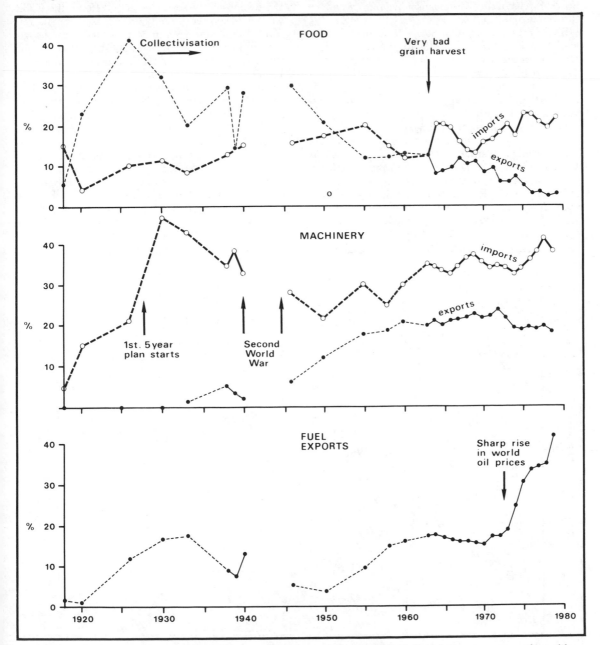

Figure 7.8 *The main types of export from and imports into the USSR during 1918–79. Data for every year are only readily available since the early 1960s. Sources: Vneshnyaya torgovlya SSSR za 1918–1940 gg Vneshtorgizdat, Moscow 1960 and various numbers of Vneshnyaya torgovlya SSSR*

At the same time it has lagged behind major Western industrial countries in some aspects of technology and continues to require various sophisticated producer goods. In the 1960s, for example, it imported much equipment to expand its inadequate chemicals industry. Space does not allow the presentation of more than a small amount of data on Soviet foreign trade. The graphs in *Figures 7.8* and *7.9* illustrate some key features.

Two distinct periods of Soviet foreign trade may easily be distinguished, 1918–40 and 1945 to the present. In the first period (*see Figure 7.8*) the USSR exported hardly any machinery and equipment. Its exports were almost exclusively food, raw materials (such as timber and furs) and fuel. Such products were exported with the main purpose of paying for capital equipment to build up its own industries. By 1921 nearly 75 percent of Soviet imports came from three leading industrial powers of the time, the USA, UK and Germany. *Figure 7.9* shows the remarkable contribution of these three countries to the establishment of the USSR as a major industrial power by 1939, the year in which Germany and the USSR signed a non-aggression treaty. Soviet purchases of

capital goods indeed helped to create employment in the West during the depression there.

After 1945 the pattern of Soviet foreign trade changed strikingly (*see Figure 7.8*). The country continued to import machinery and equipment, now mainly from its CMEA partners. It also became an exporter of machinery and equipment, but remained a net *importer*. Whereas during 1918–41 the USSR was a net exporter of food, after the bad grain harvest of 1963 it became a net importer of food. By the late 1950s, however, it had established itself as an exporter of fuel, mainly oil, which accounted for 15–20 percent of the value of its exports until 1973. In spite of selling at lower than world prices to East European partners the USSR benefited after 1973 from the relative increase of oil prices in the world and by 1979 fuel accounted for over 40 percent of the value of all exports.

Figure 7.10 shows postwar changes in the trading partners of the USSR. In the 1950s up to 80 percent of Soviet foreign trade was conducted with 'socialist' countries, either those of CMEA or with China, so great apparently was the suspicion the USSR felt towards the Western industrial powers. The graph

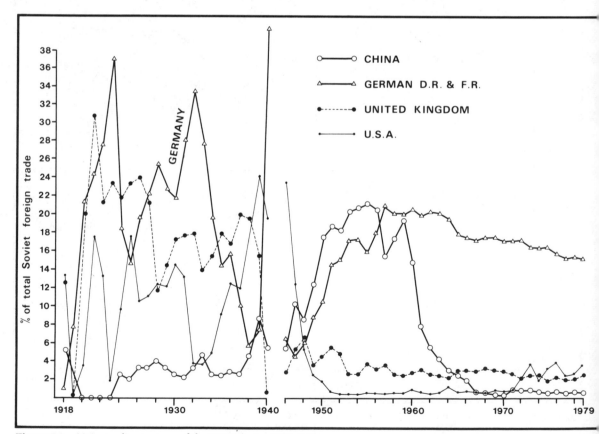

Figure 7.9   *Major trading partners of the USSR during 1918–1979*

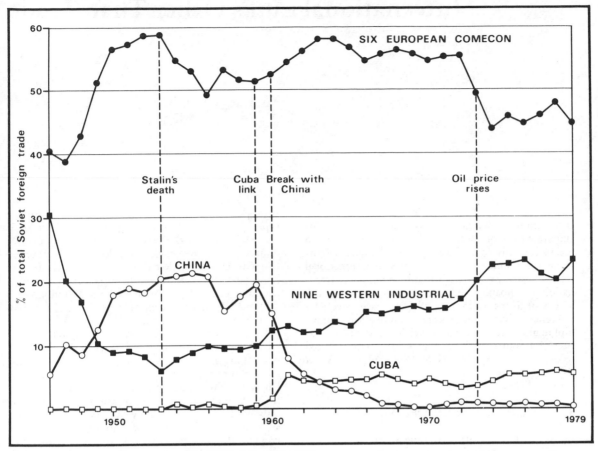

Figure 7.10 *Major trading partners of the USSR during 1946–1979*

shows the dramatic decline in trade with China after 1959 from around 20 percent of all Soviet foreign trade in the 1950s to a negligible amount after the mid-1960s. At the same time a notable trade developed with Cuba which however has only one hundredth as many people as China. The graph shows clearly how from the postwar low of about six percent in 1953 Soviet trade with nine leading western industrial countries (USA, Canada, Japan, German FR, France, UK, Italy, Netherlands and Belgium) rose gradually to 24 percent in 1979. The Soviet Union needs both food and the capital goods of these countries. It is assured of an insatiable need in West Europe and Japan for raw materials and fuels.

The geography of the USSR itself will be discussed in chapter 10 and the aims and activities in world affairs will be discussed in chapter 13. In conclusion here it may be noted that the USSR is still largely self-sufficient and any embargoes on trade imposed by Western industrial countries would only slow down the development of some sectors of the Soviet economy and could even stimulate greater effort to improve performance in some sectors like agriculture and pipeline and computer technology, which at present rely on imports.

CHAPTER EIGHT

International Links Other Than Trade

International trade is the most extensive and most elaborate form of international transactions but by no means the only one relevant to world affairs. In this chapter a number of other aspects of international relations will be dealt with: aid, investment, transnational companies, links through supranational ties and world institutions. Lack of space precludes reference to yet other aspects, including for example diplomatic links, international airline systems, migration and tourism. These topics have been covered briefly in earlier editions of *Geography of World Affairs*.

8.1 Development assistance

During and immediately after the Second World War massive international transfers of assistance were made on a scale hitherto not experienced. The USA gave or lent funds to help countries affected by the war, notably in Europe and also Japan. Such transfers were mostly from one developed country to other developed countries. Since 1950 most development assistance has gone from developed to developing countries though for example both China and Venezuela, regarded as developing countries, have also given assistance. As their colonies became independent the former colonial powers were obliged to help to put the new countries on their feet financially.

Who gives assistance?

The donors of development assistance are diplomatically subdivided in United Nations publications into two groups: developed market economies and centrally planned economies. The outflow from developed market economies is more complicated in its composition than that from centrally planned economies. Three categories of assistance from market economies are recognized: official development assistance, other official flows and private capital. In 1977 these accounted respectively for 32, 7 and 61 percent of the total of the 44 654 million US dollars that formed the net outflow of all resources from developed market economy countries. In the same year the total commitments of capital by the centrally planned economies were 3735 million US dollars, only eight percent of the amount committed by the market economies.

It is debatable whether private capital should strictly be considered as aid since it is placed in developing countries with the purpose of gaining some returns (or profit). Foreign aid, on the other hand, should include gifts from charitable institutions, though these, however well-intentioned, are only very small in quantity. Official development assistance will be taken here to represent closely enough the net outflows of assistance from the developed market economy countries. Such assistance is given either bilaterally (about two-thirds) or through multilateral institutions such as the World Bank.

The amount of assistance given by developed countries differs greatly from country to country not only according to the absolute size of the economy of the donor, but also as a proportion of total gross national product. *Tables 8.1* and *8.2* show the main donors of assistance among the developed countries. The two types of donor will be considered in turn.

All the major Western donors of assistance are included in *Table 8.1*: *Column 1* shows the great difference in the total flow of resources from countries. For its size, Switzerland is a very large foreign investor but it gives little actual aid. As would be expected, on account of the great size of its economy the USA provides the largest quantity of funds.

Column 2 shows the actual official development assistance provided in 1977. The absolute amounts must be considered against the total gross national product of each country.

Column 3 shows that Sweden, the Netherlands and Norway were the most generous of the donors. Of

Table 8.1 *The flow of official development assistance from western industrial countries in 1977 in millions of US dollars in columns 1 and 2*

Country	Total flow of resources 1	Official development 2	2 as percent of total GNP 3	2 in dollars per inhabitant 4
1 Sweden	1 445	776	0.9	95
2 Netherlands	1 957	893	0.8	64
3 Norway	495	287	0.8	72
4 Denmark	357	256	0.5	50
5 France	4 652	2 266	0.5	42
6 Australia	593	427	0.4	31
7 Belgium	865	371	0.4	37
8 Canada	2 219	985	0.4	42
9 New Zealand	67	53	0.4	17
10 UK	5 866	901	0.4	16
11 Austria	250	110	0.2	15
12 German FR	4 536	1 280	0.2	21
13 Japan	5 075	1 420	0.2	12
14 Switzerland	3 670	116	0.2	19
15 USA	10 616	4 096	0.2	19
16 Finland	65	49	0.1	10
17 Italy	1 909	166	0.1	3
Total[a]	44 654[b]	14 469[c]		

Source: UNSYB 1978 Table 203

[a] includes also Iceland, Ireland, Luxembourg, Portugal and South Africa
[b] includes 3162 in other official flows and 27023 in private capital
[c] includes 4626 through multilateral institutions

Table 8.2 *Bilateral commitments of capital by centrally planned economies in 1977*

Country	US dollars millions	Dollars per inhabitant committed	Country	US dollars millions	Dollars per inhabitant committed
Romania	275	12	Czechoslovakia	57	4
USSR	2922	11	Poland	100	3
Hungary	103	10	China	201	0.2
German DR	77	5	Total[a]	3735	

Source: UNSYB 78 Table 204
[a] Bulgaria apparently gave none

the larger industrial economies France was the most generous (or least mean).

Column 4 shows the assistance in dollars per inhabitant. The values correlate only broadly with the percentages in column *3* on account of differences among countries in gnp per inhabitant.

Table 8.2 shows that there are also marked differences in the generosity of the centrally planned economies in the provision of commitments of capital. In 1977 Romania, the USSR and Hungary gave more per inhabitant than Poland or Bulgaria but contributions change considerably from year to year. In relation to their total economies the CMEA countries (therefore excluding China) provide roughly the same proportion as the market economies do of theirs, about 0.4 percent. Being 'socialist' does not make CMEA any more ready to share its wealth with the poor countries. One Soviet argument is that the CMEA countries have not in the past exploited the present developing, former colonial countries of the rest of the world, and do not do so now, so they have no obligation to make amends for past exploitation.

Why is aid given?

The justification for giving aid to poor countries rests on two distinct lines of argument: altruism and self-interest. The first justification rests on the moral, ethical or philanthropic grounds that the world is socially unjust. The counter arguments are many: that there are plenty of poor in need of help in the rich countries themselves, that the poor countries are poor because their population is lazy or inept (not because they have been run as colonies). It is self-defeating to help beggars.

Self-interest seems a better justification for helping the poor than philanthropy in the cynical world we live in. Explicitly or implicitly, the following reasons may be found:

(1) Countries that are politically or ideologically sympathetic to one's own system should be helped (sometimes militarily). Conversely, undermine the political regime in a country where a change could favour you.
(2) Countries that are of strategic importance to one's own defence systems should be helped

(Cuba to the USSR, South Viet Nam to the USA in the 1960s).
(3) Countries with particular natural resources vital to one's own economy should be helped (Niger with its uranium, Indonesia with its oil).
(4) Save the developing countries from very serious poverty or (as in the French Revolution) the poor of the world will rise up and set about acquiring the wealth of the rich by force.
(5) Help to expand the economies of the poor countries because if they produce and consume more they will need to trade more with the rich countries, thereby creating employment, a view expressed in the Brandt Report, 1980, a worthy, apparently reasonable idea, but perhaps with flaws that are difficult to state clearly.

Who gets the aid?

Tables 8.3 and *8.4* show selected recipients of assistance from market and from centrally planned economies respectively. They will be considered in turn.

Table 8.3 *Transfers of official development assistance from developed market economies to selected developing countries, average during 1975–1977*

The 20 'poorest'				More 'affluent'			
Country	1	2	3	Country	1	2	3
Bhutan	70	3	2	French Guiana	na	65	1087
Laos	90	33	10	Reunion	1920	297	583
Ethiopia	100	134	5	Martinique	2350	190	512
Mali	100	101	17	Guadeloupe	1500	160	443
Rwanda	110	83	19	Djibouti	1940	32	294
Bangladesh	110	690	9	Surinam	1370	82	187
Upper Volta	110	92	15	Israel	3920	637	184
Somalia	110	78	24	Netherlands Antilles	1680	42	176
Burma	120	77	2	Seychelles	580	9	143
Burundi	120	49	13				
Nepal	120	56	4	Uganda	240	19	1.6
Chad	120	66	16	Argentina	1550	28	1.1
Benin	130	54	17	Brazil	1140	119	1.1
Zaire	140	215	8	Nigeria	380	58	0.9
Malawi	140	68	13	Southern Rhodesia[a]	550	6	0.9
Guinea-Bissau	140	23	43	Mexico	1090	58	0.9
Guinea	150	14	3	Saudi Arabia	4480	8	0.8
India	150	1317	2	Venezuela	2570	6	0.5
Niger	160	121	26	Hong Kong	2110	2	0.4
Afghanistan	160	63	3	Iran	1930	4	0.1
				World	1650	13 594	7

Main Source: UNSYB 78, Table 206.

1 = GNP in US dollars per inhabitant; *2* = Total assistance received in millions of US dollars; *3* = Assistance received in US dollars per inhabitant.

[a] Now Zimbabwe.   na not available.

Table 8.4 *Recipients of the largest amounts of commitments of capital by centrally planned economies during 1954–1973 in millions of US dollars*

Country	Total 1954–73	1974	1975	1976	1977
1 Egypt	2 798	218	125	7	20
2 India	2 330	–	32	–	340
3 Iran	1 603	3	35	–	1600
4 Iraq	1 317	40	88	10	–
5 Algeria	1 103	–	294	265	–
6 Pakistan }	1 148	230	27	–	55
Bangladesh }		102	86	–	50
7 Syria	828	875	248	145	–
8 Indonesia	794	–	100	–	–
9 Chile	631	–	–	–	–
10 Afghanistan	607	6	670	34	21
11 Brazil	372	–	180	100	600
12 Sudan	303	53	62	–	–
13 Turkey	282	–	706	1715	138
14 Tanzania	280	78	4	–	47
15 Somalia	236	1	62	–	–
Total[a]	18 265	2941	3046	2876	3735

Source: UNSYB Table 204.

[a] totals are for 73 countries listed altogether.

Table 8.3, left hand columns *1–3*, shows the 20 poorest countries of the world, excluding some very small political units, all with a gross national product per inhabitant less than one-tenth of the world average of 1650 dollars in 1977. During 1975–77 the developed market economy countries (*Table 8.1*) provided each year on average seven US dollars per inhabitant to the developing countries (excluding China). Eight of the twenty poorest listed received less than the world average.

The right hand side of *Table 8.3* shows that quite affluent developing countries received very generous amounts of assistance per inhabitant. All but Israel have less than one million inhabitants and a fairly small absolute amount of aid goes a long way in a small country. Five of the nine countries are actually small islands and all but Israel have been or are colonies of France, the Netherlands or the UK. Ten countries receiving only very small amounts of assistance per inhabitant are also shown in *Table 8.3* (lower right). These countries also mostly have high gross national product values per inhabitant for developing countries. Some of them are large in population, while several are exporters of oil and in OPEC or are fairly highly industrialized.

Without examining assistance data for all countries for a considerable period it is not possible to give a full picture of the situation, but a number of tendencies are evident:

(1) The amount of aid received per inhabitant tends to be related negatively to the absolute size of the country. In an extreme situation a donor will make more political impression and gain more publicity spreading assistance among 70 small countries averaging 10 million inhabitants each then by placing it all in one country, India, with 670 million. India gets two to three dollars of assistance per inhabitant, including a pittance from the socialist countries.

(2) Before the 1970s there had been little attempt to discriminate betwen the richer and poorer developing countries in placing assistance. Indeed it could be argued that even five to ten dollars per inhabitant is helpful in a country where annual gross national product per inhabitant is only 100 to 150, whereas five to ten dollars per inhabitant in a country with a gross national product per inhabitant of 1000–1500 is of little consequence.

(3) One can easily conclude that the preoccupation with small countries, including many in the Pacific not shown in *Table 8.3* is indeed strategic.

The distribution of aid from the centrally planned countries seems to be even less attributable to a plan or pattern than aid from the market economies. *Table 8.4* shows the countries favoured with the largest absolute amount of socialist aid (excluding China) during 1954–73, though amounts given *per inhabi-*

tant would bring other countries to the fore. The point of showing the yearly transfers in the 1970s is to indicate how suddenly the fortunes of the recipients can change. Chile, for example, was helped extensively during the presidency of Allende (1970–73) but was dropped at once when he was assassinated in 1973. China was similarly 'dropped' around 1960. For no apparent reason at the time, Afghanistan received a massive quantity of Soviet aid in 1975, Turkey in 1975–76 and Iran in 1977, perhaps some Soviet plot to start things moving in the Middle East. It remains a matter of speculation with regard to the extent of collaboration among CMEA countries in the choice of recipients of their aid. Romania may take an independent line but broadly the distribution of aid has presumably been determined by the USSR.

Table 8.5 *US foreign aid commitments for economic assistance 1970–1979*

Country	US Dollars total in millions	US Dollars per inhabitant	Country	US Dollars total in millions	US Dollars per inhabitant
Israel	3575	917	Kampuchea	322	59
Egypt	3344	77	Bolivia	262	48
South Viet Nam	2032	68	Laos	256	71
Indonesia	772	5	Turkey	247	5
Pakistan	680	8	Brazil	230	2
Bangladesh	676	7	Korean R	207	5
Jordan	655	198	Peru	162	9
India	639	1	Thailand	157	3
Syria	439	47	Panama	154	81
Portugal	435	44	Guatemala	147	20
Philippines	430	9	Nicaragua	135	54
Colombia	406	15			
			World total	26 910	10 approx

Source: Statistical Abstract of the United States 1980 Table 1534

Table 8.6 *The direction of US foreign military aid in millions of US dollars*

Country	Millions of US dollars 1970–79	Millions of US dollars 1977–79	Percentage 1970–79	Percentage 1977–79
Israel	11 665	6 000	30	53
Viet Nam	10 727	0	28	0
Korean R	3 452	670	9	6
Turkey	1 765	480	5	4
Greece	1 242	503	3	4
Kampuchea	1 182	0	3	0
Taiwan	1 009	60	3	0.5
Laos	1 008	0	3	0
Jordan	827	368	2.1	3.2
Thailand	693	113	1.8	1.0
Spain	672	438	1.7	3.9
Indonesia	335	134	0.9	1.2
Philippines	332	107	0.9	0.9
Brazil	214	0	0.6	0
Morocco	213	121	0.6	1.1
Portugal	105	90	0.3	0.8
Total	38 812	11 268	100	100

Source: Statistical Abstract of the United States 1980 Table 1533

In view of the above description of the origin and destination of foreign assistance it is not surprising that the donors of foreign aid have been criticized both from within and by the recipients. Only 0.3–0.4 percent of total gross national product is given. Much is given with conditions and in the end most is in the form of loans that have to be repaid, not in the form of pure gifts. It is not distributed fairly according to the needs of the recipient countries and is used for political and ideological influence in some countries. Inside the receiving countries much is wasted and much reaches the 'wrong' hands. It helps to keep the wealthiest five to ten percent at the top. Aid should be given to individuals (or families) not to whole countries or regions. The development gap might be attacked more successfully if foreign trade were to be modified, giving more favourable prices for both the primary and the manufactured products of the developing countries.

8.2 US and Soviet foreign assistance and US investments

Since the USA and USSR are the two countries most likely to want to influence world affairs to their own ends it is appropriate to give more details on the direction of their assistance.

The 23 countries receiving the largest absolute amount of US foreign aid during 1970–79 are given in *Table 8.5*. Very marked differences can be seen among the recipients in the amount received per inhabitant. India is penalized through its great size while very poor countries such as Indonesia and Bangladesh also get very little. Assistance to Viet Nam, Kampuchea and Laos had virtually ceased by the mid-1970s.

The 16 countries receiving the largest amounts of US foreign military aid are given in *Table 8.6*. The US withdrawal from Indo-China comes out very

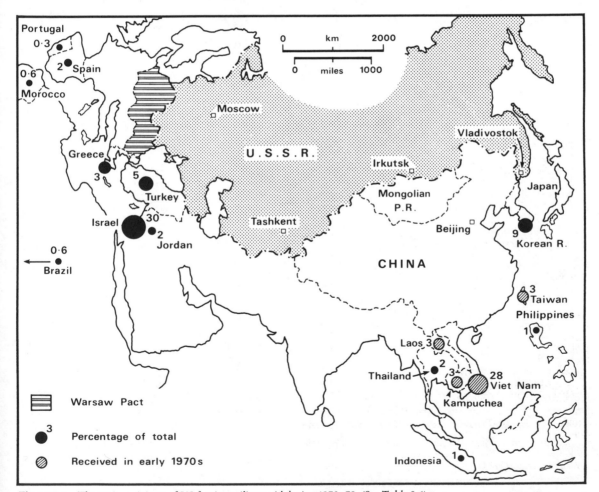

Figure 8.1   *The main recipients of US foreign military aid during 1970–79. (See Table 8.6)*

clearly. The location of the receiving countries is shown in *Figure 8.1*. Their positions in relation to a perceived or expected Soviet threat are evident.

The worldwide distribution of US direct investments in 1979 is shown in *Table 8.7*. The main sectors among which US investment was distributed in 1979 were manufacturing (43 percent), petroleum (22 percent), and finance and insurance (14 percent). The preference for developed countries is evident from the data. With less than one-quarter as many inhabitants as the developing host countries for US investments, the developed countries have nearly three-quarters of the total invested. Returns from investments in developing countries are on average much higher than they are from developed ones, but

on the whole the risk is much greater. When returns on US investments are measured against the number of people in each host country, Canada and Switzerland emerge as the largest contributors. Among developing regions, Latin America is much more attractive to US investments than most of Africa or Asia.

The direction of that part of Soviet foreign assistance referred to as projects is shown clearly by sets of data published periodically in the USSR. *Table 8.8* shows the number of Soviet projects in various countries in selected years. The data in columns *1* to *4* show the total number of projects, complete or incomplete, in four different years. Column *5* shows that in some countries all projects

Table 8.7 *The distribution of US direct investments in 1979*

	Investment 1	Income 2	Return 3	Per inhabitant 4
Total	192.6	37.8	20	70
Developed countries (popn 520 million)	137.9	24.4	18	250
Developing countries (popn 2250 million)	47.8	12.7	27	20
International and unallocated	6.9	0.7	10	–
Individual areas				
Developed				
Canada	41.0	5.3	13	1710
Belgium and Luxembourg	5.7	1.1	19	550
Denmark and Iceland	3.1	0.5	16	370
France	7.7	1.0	13	140
German FR	13.5	2.9	21	220
Italy	4.2	0.9	21	70
Netherlands	6.4	1.9	30	450
UK	24.3	5.3	22	440
Switzerland	8.6	1.9	22	1370
Other West Europe	10.0	1.1	11	120
Australia, New Zealand, South Africa	9.7	1.5	15	210
Japan	5.8	0.9	16	50
Developing				
Mexico and Central America	6.9	1.3	19	70
Argentina	2.3	0.6	26	80
Brazil	7.5	0.4	5	60
Chile	0.3	0.04	13	30
Colombia	0.9	0.03	3	30
Peru	1.8	0.5	28	100
Venezuela	2.2	0.2	9	140
Other Western hemisphere	12.5	2.9	23	320
Africa except South Africa	3.6	1.4	39	8
Other Asia and Pacific	7.8	2.4	31	6

Source: Statistical Abstract of the United States 1980, Table 1529.

1 = US direct investment position, thousands of millions of dollars in 1979; *2* = Preliminary income from *1*, thousands of millions of dollars; *3* = Return on investment in 1979, *2* as a percentage of *1*; *4* = US direct investment in dollars per inhabitant in host area.

[a] Excluding China, Viet Nam and other Socialist countries.
[b] Middle East is not included. It had negative US investment.

Table 8.8 *Soviet projects of various kinds in foreign countries since 1945 (as on 1 January of year shown)*

	1967 1	1974 2	1977 3	1980 4	Comp 1980 5	% Comp 1980 6	1967–80 7
Total							
All projects	2018	2865	3551	3965	2435	61	196
Socialist	1413	1984	2541	2793	1782	64	198
Non-socialist	599	871	998	1157	643	56	193
European Socialist							
Albania	45	45	45	45	45	100	100
Bulgaria	166	238	307	307	191	62	185
Czechoslovakia	27	39	40	42	26	62	156
German DR	26	36	83	82	35	43	315
Hungary	76	85	114	122	86	70	161
Poland	108	150	202	218	136	62	202
Romania	101	123	158	159	114	72	157
Yugoslavia	46	98	108	130	70	54	283
Developing Socialist							
China	256	256	256	256	256	100	100
Cuba	104	244	342	438	184	42	421
Korean DPR	58	73	70	70	58	83	121
Laos	nm	nm	nm	34	3	9	na
Mongolian PR	257	378	571	621	389	63	242
Viet Nam	143	219	245	269	189	70	188
Non-Socialist							
Afghanistan	59	80	115	147	73	50	249
Algeria	74	90	98	100	62	62	135
Bangladesh	na	10	16	15	11	73	na[a]
Burma	5	8	7	7	7	100	140
Cambodia	4	nm	nm	nm	nm	na	na
Egypt	102	148	107	107	94	88	105
Ethiopia	6	9	9	59	4	7	983
Ghana	20	nm	nm	nm	nm	na	na
Guinea	31	27	28	31	25	81	100
India	45	68	67	74	53	72	164
Indonesia	20	nm	nm	nm	nm	na	na
Iran	21	79	108	118	66	56	562
Iraq	49	86	104	98	61	62	200
Mali	13	14	14	15	12	80	115
Morocco	nm	3	3	6	4	67	na
Nepal	6	7	7	8	6	75	133
Pakistan	21[a]	13	12	13	7	54	133[a]
Somalia	17	25	36	nm	nm	na	na
Sri Lanka	11	7	11	10	9	90	91
Sudan	14	15	15	15	8	53	107
Syria	19	29	43	56	30	54	295
Turkey	nm	9	12	14	8	57	na
Yemen AR	13	13	12	11	11	100	85
Yemen PDR	nm	17	24	32	14	44	na

1–4 = Cumulative total numbers (size of project not given); *5* = total completed by 1980; *6* = percentage completed by 1980; *7* = 1967–1980 (1967 = 100).
nm = country not mentioned
na = not available or not calculable

[a] Bangladesh included in Pakistan in 1967

had been completed by 1980 while in others many projects were still incomplete. Unfortunately the size of projects in terms of their cost is not given and it cannot be assumed that all countries have the same mix of small, medium and large projects. *Figure 8.2* shows the location of the projects.

Special significance may be given to projects as a form of assistance because presumably they require the presence of Soviet personnel in the receiving country. Over 70 percent of all projects referred to in *Table 8.8* are in countries defined by the USSR as socialist. More than half of the projects were or are in the developing socialist countries of Asia, together with Cuba. In Europe, Bulgaria is the most favoured while Albania has not been the scene of Soviet activity since the 1960s. China also received no Soviet assistance after about 1960 and it is perhaps striking to find that the bitter reaction to China's policy around 1960 to become independent from the influence of support from its powerful neighbour

means that the Mongolian People's Republic, with fewer than two million people, had 621 Soviet projects by 1980 compared with 256 in China, which has about 1000 million people.

Listed recipients of Soviet assistance in the form of projects in non-socialist developing countries are all in Africa and Asia. Most Soviet projects outside the socialist sphere are in countries in North Africa, Southwest or South Asia. A comparison of numbers of projects in the four years given in columns *1* to *4* in *Table 8.8* shows that the fortunes of receiving countries fluctuate through time. Thus in Egypt many Soviet projects were cancelled between 1974 and 1977, whereas Ethiopia had only 9 Soviet projects in 1977 but 59 by 1980. Soviet projects do not seem to be distributed closely according to the level of poverty of different developing countries nor to the size of their populations. Afghanistan, for example, has had twice as many projects as India yet has only one-fortieth as many inhabitants.

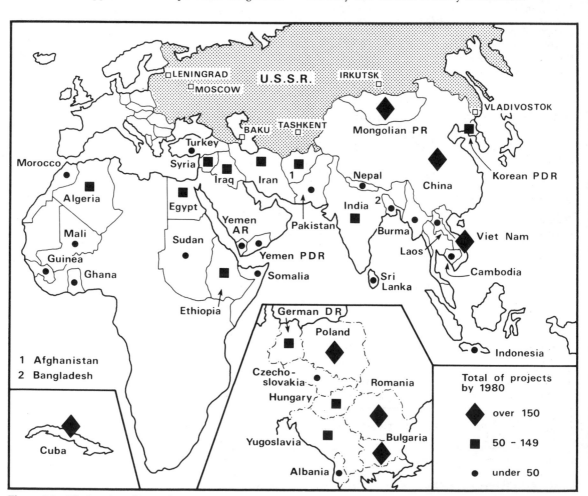

Figure 8.2   *The location of Soviet foreign projects since 1945. (See Table 8.8)*

8.3 Transnational companies

Although the large companies of the Western industrial countries usually attempt to keep a low profile in international affairs and disputes their presence is undoubtedly very influential and at times they are the subject of criticism from various quarters. In this section three aspects of transnational companies will be noted, the scale of their financial transactions, the worldwide nature of their operations and the arguments for and against them.

Although Canada, Australia and South Africa have some companies of giant size in relation to their economies the largest are based in three areas, the USA, West Europe and Japan. The size of a company may be judged in various ways including capital employed, the labour force and income (before or after tax). From the data readily available the most

Table 8.9 *The largest companies. All values are in thousands of millions of US dollars*

US companies	Turnover	Capital	Income after tax
1 Exxon	84	33.7	4.3
2 General Motors	66	22.4	2.9
3 Mobil	48	15.9	2.0
4 Ford Motor	44	14.0	1.2
5 Texaco	38	16.2	1.8
6 Standard Oil of California	31	11.9	1.8
7 Gulf Oil	26	12.1	1.3
8 International Business Machines	23	18.1	3.0
9 General Electric	22	9.8	1.4
10 Standard Oil (Indiana)	20	12.4	1.5
Total	402	166.5	21.2

European industrial groups	Headquarters	Sales
1 Shell/Royal Dutch Petroleum	UK/Netherlands	72
2 British Petroleum	UK	54
3 Unilever/Unilever N V	UK/Netherlands	23
4 Veba (Heavy industry)	German FR	21
5 Compagnie Française des Pétroles	France	18
6 Fiat	Italy	18
7 Peugeot-Citroen	France	18
8 Volkswagonwerk	German FR	17
9 Philips' Lamps	Netherlands	17
10 Renault	France	17
Total		275

Japanese companies	Head Office	Sales
1 Nippon Steel	Tokyo	13
2 Toyota Motor	Toyoda	13
3 Nissan Motor	Tokyo	12
4 Nippon Oil	Tokyo	12
5 Tokyo Electric Power	Tokyo	9
6 Masushita Electric Industrial	Osaka	8
7 Hitachi	Tokyo	8
8 Maruzen Oil	Osaka	7
9 Toshiba	Kawasaki	7
10 Mitsubishi Heavy Industries	Tokyo	7
Total		96

Source: Derived from data in The Times 1000, 1980–1981, Times Books Ltd., 1980, pp. 119, 122.

convenient way of assessing the size of large companies is to note the value of their turnover or sales in a given period. *Table 8.9* shows the ten largest companies or industrial groups in each of the three major industrial areas of the Western world.

The data show clearly that the ten largest companies of the USA are, company for company, considerably larger than the ten largest in West Europe. The ten largest Japanese companies are much smaller than the ten largest of the other two areas.

A rough idea of the scale of activities of the largest companies can be obtained from a comparison of their sales with the gross national product of whole countries (*see Table 5.17*). The largest US Corporation, Exxon (oil), has a turnover roughly equal to one-thirtieth of the whole gross national product of the USA. Its turnover exceeds the whole gross national product of Switzerland or Austria, and even that of Indonesia, which has about 150 million people. The sales of the whole Shell group are roughly equal in value to half the total gross national product of the Netherlands and exceed the total gross national product of South Africa. Each of the top ten Japanese companies has sales in excess of the total gross national product of Sri Lanka or Burma.

Many of the 30 companies listed in *Table 8.9* are familiar names and the main products of most are well known. They are concerned mostly with the handling of minerals and the manufacture of steel and engineering products. Such industries require large amounts of capital and their individual plants often benefit from economies of large scale. Other goods and services widely associated with large companies include food products, chemicals and retailing.

Most of the large companies of the developed market economy countries have operations in more than one country. Some, such as the major oil companies, may be based primarily on the natural resources of other countries. In the 1920s, for example, large oil deposits were found in the Venezuelan concessions of Esso and Shell. Many industrial companies operate factories in various parts of the world. Thus General Motors and Ford manufacture or assemble vehicles outside the USA in both developed and developing countries.

For simplicity one large company has been taken to illustrate the extent of operations of a large transnational company. The 'Shell' Transport and Trading Company has its head office in London. It is part of the Royal Dutch/Shell Group of Companies. *Table 8.10* shows the number of countries, large and small, in which Shell operates. The list includes all but a few non-socialist countries. Altogether in 1980 there were about 280 operating units, about 160 entirely owned by Shell and the rest under joint ownership. Most of the host countries had only one or two units but some, notably the UK, Netherlands,

Table 8.10 *Operating units and interests of Shell*

Area	Countries	Operating units
Europe	18	82
Africa	37	56
Middle East[a]	9	16
Far East	13	50
Australasia	6	20
North America	2	9
Latin America	31	46
Total	116	279

Source: The 'Shell' Transport and Trading Company, Ltd., Annual Report 1980

[a] Including Egypt and Libya

Oil and natural gas: exploration, production, refining

Agriculture	Metals
Chemicals	Nuclear energy
Coal	Pipelines
Marketing	Research
Marine	Storage

the German FR, Japan and Australia had at least ten. *Figure 8.3* shows the worldwide distribution of the operations of Shell. Outside the socialist or centrally planned countries Shell is represented widely, but its absence may be noted in some African countries (eg Madagascar) and Asian ones (eg Bangladesh, Burma, Iraq, Iran).

One does not have to read far in a report or company magazine of the advantages of the large transnational companies. Given their large capital resources they are able to benefit from economies of scale and to spread their activities internationally, extending where prospects are good and retracting where they are deteriorating or where raw materials have been used up. They have great flexibility and high survival value.

Transnational companies are criticized, however, for many reasons. It can be argued that they are not necessarily concerned about the countries in which they operate and they may stop activities in a given country if it suits their plans regardless of the local consequences. In some cases they have been accused of trying to market products that are not really needed, especially in developing countries, or even, as with milk products to replace breastfeeding of babies, harmful to local communities. Many multinational companies are engaged in extracting or growing raw materials in developing countries with the main purpose of exporting these to industrial countries. Multinational companies may be so large and diffuse that individuals become forgotten and

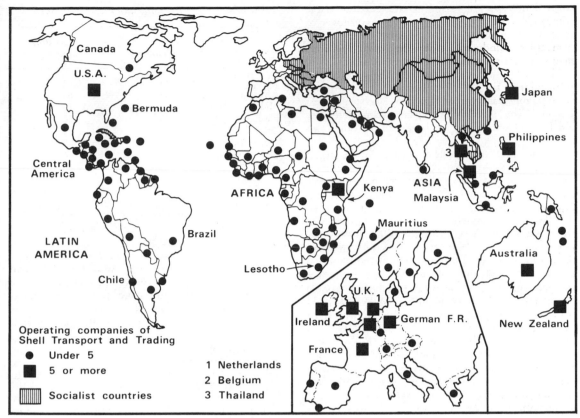

Figure 8.3 *The location of operating companies of Shell Transport and Trading*

different sections may even compete among themselves.

The impersonal nature of production in a large company and other faults may be found on a still larger scale in a country like the USSR in which the whole economy is in effect run by the state as a kind of super-company. Whatever the objections to transnational companies they seem likely to be prominent in world affairs for a good time to come, keeping as ever a fairly low profile and adapting to the changing attitudes and needs of their various host countries.

8.4 Supranational groupings

Since the Second World War, two contradictory processes have affected the number of sovereign states in the world. Firstly, many new independent countries came into being as a result of decolonization. Secondly, attempts have been made with varying success to create supranational units. Many groupings of countries associated through various kinds of interest now exist. The strength of association varies greatly among groups. The EEC aims eventually to establish a federal union between its members. OPEC on the other hand is a gathering of convenience by petroleum exporting countries. Some groupings are regional associations containing only countries that are neighbours or are close to one another while others have members scattered round the world. Some of the major groupings will be described below and their importance in world affairs noted. The two military alliances, NATO and WTO were referred to in chapter 3.

The following groupings will be discussed:

(1) Worldwide
    (a) OECD
    (b) OPEC
    (c) The British Commonwealth

(2) Regional
    (a) EEC and EFTA
    (b) CMEA
    (c) LAFTA
    (d) The Colombo Plan
    (e) The Arab League

Table 8.11 *OECD members*

EEC	EFTA	Other European	Outside Europe
Belgium	Austria	Spain	Australia
Denmark	Finland	Turkey	Canada
German FR	Iceland		Japan
France	Norway		New Zealand
Greece	Portugal		USA
Ireland	Sweden		
Italy	Switzerland		
Luxembourg			
Netherlands			
UK			

OECD

After the Second World War the Organization for European Economic Co-operation (OEEC) was set up to coordinate the reconstruction of war-damaged Europe. In 1961 the membership of OEEC was extended to bring in developed countries outside West Europe and the name of the body was changed to OECD (Organization for Economic Co-operation and Development). The 24 members of OECD in the early 1980s are listed in *Table 8.11*. Nineteen including Turkey are in Europe. These countries had about 750 million inhabitants in 1981, about 17 percent of the total population of the world, but they account for more than 60 percent of the total production of goods and services in the world.

OPEC

The Organization of Petroleum Exporting Countries was founded in 1960. The founder members, inspired by Iraq, were five of the major oil exporters of the time. They and later joiners are listed in *Table 8.12*.

In 1980 OPEC countries accounted for 44 percent of total world oil production and 58 percent of world oil production excluding that of the USSR, its CMEA partners and China. More than half of the oil produced in the world in that year was exported and nearly 90 percent of the exports came from OPEC members and a few smaller developing world exporters. Much of the rest was from the USSR. Ironically the Headquarters of OPEC is in Vienna,

capital of an oil importing country, presumably a more 'central' and more 'neutral' place to be located than in one of the member countries.

OPEC countries have been successful in pushing up the price of oil many times since the 1960s, and particularly during 1973–74. As a result, oil is more highly priced relative to most other products than it was a decade ago though the actual prices of most products have also gone up during the same period. OPEC members have undertaken to provide aid through a Special Fund to the many developing countries with little or no oil. In practice much of this aid has gone to selected developing countries, particularly those in the Arab cultural region.

The British Commonwealth

Membership is confined to those countries that were members of the British Empire at some time in their history. Ireland, Burma, South Africa and Pakistan do not now belong to the Commonwealth, but virtually every other part of the former British Empire does, and the total membership is substantial. Through accident rather than intention the British Commonwealth is one of the few supranational groupings of the world that accepts indiscriminately both developed and developing countries. The assistance provided by the richer members to the poorer ones is considerable and includes not only material assistance but also cultural exchanges and even a limited flow of migrants.

The British Commonwealth is shown in *Figure 8.4*. Pacific island members are not shown on the map: Fiji, Western Samoa, Tonga, Nauru, Kiribati and Tuvalu (formerly the Gilbert and Ellis Islands). Currently the countries mainly with people of European origin are referred to as the 'Old' Commonweath (Canada 1867, Australia 1901, New Zealand 1907) while the more recently independent countries belong to the 'New' Commonwealth (India 1947, Sri Lanka 1948, Ghana 1957 and many others). Some units, including Hong Kong, remain Crown Colonies.

Table 8.12 *OPEC countries*

Founder countries	Others	Others
Iran	Algeria	Libya
Iraq	Ecuador	Nigeria
Kuwait	Gabon	Qatar
Saudi Arabia	Indonesia	United Arab Emirates
Venezuela		

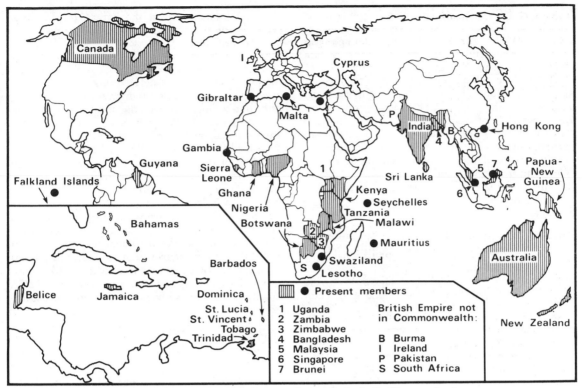

Figure 8.4 *Members of the British Commonwealth. Some small countries in the Pacific are omitted*

EEC and EFTA

The countries of Europe not under Soviet control have not been able to form a bloc including all theoretically eligible members. For various reasons several countries chose not to join the European Community as it evolved in the 1950s and others were not acceptable. The non-socialist countries of Europe may now be divided into three categories: members of EEC, members of EFTA and members of neither. In *Table 8.11* the three categories are given separately.

The origins of the European Economic Community go back to the establishment of the ECSC (European Coal and Steel Community), in which the steel-producing countries France, West Germany, Belgium and Luxembourg, associated through the sharing of their coal and iron ore, agreed to cooperate in the integration of the iron and steel industry. In 1957 the signing of the Treaty of Rome led to the establishment of more extensive and closer cooperation between the six founder countries of EEC. In 1958, Euratom was set up to deal with nuclear power and in 1967 the three bodies were merged. In 1973 the UK, Ireland (the Irish Republic) and Denmark joined and in 1981 Greece followed. After a referendum in the country, Norway decided not to join (1973). It is expected that Spain will be a member by 1984.

Finland and Austria are virtually excluded from joining a body such as EEC through the neutral status required of them when Soviet forces withdrew from each soon after the Second World War. Switzerland regards its neutrality as more important than membership of EEC and Sweden also professes to have a neutral position in world affairs. The European Free Trade Association (EFTA) to which the UK and Denmark also belonged until 1973, is, therefore, a group of leftovers established in 1960 to try to make up for the disadvantages of exclusion from EEC. The members are listed in *Table 8.11*. *Figure 8.5* shows the location in Europe of EEC and EFTA members.

CMEA

In addition to the USSR the Council for Mutual Economic Assistance (also COMECON) includes six countries of 'Eastern' Europe, Poland, the German DR, Czechoslovakia, Hungary, Romania and Bulgaria (*see Figure 8.5*) and three outside Europe, the Mongolian People's Republic (1962), Cuba (1972) and Viet Nam (1978). All members of

CMEA are defined as 'socialist' in Soviet publications but China, the Korean DPR, Yugoslavia (observer since 1964) and Albania (1949–61), though also socialist are not full members. New observers are Angola, Ethiopia and Laos. The 10 full members of CMEA had a combined population in 1981 of about 445 million, almost exactly 10 percent of the total population of the world. The bloc accounts for about 17–18 percent of the total production of goods and services in the world. The member countries are economically interdependent, with specialization in certain sectors of production by each country and synchronization and coordination of state plans. In the late 1970s over half of the value of all Soviet foreign trade was with its CMEA partners.

LAFTA

The late 1950s was a time when supranational units were fashionable and it even appeared that some might gradually become enlarged superstates of the future. In 1961 most of the major countries of Latin America joined to form the Latin American Free Trade Association. The seven founder members were Mexico (in North America) and (all in South America), Argentina, Brazil, Chile, Paraguay, Peru and Uruguay (*see Figure 8.5*). Colombia, Ecuador and Venezuela joined later. During the 1960s there was some increase in trade between member countries but the plan to help the smaller (and poorer) countries such as Paraguay and Ecuador was not carried out to any extent. One possible advantage of the establishment of LAFTA has been the diversion of much of the oil exported by Venezuela to other South American countries. The rapid growth of the manufacturing sector in Brazil since the 1950s may also owe a little to the presence of a sympathetic if captive market for its products in other LAFTA countries.

Dissatisfied with the slow application of LAFTA principles and exchanges, a subgroup of the members, joined also by Bolivia, formed the Andean Pact in 1969 (*see Figure 8.5*). In 1981 the ten LAFTA countries listed above had about 305 million inhabi-

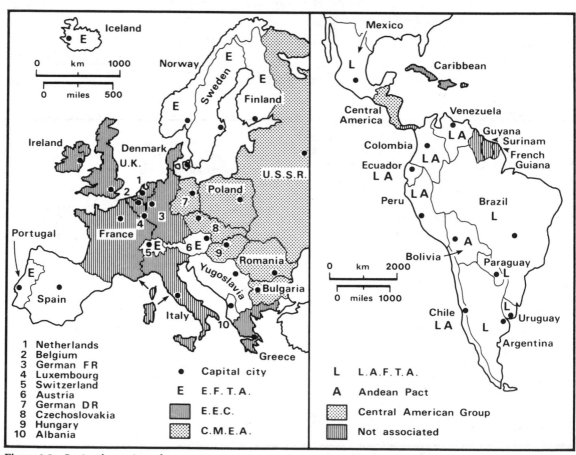

Figure 8.5 *Regional groupings of countries in Europe and in Latin America*

tants, a little under 7 percent of the total population of the world. They account for about 5 percent of the total gross national product of the world.

Many smaller Latin American countries are not members of LAFTA. In the late 1950s five small countries on mainland Central America formed a Common Market (ODECA — Organización de Estados Centro-Americanos). The members are Guatemala, El Salvador, Honduras, Nicaragua and Costa Rica. Cuba, as already noted, is linked to CMEA and its economy is virtually controlled by the USSR. Various attempts have been made to unite the British or former British possessions of the Caribbean area and in the early 1980s the Caribbean Community (Caricom) was the current body.

The League of Arab States

The League of Arab States came into existence in 1945 in an area either dominated by the Turks in their Ottoman Empire until the First World War or colonized or strongly influenced by France, the UK and Italy. The seven founder members were: Egypt, Iraq, Saudi Arabia, Syria, Lebanon, Jordan and Yemen (now AR). All these countries were effectively independent at the end of the Second World War. The 14 later joiners were still colonies or were 'protected' by outside powers at the time. Strictly the Palestine Liberation Organization (PLO) does not have a sovereign *de facto* territorial base. The

locations of the member states of 1980 are shown in *Figure 8.6*. The headquarters was in Cairo until 1979 when the establishment of friendly relations between Egypt and Israel led to its transference to Tunis.

The members of the Arab League are held together by common cultural ties through their language and also through their association with Islam. They differ among themselves economically, however, through the presence in some of very large oil deposits and the virtual absence of oil in others. The production and export of oil has given some very high gross national product values per inhabitant. Politically, too, the Arab League countries are divided, with some openly pro-Soviet and some pro-USA. In 1981, for example, Egypt, Sudan, Somalia and Oman were involved with the USA in a military exercise (Bright Star). Algeria, Libya, Syria, Iraq and Yemen PDR (formerly the British Aden Protectorate) were notably pro-Soviet (*see* Soviet aid in section 8.2). In 1981 the countries shown in *Figure 8.6* had a combined population of about 170 million, somewhat less than 4 percent of the total population of the world. More than two-thirds of the population is in the African countries of the group.

Colombo Plan

An economic association consisting of both developed and developing countries, the Colombo Plan, was established to provide assistance for Asian

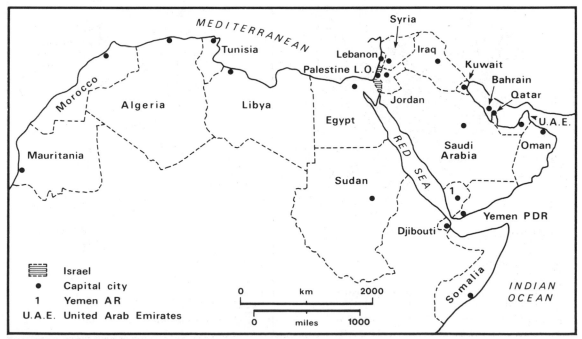

Figure 8.6   *The League of Arab States*

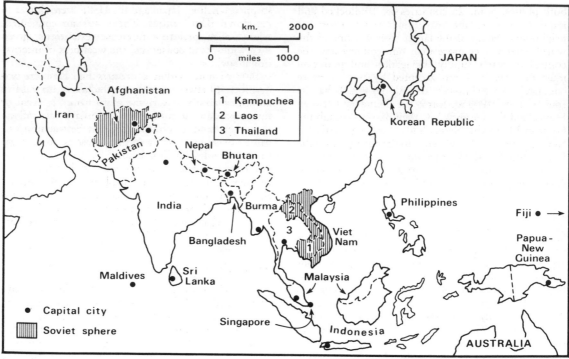

Figure 8.7 *the 'developing' members of the Colombo Plan*

countries. The five non-European OECD countries (USA, Canada, Japan, Australia and New Zealand) together with the UK are the 'rich' members. *Figure 8.7* shows the location of the 'poor' member countries and of Japan and Australia. South Viet Nam, Cambodia (Kampuchea), Laos and Afghanistan were once members of the Colombo Plan but Soviet penetration since the 1970s has been moving them into the Soviet bloc and eventually (as already with Viet Nam) they may join CMEA. Iran, too, seems to be heading for a new alignment in world affairs. The developing Colombo Plan countries shown in *Figure 8.7*, excluding the ones under Soviet influence (shaded), had a combined population in 1981 of about 1285 million, 28.6 percent of the total population of the world but they account for only about 4 percent of total world gross national product.

8.5 World institutions

One eventual prospect for the world is world government. Under such a system a major world conflict would presumably no longer be possible. On the other hand central control by a particular group could lead to oppression of the rest of the world. At least now it is possible theoretically to escape from one country to another.

The number of independent countries in the world has increased more than three times since 1945 largely due to the achievement of independence by many comparatively small colonies. On the other hand, as shown in the previous section, many countries of the world now belong to supranational groups of some sort, whether military, economic or potentially political. One could envisage in the future a world in which most countries belong to any one of 10–12 supranational groups. It is more difficult to envisage the world organized as one political unit. Nevertheless in the 20th century there have been two serious attempts to create a world body to regulate or at least discuss international transactions and conflicts, the League of Nations and the United Nations. The former collapsed with the Second World War. The latter has at least survived twice as long as its predecessor.

In a world in which some 200 sovereign states (excluding micro-states like San Marino) of equal status though very unequal size co-exist and transact, international agreement is needed on many matters, whether on land, sea, in the air or in space beyond. Navigation on the oceans, the sharing of wavelengths for broadcasting and the organization of postal links are three examples. Several hundred different pairs of countries share a stretch of international boundary, many not demarcated entirely to the satisfaction of both parties. Some 25 countries are landlocked and require access to the seas across the territory of neighbours. Many of the larger river systems of the world extend over more than one country and the control of water supply, navigation and pollution must be organized and accepted by two or more countries. An example is the Rhine, which has its source in Switzerland, forms the boundary between France and the German FR, then passes through the German FR to the Netherlands, carrying traffic for all four countries and water (with impurities) into the water supply of the Netherlands. Over the centuries various international agreements have developed to cope with problems of the kind noted above.

International law

Some centuries ago international co-operation was very limited and in practice in Europe after the 15th century war was the normal state of relations between countries, even fellow Christian countries. Foreigners could be treated in any way by rulers unless safe conduct had been agreed within their domain. As European national states became more distinct and powerful in the 15th and 16th centuries they were frequently in conflict in Europe and as they extended their influence elsewhere in the world, they attacked each other's shipping at sea and fought on other continents. The high seas were under no jurisdiction or agreement. In the 16th century English heroes or pirates, depending on how you regard them, regularly attacked Spanish convoys taking silver from the New World.

The Treaty of Tordesillas (*see* chapter 2) proposed by the Pope and accepted by Spain and Portugal in 1493 was an early example of an agreement (not a law) to reduce conflict. Modern international law is conventionally regarded as the creation of a Dutch lawyer, Hugo Grotius (1583–1645). Gradually European countries came to accept the need for some kind of mutual trust and respect, but these attributes were not extended to the rest of the world, where colonies had no sovereignty and uncolonized areas were not in the system. International law could be seen by the rest of the world as a concept designed to assist European expansion.

International agreements have often led to desirable changes in spite of the inherent problem of lack of machinery to enforce any laws. It was accepted by the Treaty of Paris in 1815 that the slave trade (not necessarily slavery itself) should not be permitted to continue. Such a change required the agreement of several influential states. The trade was gradually run down only through the pressure and enforcement of individual states, particularly the UK, with its increasing world role. In the 1960s it was becoming evident that some species of whale were threatened with extinction and many countries voluntarily stopped whaling. Japan and the USSR were slower to cut down their catches. There was no convenient ocean police force to enforce the restrictions agreed by a widespread consensus, the weakness of international law.

In a dispute within a country, as between two districts or between a company and a city, there is an authority above to sort the matter out. Internal or municipal law is enforceable. International law, on the other hand, is only by consent between two or more sovereign states and the world is an unorganized society. Thus, for example, if a treaty between two or more countries is subsequently broken by one of them, only good faith is broken, no law, because there is no independent judicial authority to enforce the agreement and sanction (or punish) an offending party. International laws are only guidelines and are potentially unenforceable.

The United Nations

At the end of the Second World War the ineffective and discredited League of Nations was succeeded by the United Nations. Many aspects of international relations both military and civil came under the general umbrella of the new global, multipurpose institution. In recognizing the right of peoples and countries to self-determination the United Nations by its charter excludes, however, not only the intervention of one member country in the internal affairs of another country but also its own intervention. Broadly, then, the United Nations is concerned with international transactions and, as previously noted, it cannot itself enforce sanctions, though all or some of its members may agree to do so, as on trade with Rhodesia in the 1970s.

The United Nations includes resolutions about which no self-respecting country could argue. It bans crimes against peace, war crimes, crimes against humanity and genocide. The Universal Declaration of Human Rights protects peoples in a moral though not legal sense. There is general protection for nationals abroad and in 1961 the Vienna Convention on Diplomatic Relations ensured special immunity to diplomats, an agreement conspicuously violated by the government of Iran when the US Embassy in Tehran was occupied in 1979. The weakness of international control was very evident here. An 'illegal' act had breached an international obligation, not a law that could be enforced. The United Nations is also concerned with such questions as the granting of asylum and of extradition, denuclearization, the Law of the Sea, and international labour.

The right of peoples and nations to self-determination is one of the key themes in the United Nations yet it is bitterly ironical that there are far more 'nations', nationalities, linguistic groups, people, tribes and such like entities without sovereignty

Table 8.13 *Members of the United Nations showing yearly contributions to the body in millions of US dollars*

Europe and USSR		Middle America		Africa (contd.)		Southwest Asia	
Albania	2	Bahamas	2	Comoros R	2	Afghanistan	2
Austria	63	Barbados	2	Congo	2	Bahrain	2
Belgium	107	Costa Rica	2	Djibouti	2	Cyprus	2
Bulgaria	13	Cuba	13	Egypt	8	Iran	43
Belorussia[a]	40	Dominican Rep	2	Eq. Guinea	2	Iraq	10
Czechoslovakia	87	El Salvador	2	Ethiopia	2	Israel	24
Denmark	63	Grenada	2	Gabon	2	Jordan	2
Finland	41	Guatemala	2	Gambia	2	Kuwait	16
France	566	Haiti	2	Ghana	2	Lebanon	3
German DR	135	Honduras	2	Guinea	2	Oman	2
German FR	774	Jamaica	2	Guinea-Bissau	2	Qatar	2
Greece	39	Mexico	78	Ivory Coast	2	Saudi Arabia	24
Hungary	34	Nicaragua	2	Kenya	2	Syria	2
Iceland	2	Panama	2	Lesotho	2	Turkey	30
Ireland	15	Trinidad[b]	2	Liberia	2	United Arab E	8
Italy	330			Libya	17	Yemen AR	2
Luxembourg	4			Madagascar	2	Yemen PDR	2
Malta	2	*South America*		Malawi	2		
Netherlands	138	Argentina	83	Mali	2	*South Asia*	
Norway	43	Bolivia	2	Mauritania	2		
Poland	140	Brazil	104	Mauritius	2	Bangladesh	4
Portugal	20	Chile	9	Morocco	2	Bhutan	2
Romania	26	Colombia	11	Mozambique	2	Burma	2
Spain	153	Ecuador	2	Niger	2	India	170
Sweden	120	Guyana	2	Nigeria	13	Indonesia	14
Ukraine[a]	150	Paraguay	2	Pakistan	6	Kampuchea	2
USSR	1133	Peru	6	Rwanda	2	Laos	2
UK	444	Surinam	2	Sao Tomé	2	Malaysia	9
Yugoslavia	38	Uruguay	4	Senegal	2	Maldives R	2
		Venezuela	40	Sierra Leone	2	Nepal	2
				Somalia	2	Philippines	10
Australasia				South Africa	40	Singapore	8
		Africa		Sudan	2	Sri Lanka	2
Australia	152			Swaziland	2	Thailand	10
Fiji	2	Algeria	10	Tanzania	2	Viet Nam	2
New Zealand	28	Benin	2	Togo	2		
Papua-NG	2	Botswana	2	Tunisia	2	*East Asia*	
		Burundi	2	Uganda	2		
		Cameroon	2	Upper Volta	2	China	550
North America		Cape Verde R	2	Zaire	2	Japan	866
		Central African R	2	Zambia	2	Mongolia	2
Canada	296	Chad	2				
USA	2500						

[a] part of USSR
[b] and Tobago
R Republic

than with it. Groups that might like a voice in the United Nations include such submerged peoples as the Basques in Spain, the Welsh in the UK, the Latvians in the USSR, the Kurds (*see* section 3.2), the Palestinians, the Tibetans in China, the Ibos in Nigeria, and many others. How representative is the United Nations? *Table 8.13* is included for reference.

It lists the members of the United Nations in 1980 and their yearly contributions to the body in millions of US dollars. *Table 8.14* includes smaller political units. *Figure 8.8* shows the location of all UN members not among the 100 largest countries of the world in population already mapped in *Figures 2.9* and *2.10*.

Table 8.14 *'Countries' included in the mailing price list of the British Post Office in 1981 but not yet members of the United Nations*

Anguilla	Falkland Is	Montserrat	Seychelles
Antigua	Fr Guiana	Namibia	Solomon Is
Ascension	Fr Polynesia	Nauru Is	Ceuta etc
Azores	Fr West Indies	Neth. Antilles	Taiwan
Belize	Gaza	New Caledonia	Tibet
Bermuda	Gibraltar	Norfolk Is	Tonga[b]
Br Virgin Is	Grenada	Pitcairn Is	Tristan da Cunha
Brunei	Hong Kong	Puerto Rico	Turks and Caicos
Canary Is	Kiribati[a]	Reunion	Tuvalu[c]
Caroline Is	Korea R	St Helena	Vanuatu[d]
Cayman Is	Korca DPR	St Kitts-Nevis	Virgin Is (US)
Christmas Is	Macao	St Lucia	Wake Island
Cocos Is	Madeira	St Pierre et Miquelon	Western Samoa
Dominica	Mariana Is	St Vincent	
East Timor	Marshall Is	Samoa (US)	

[a] formerly Gilbert Is.
[b] formerly Friendly Is.
[c] formerly Ellice Is.
[d] formerly New Hebrides

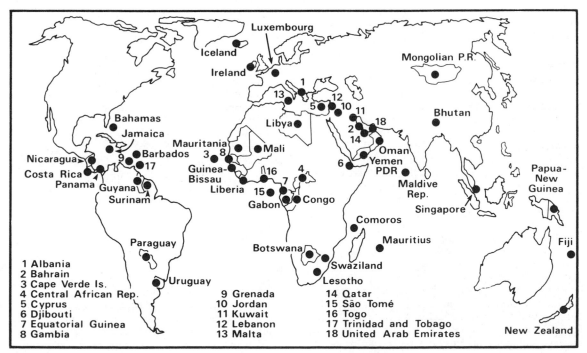

Figure 8.8 *Smaller members of the United Nations (see Figures 2.9 and 2.10 for the location of the hundred largest countries)*

8.6 Some international problems

The acuteness of an international dispute is not necessarily measured by the length of a common boundary or the size of a particular mountain range or river basin. Israel and Syria share a very short common boundary but the displacement by the Israelis of the boundary in 1967 and their annexation of the Golan Heights in 1981 are a major problem in world affairs. Many boundary disputes have been settled by mutual consent or by arbitration through a third country (eg in the late 1970s a dispute between Argentina and Chile was referred to external arbitration). Many boundaries are still disputed and some are undefined. Some types and examples of sources of friction and conflict are noted:

(1) Rivers as international boundaries (eg disputes along the Amur and Ussuri Rivers between the USSR and China).
(2) Rivers flowing from one country into another (eg tributaries of the Amazon from several Andean countries into Brazil).
(3) Boundaries disputed still after decolonization (eg part of Guyana claimed by Venezuela, Spanish Sahara claimed by Morocco, Falkland Islands claimed by Argentina).
(4) Attempts at internal secession and independence movements put down by force (eg Iboland in Nigeria, Eritrea in Ethiopia).
(5) Invasion across *de jure* or *de facto* international boundaries (eg North into South Korea, South Africa into Angola, China into India).

With regard to the seas and oceans, international law or agreement has been no less necessary than on land, but more effective. Outside coastal waters, anywhere in the seas and oceans is theoretically common to any country whereas on the land virtually everywhere is the undisputed territory of only one country or a mutual zone between a pair of countries.

The idea of laying claim to an area of sea in addition to territory on land is not a new one. At the height of its power some centuries ago Venice effectively controlled the Adriatic. Between the two World Wars Mussolini's aspiration was control of the Mediterranean, a dream shattered with the sinking of his inadequate navy early in the Second World War. In the same period the USSR claimed as its territorial waters all the Arctic Ocean to the north of its shores as far as the North Pole itself, in effect a triangular area bounded by meridians 30/35 degrees East and roughly 170 degrees West. Other countries, including the UK and USA, tended to advocate jurisdiction by states over only a very narrow zone of territorial waters. In 1981 and again in 1982, for example, the USA recognized only a 12 mile limit to the territorial

seas of Libya and carried out naval exercises just outside the limit, but still in an area claimed by Libya itself.

One of the achievements of the League of Nations was the Codification of International Law. In 1930 at The Hague in the Netherlands some aspects of the Law of the Sea were clarified. There should be a uniform three-mile limit to the width of a territorial sea while within a broader zone of twelve miles in width a coastal state could exercise jurisdiction in customs, sanitary and security matters. Of the natural resources in the seas and oceans, fisheries were the main aspect of concern at that time.

In 1945 President Truman proclaimed that without restricting the navigation of other countries on the high seas the USA would regulate fisheries in sea areas contiguous to its coasts and that it also had jurisdiction over the natural resources of its continental shelf, where oil and natural gas discoveries offered the main prospects. The terms (Glassner 1978) were:

'the Government of the United States regards the natural resources of the subsoil and sea bed of the continental shelf beneath the high seas, but contiguous to the coasts of the United States as appertaining to the United States, subject to its jurisdiction and control.'

Soon after 1945 Mexico and Argentina made similar claims on their continental shelf areas while in 1947 Chile and Peru extended not only their jurisdiction but also their sovereignty over a 200 mile wide zone of the Pacific in which there happened to be extensive fisheries. Clearly new developments were taking place in the seas and oceans and eventually in 1958 a Conference was held in Geneva to modify, update and give proper standing to the Law of the Sea.

Virtually every piece of continental shelf, accepted to reach to a depth of 200 metres below sea level, and every area of sea within 200 miles of any piece of land, even ridiculously small islands, no more than the tops of 'mountains' projecting through the surface of the ocean, form the basis for claims. When the UK 'occupied' tiny uninhabited Rockall, which lies to the west of Scotland, it staked a claim over a generous portion of the Atlantic Ocean. *Figure 8.9* shows an unfamiliar shaped zone of 'coastal' seas between the land areas of the world and the rest of the oceans. The exploitation of natural resources, the prevention of pollution and the policing of navigation in busy shipping areas are three of the many matters of dispute and subsequent agreement connected with the seas adjacent to land. Most of the best sea fisheries are fairly close to the land, not far out in the oceans. Many disputes have eventually been resolved with little or no risk of a major conflict.

There remains the rest of the oceans, beyond the continental shelf and mostly very much deeper than 200 metres. While even the most humble small developing countries are credited with their offshore

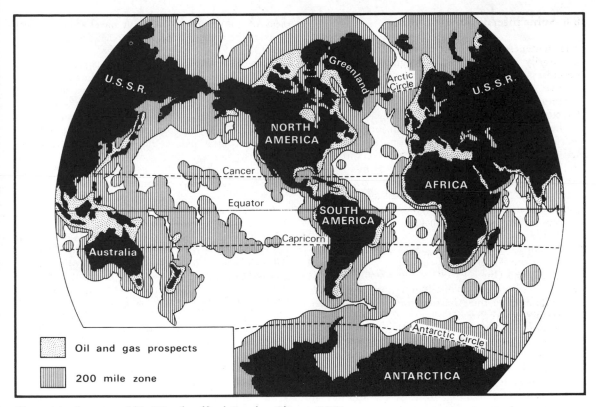

Figure 8.9 *Sea areas within 200 miles of land. Based on Glassner (1978)*

zones, even if they cannot themselves exploit the resources yet, the resources of the oceans are too deep to be reached by any but the most sophisticated methods of extraction. It is not technically possible at present to drill into the sea floor where this lies more than a few hundred metres below the surface of the sea. The riches of the oceans are the minerals lying on the sea bed. In the form of nodules, created by precipitation of minerals out of the water itself, large quantities of manganese, nickel, copper and cobalt are available, as well as smaller quantities of many other elements. Realizing that only the large, industrially advanced countries (in effect the USA, USSR, Japan or EEC) could hope to find ways of extracting the nodules commercially in the fairly near future, a so-called 'Group of 77' developing countries argued in 1967 that, unlike the continental shelf areas, the seabed and ocean floor should be reserved exclusively for peaceful purposes and that the resources should be used in the interests of all mankind.

While landlocked countries could hardly expect to lay claim to territorial seas (though some have navies) they in particular, like the poorer countries in general, faced the prospect of losing any chance of benefiting from the natural resources of the oceans, the 'commons' of the world. One proposal in the late 1960s was that any country that succeeded in extracting minerals from the ocean floor should at least pay a royalty to some fund held by the United Nations or to be used as aid for poorer countries.

In August 1981 yet another conference was held (in Geneva) on the international law of the sea. By then the USA, UK and German FR had already passed national legislation to enable mining companies to go ahead with the exploitation of deep sea minerals before the convention's decisions were made.

Attempts to establish rigid international law (or laws) and to apply them have been frustrated on many occasions in the 20th century. In the last resort a major world power will act unilaterally or against international feeling and pressure if it feels the need to do so. One example has been Soviet reluctance to allow international inspection of its military capacity. Another is US insistence on maintaining free access to the minerals anywhere in the ocean, a policy quite inconsistent with its earlier claim to the territorial continental shelf. Antarctica (1959) and outer space (1967) are two other regions in which international co-operation is essential.

CHAPTER NINE
Developed Regions Poor in Natural Resources

9.1 Japan

Introduction and historical background

In 1981 Japan had a population of 118 million, 2.6 percent of the world total. Its area is 378 000 square kilometres, about 0.25 percent of the total land area of the world. It consists of four main islands (*see Figure 9.1*) together with many smaller ones. These give it a very long shoreline (30 000 kilometres), but nowhere is more than 110 kilometres from a coast. It does not share a land boundary with any other country but its four closest neighbours, the USSR, the two Koreas and China, are not far away across the Sea of Japan.

Japan became unified as a nation about 1500 years ago but culturally it has always been strongly influenced by China. It was little influenced by the Portuguese, Spanish and Dutch as they colonized parts of eastern Asia from the 16th to the 18th centuries. In the middle of the 19th century Japan was still a deeply rural and agricultural country, with about 85 percent of its population engaged in agriculture and associated activities.

In 1854 Japan was forced to sign a treaty to open its ports and to trade with the USA and subsequently with European countries. It quickly adopted a positive attitude to economic change and development and in the Meiji era (from 1868) borrowed and developed the techniques of existing industrial countries. Since the country has few fossil fuels and raw materials it sought to acquire control of the natural resources of other countries in the region, taking Formosa (held 1895–1945) and Korea (1910–1945), occupying Manchuria in the early 1930s and eventually embarking on the conquest of the rest of China in 1937. As a result of its early conquests in the Second World War, aimed particularly to gain control of oil fields in the Dutch East Indies and Burma, Japan occupied a large part of Southeast Asia, reaching to the confines of India and Australia.

Between 1875 and 1945 the population of Japan had roughly doubled, from 36 million to 72 million

and the country had become highly industrialized. As a result of the Second World War, however, it lost all its colonies, was left with much of its industrial capacity and housing in ruins, and was humiliated by being the recipient of two nuclear bombs (Hiroshima and Nagasaki). Gross national product per inhabitant just after the war was about US 100 dollars (equivalent to several hundred dollars in 1980). Around 1980 gross national product per inhabitant had reached about 9000 US dollars.

If the industrial progress of Japan was impressive between the 1870s and the Second World War it has been almost unprecedented since its fresh start in the late 1940s. By the 1970s Japan had compressed into a century what Britain had achieved in two centuries. Japan's development is especially remarkable because in relation to the size of its population and its high level of technology and production the natural resources are very limited. As a result of almost complete dependence on other countries for fossil fuels and raw materials the Japanese economy is particularly difficult to plan far ahead.

Natural resources and production

The four main islands of Japan are situated between parallels 46 degrees North and 31 degrees North. The same parallels in the USA pass through Maine and northern Florida while in Europe they pass through central France and North Africa. Japan covers about one and a half times the area of the UK but has more than twice as many inhabitants. In *Figure 9.1* Japan is compared in scale with the British Isles and with California.

All the islands of Japan are very rugged. Less than one-sixth of the land is flat or gently sloping. Almost all the cultivated land is concentrated here as well as most of the urban land. In spite of its comparatively low latitude much of Japan has severe winters, with heavy falls of snow. The whole of the country is hot in summer, the period when most of the heavy monsoon rain comes. The position of Japan along the

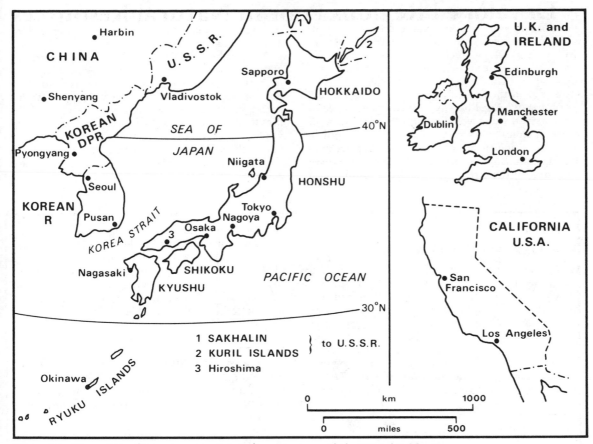

Figure 9.1 *The position of Japan in eastern Asia and a comparison with the UK and California*

junction of converging tectonic plates causes frequent earthquakes (some 5000 annually) and volcanic activity (about 50 out of 450 volcanoes in the world active in historical times) while heavy rains and strong winds also bring frequent disasters.

Land over about 1000 metres above sea level is shown in *Figure 9.2* together with the larger of the numerous plains, five of them named. Many peaks in central Honshu exceed 3000 metres and the steep well-watered slopes, with frequent river valleys and gorges, provide sites for hydro-electric stations (*see Figure 9.2*). The land use of Japan in percentages is as follows:

About *18 percent* consists of the coastal plains, flat interior valley floors, or gentler slopes, and is either cultivated or covered by urban uses.
About *67 percent* is forest, mainly planted or replanted in the last hundred years.
About *15 percent* is poor pasture or waste.

Off the coast of Japan there are good fisheries.

Japan has no major deposits of minerals. Only token quantities of oil and natural gas have been found. There are several coalfields but the coal mostly contains excessive volatile material and ash, the coal seams are thin, and they are extensively folded and faulted. The coal deposits are mostly located at some distance from the main industrial areas (*see Figure 9.2*). Among the non-fuel minerals found in Japan only zinc, manganese ore, sulphur and limestone make important contributions to the economy. Small reserves of copper, tin, mercury, tungsten, molybdenum, chromium and antimony are exploited, but there is virtually no iron ore, bauxite, phosphates or nickel.

Japan is forced to import the materials it lacks from many parts of the world and as a result it makes a great impact on world trade. It has even been argued that Japan is at an advantage *because* it lacks fuel and raw materials and therefore does not have to invest at home in developing the production of these items. As a major importer it is therefore in a good position to obtain good trading conditions round the world. On the other hand Japan is vulnerable to sudden price increases on world markets and to threats to cut off sources of materials. Moreover countries at present with a surplus of fuel and raw materials may eventually themselves need to use all of their own materials as they themselves industrialize.

In spite of the limited extent of its cultivable land Japan has depended heavily on its agricultural sector until a few decades ago. Even in 1914, 55 percent of the economically active population was engaged in agriculture and associated activities, 20 percent in mining and manufacturing and 25 percent in tertiary activities. In the early 1980s the figure had dropped to about 10 percent in agriculture, with 35 percent in mining and manufacturing and 55 percent in tertiary activities.

The number of persons working in a given area of cultivated land is about 100 times as great in Japan as in Australia. Japanese farms are very small and often do not support the families that cultivate them, forcing many farmers to find other work as well. In the 1970s it was estimated that 1.5 hectares was needed to support a farming family but only one-sixth of all families have this amount while the average size is only about 0.9 hectares. Fields are generally small and mechanization, though wide-

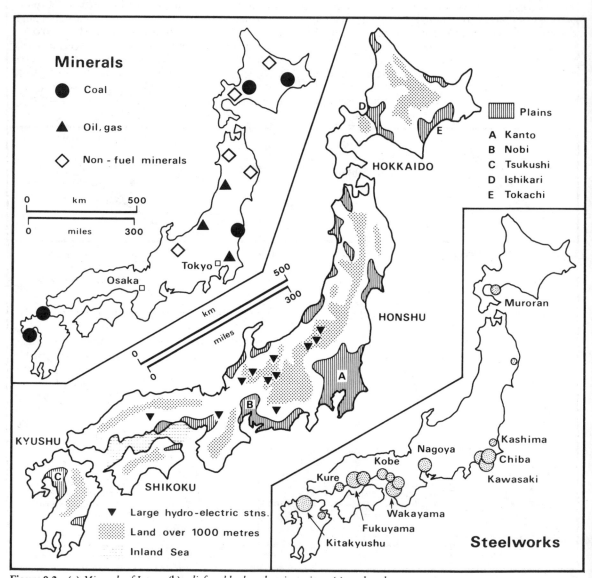

Figure 9.2   (a) *Minerals of Japan,* (b) *relief and hydro-electric stations,* (c) *steelworks*

spread, is characterized by tractors and other machines that are appropriately small in size. Almost all the land under field and tree crops (about 5 million hectares) is devoted to the production of food and beverages for human consumption. Little fodder is grown for livestock and Japan does not produce either much meat or many dairy products or many plant and animal raw materials. Pigs and poultry are the main forms of livestock and they tend to be kept in arable farms. The livestock sector accounts for about 25 percent of the total value of agricultural production.

Roughly half of the cultivated land of Japan is under rice, almost all from wet lowland paddy fields, and the crop accounts for about half the value of output from all crops. Though very high yields of rice are produced per unit of area thanks to the heavy use of fertilizers, it is still necessary to protect and subsidize rice growers. Other cereals are also grown and in the warmer areas they can be double cropped with rice. Other crops include sugar beet and vegetables as well as bush and tree crops, especially tea, the mulberry (the leaves are fed to silkworms) and various fruits. Although the cultivated land of Japan is used intensively, several million tonnes of wheat may be imported each year. The large fish catch is the source of about half the animal protein consumed, and meat is not consumed in large quantities.

The extensive forests of Japan provide some compensation for the lack of cultivated land. About two-thirds of Japan is forested, though only about 40 percent of the forest is commercially useful. The natural tree species of the country have been widely replaced by artificially planted larches and spruces in the north and by cedars, pines and cypresses elsewhere. These species provide wood for pulp, construction timber and other uses, but not enough to use as fuel; some timber is imported.

Japan depends more on imported fuel than any other major industrial country. Most of its hydro-electric potential has now been harnessed. Coal production has actually declined drastically since a peak of production in the 1960s (*see Figure 9.3*).

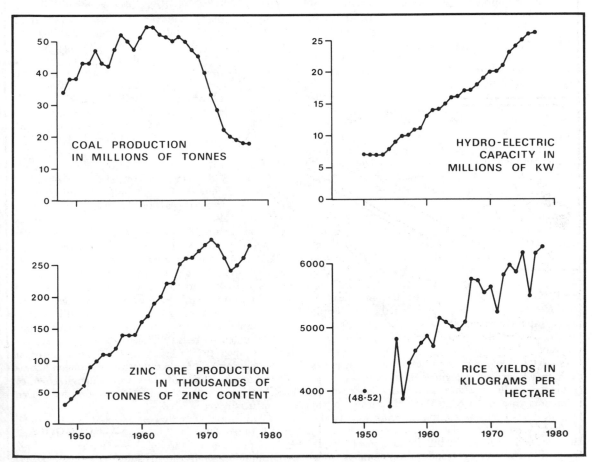

Figure 9.3   *Selected aspects of Japanese production 1948–1977*

Between 1960 and 1971 the number of pits in use dropped from 620 to 70. Japan imports large quantities of coking coal but imported oil accounts for about three-quarters of all its energy requirements. The largest supply comes from Southwest Asia while some comes from Indonesia, the USSR, China and Latin America. There are many large oil refineries along the coasts of Japan, most in the industrial belt of South Honshu. The importance of liquefied natural gas is increasing. Nuclear power has also been developed since 1966 but safety problems connected with the occurrence of earthquakes, and emotional opposition arising from the atomic bombs dropped on Japan in 1945 appear to have delayed progress until recently but between 1975 and 1980 nuclear generated electricity increased more than three times. It is hoped that electricity will be produced commercially from the safer method of nuclear fusion by the end of this century.

Japan's economic success depends essentially on the ability of industry to produce goods greatly in excess of the value of the imported fuel and crude materials it imports. A certain proportion of industrial production is exported to pay for these items and for food and essential capital goods. Japan imports very few consumer manufactures.

Since so many of the ingredients of Japanese industry are imported and therefore arrive by sea it is not surprising that most of the major industrial establishments are located on or very near to the coast. Approximately six-sevenths of Japanese industrial capacity is located in a zone from Hitachi a little way north of Tokyo in the east to Kitakyushu in the west (*see Figure 9.2*). Only about one-quarter of Japanese industrial workers are however employed in very large, high productivity plants while at the other extreme there are many small factories and workshops.

One of the earliest modern industries to be established in Japan was the manufacture of textiles and clothing. This sector has declined relatively, but it accounts for about 10 percent of the labour force. It is concentrated especially in Osaka-Kobe. In spite of the lack of iron ore and coking coal, except during the 1930s in Manchuria, Japan has become a leading producer of steel. *Figure 9.2* shows the location of major Japanese steel plants and their distribution reflects broadly the concentration of all industry in the country.

On the basis of its iron and steel industry Japan has developed a successful engineering industry, a sector employing about one-third of the total labour force in manufacturing. Early in the 20th century Japan was producing railway equipment and ships. In the 1960s it began to manufacture motor vehicles on a large scale. Its success in the production of a wide range of engineering goods using limited raw materials but of high value (and good quality) is well

known. Another sector of industry to be developed in Japan since the Second World War has been oil refining and chemicals.

The rises in the price of oil since the early 1970s have been one reason for a series of setbacks in Japanese industry. Steel producing capacity has not been used fully and shipbuilding production has declined, being in 1979 only about one-third what it was in 1974. The Japanese government encourages the development of new types of industry and in the 1970s and 1980s interest has grown in electronics and computers, aerospace, nuclear power and agricultural equipment suitable for developing countries.

Internal features and problems

Between about 1700 and 1850 the population of Japan was stable at about 30 million. It reached 60 million in 1925, 90 million in 1956 and should be about 120 million by 1984–85. In spite of the fourfold growth, the annual increase of population in Japan since 1850 has never been as high as that in many developing countries at present. The average citizen of Japan consumes perhaps 25 times the quantity of goods and services consumed by his predecessor around 1850 and pressure has therefore grown enormously on the land, coastal waters and natural resources of Japan. In order to maintain its high level of production and considerable living standards, Japan has to perform a delicate balancing act. Many problems have arisen as a result of the way it has developed, some of which may be noted:

(1) Japan depends very heavily on imported goods and it must, therefore, retain the good will of other countries. Though militarily allied to the USA it attempts to keep a neutral position in world affairs. Its push to find outlets for its manufactured goods, however, has caused great friction with Western industrial countries.
(2) Lack of space has meant that much of Japanese industrial capacity has had to be concentrated in or close to urban areas, often using up good agricultural land. Various ways have been proposed to reclaim offshore areas, sometimes conflicting with inshore fishing.
(3) So much waste is produced from various sources in Japan that it has been difficult to prevent pollution. There have already been many victims of mercury and cadmium poisoning while tens of thousands of victims of air pollution have been officially recognized.
(4) Population is highly concentrated in a few small regions of Japan. About 20 percent of the total population of Japan is concentrated in three adjoining prefectures (Chiba, Tokyo and Kanagawa), which cover only about 2.5 percent of the

area of Japan. About 12 percent is concentrated in two others (Osaka and Hyogo), also covering only about 2.5 percent of the area of the country. *Figure 9.4* shows areas losing and gaining population in the period 1965–70, a period of particularly fast economic growth in Japan. Most of the areas with an increase correspond to the major urban centres of Japan. The larger Japanese cities are characterized by bad urban planning, congested housing, lack of parks, traffic problems, long journeys to work and heavy pollution.

(5) Although by world standards Japan is a comparatively homogeneous country, marked inequalities occur. Contrasts exist, particularly, between town and country, with rural settlements tending to lose their more able inhabitants. *Figure 9.4* shows that one measure of prosperity, car ownership, is by no means even over the national area of Japan. Other indicators also show regional disparities. Japan also has a class of untouchables or outcastes, the Burakhamin.

(6) The fragmented and elongated layout of the land of Japan has caused problems of cohesion. Rapid land transport is assured between Tokyo and the southwest of Japan by a special railway while Kyushu and Hokkaido are linked by tunnels with Honshu. Distances are however considerable and roads heavily congested.

(7) Educational and health levels in Japan are high and crime low. Illiteracy is almost non-existent, a quarter of school leavers go on to higher education, and newspaper circulation and television ownership are high. Hospital facilities are widely available and Japan has one of the lowest levels of infant mortality in the world. On the negative side it is not surprising that in such a highly competitive society tension is generated in the education system.

Japan, the outside world and the future

In 1978 Japan took six to seven percent of the imports in world trade but provided eight to nine

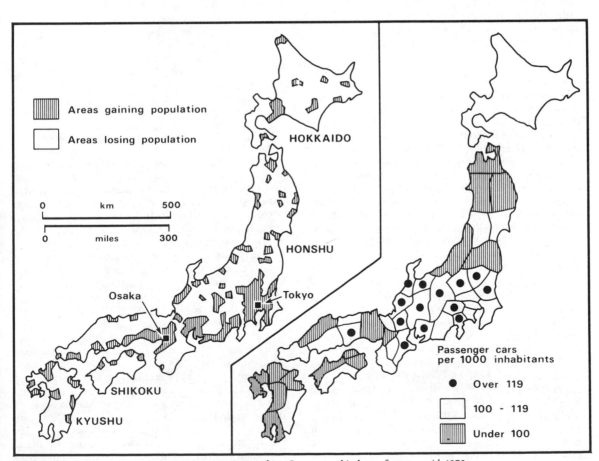

Figure 9.4(a) *Population change in Japan 1965–70*   **(b)** *Car ownership by prefectures, mid-1970s*

Table 9.1 *Japanese foreign trade*

Commodity groups in Japanese foreign trade

Commodity group	Percentage imports	Percentage exports
Foodstuffs	13.0	1.2
Raw materials and fuels	61.0	1.0
Heavy industry products	16.5	85.1
Light industry products	8.1	11.6
Other	1.4	1.1

Principal trading partners of Japan 1979

Imports from	Percentage	Exports to	Percentage
USA	18.5	USA	25.6
Saudi Arabia	11.0	Korean R	6.1
Indonesia	7.9	Taiwan	4.2
Australia	5.7	German FR	4.1
Kuwait	3.9	Saudi Arabia	3.7
Iran	3.9	China	3.6
Canada	3.7	Hong Kong	3.6
United Arab E	3.3	UK	3.0
Korean R	3.0	Singapore	2.6
Malaysia	2.9	Australia	2.5
Other	36.2	Other	41.0

Source: Statistics Bureau, Prime Minister's Office, 1980 Japan, pp 13–14

percent of the exports. In 1979, however, it had an untypical balance of payments deficit which in the early 1980s was being 'corrected'. The fact that 75 percent of the value of Japan's imports consists of food, fuel, and raw materials underlines clearly the lack of agricultural land and mineral resources in the country. To pay for its imports Japan depends almost exclusively on manufactured goods. *Table 9.1* shows that most of Japan's suppliers are either providing oil (eg Saudi Arabia, Kuwait), food, or raw materials (eg Australia, Canada). These countries are either unable or unwilling to import an equivalent value of Japanese manufactures. Japan is therefore forced to search for other trading partners, many of them also industrialized countries such as the USA and the UK. Through these countries Japan must achieve a positive (to itself) overall balance of trade. Such is one of the underlying reasons for the penetration by Japan of the US and West European markets.

In order to achieve the above aim, Japan has to export manufactured goods that are either cheaper or of better quality (or both) than those available in other industrialized countries. How does Japan do all this? Many reasons have been proposed and the following are only a summary of some of the main ones.

The Japanese economy is largely a market economy, with private ownership of many of the means of production. Competition between companies within Japan is encouraged and is fierce. Nevertheless the government and the banks have considerable influence in guiding the development of various sectors of the economy. Indeed indirectly the banks own many Japanese enterprises and therefore the many savers in Japan have a share in ownership.

The Japanese generally participate willingly in economic development. They save about 20 percent of their income (twice to three times the percentage in the USA and West Europe), thus forgoing consumption in order to build for the future. Workers participate with management in maintaining good labour relations. Economic patriotism is either an indigenous characteristic of the Japanese or has been instilled into them. As with Israel, perhaps Switzerland, and a few other countries, Japan is an extended family. The Japanese generally do not like foreigners to settle in their country and do not like Japanese to marry foreigners.

Early Japanese development was associated with the imitation of methods used in more advanced industrial countries. Japan now devotes a large part of its income to research and development (six percent compared with one percent in the USA). Although Japanese researchers and industrialists are not regarded as very creative or original themselves, they are quick to adopt and apply the innovations of others.

In order to retain a favourable balance of trade Japan resorts to a number of devices that are not regarded as all that fair. The Japanese themselves (their patriotism?) prefer to buy Japanese manufactures rather than foreign ones. Foreign capital investment is discouraged although Japan itself has become a large investor in many other countries. Obstacles are put in the way of exporters of manufactures looking for a market in Japan.

Japan goes deliberately for the latest technology. The Industrial Revolution has so far been characterized by the use of inanimate sources of energy to supplement (enormously) or replace human (and animal) muscles. The new Industrial Revolution, seriously possible only since the successful development of electronic computers in the 1950s, allows the human brain to be supplemented or even replaced. To be the world innovator in the robotization of industrial and other processes would benefit Japan. Suitable labour for the industrial sector is not lacking but industrial goods can be produced more cheaply and more reliably by robots. Knowhow and equipment could be exported by Japan and the country could therefore exercise considerable influence on the world economy in the next decades in the way that the UK did a century and a half ago when it provided a large part of the exports of industrial goods. Many futurists have feared that people would be turned into 'robots'. The Japanese are turning robots into 'people'.

Japan has been considered here in some depth for a number of reasons. Its economic development has lessons for developed and developing countries alike. Its attributes, achievements and problems may serve as a checklist for comparison with the rest of the regions of the world to be considered next.

Of the developed regions of the world Japan is the one with the most unfavourable ratio of natural resources to population. West Europe and East Europe, though better off than Japan, face similar problems. The USSR, USA, Canada and Australia, on the other hand, are much better endowed and more self-sufficient. They feel less urgency in innovating in industry.

The developing regions of the world may see the path of Japan since the 1860s as one to be followed in the future. If, however, many developing countries that lack natural resources (eg India, Bangladesh, China) industrialize to the extent that Japan has done, then they would be competing with Japan to obtain fuel, raw materials and food elsewhere in the world.

9.2 West Europe

Introduction

West Europe is understood here to include all the non-socialist countries of the continent. At any given time several of the countries may have a self-styled socialist party in power but the economies of the countries are 'mixed' economies, with a public sector that has developed mainly since the Second World War. The economic role of the state tends to be to support declining or unsuccessful sectors of the economy and to guide economic development rather than to enforce national plans.

West Europe as defined here had a population of about 350 million in 1981, a little under 8 percent of the total population of the world. With an expected 360 million inhabitants in the year 2000 it would then have less than six percent of the world's population. It has less than three percent of the earth's land surface and is therefore less than half as large as Australia and about one-sixth the size of the USSR. Its land is stretched and fragmented into peninsulas and islands. The scene of innumerable wars between member nations until 1945, happily since that year there have been no serious military conflicts. The presence of the Soviet Union and its East European allies so near has produced a feeling of unity in the region not experienced for centuries.

Politically West Europe is subdivided into 19 sovereign states (counting Iceland, Luxembourg and Malta, but not the very small units of Andorra, Liechtenstein, San Marino and The Vatican). Their location is shown in *Figure 9.5* and economic demographic and social data are given in *Table 9.2*. All the countries except Spain and Portugal belonged in the early 1980s either to the European Community (*see* chapter 8) or to the European Free Trade Association. West Europe has some of the richest countries of the world but also, in some parts of southern Europe, regions that in some respects could belong to developing countries. Militarily most of West Europe is directly associated with the USA through the North Atlantic Treaty Organisation, but Austria, Finland, Switzerland and Sweden maintain a neutral policy. Spain has had US military bases since the 1950s but, like Portugal, it was regarded as politically undesirable as a prospective member of NATO. France is associated with NATO but does not allocate forces specifically to it while Spain is expected to be admitted by the mid-1980s.

In spite of the large number of sovereign states in Europe the existing boundaries do not exactly fit underlying cultural features. Most countries contain more than one language group and several contain a mix of Roman Catholics and Protestants. Movements aimed at complete independence or at least a degree of regional autonomy within rigidly unitary states

Figure 9.5 *The countries and main cities of West Europe*

Table 9.2 *Data for the countries of West Europe*

Country	Area 1	Popn 2	% crops 3	% forest 4	Popn/ arable 5	Popn/ change 6	Non- agric 1960 7	Non- agric 1979 8	Energy cons 1979 9	Infant mort 10	Pupils/ teacher 11	GNP 12
Finland	337	4.8	7	70	185	0.4	64	86	5210	8	24	8 260
Sweden	450	8.3	7	59	223	0.1	86	95	5950	7	17	11 920
Norway	324	4.1	3	26	455	0.3	80	92	5570	9	20	10 710
Austria	84	7.5	20	39	202	−0.1	76	90	4050	15	28	8 620
Switzerland	41	6.3	10	25	314	0.3	89	95	3690	9	23	14 240
Portugal	92	10.0	39	40	244	0.7	56	73	1030	39	34	2 160
Spain	505	37.8	41	30	120	0.8	58	82	2410	13	35	4 340
Denmark	43	5.1	61	11	175	0.1	82	93	5420	9	20	11 900
Ireland	71	3.4	14	4	60	1.0	64	79	3290	15	34	4 230
UK	245	55.9	29	8	304	0.1	96	98	5210	13	23	6 340
German FR	249	61.3	32	29	465	−0.2	86	96	6020	15	30	11 730
Netherlands	37	14.2	23	8	694	0.4	89	94	5330	8	35	10 240
Belgium	33	9.9	27	21	682	0.1	92	97	6080	12	23	10 890
France	547	53.9	35	27	169	0.4	78	91	4370	10	24	9 940
Italy	301	57.2	41	21	326	0.2	69	88	3230	15	28	5 240
Greece	132	9.6	29	20	105	0.7	44	62	1930	19	43	3 890

Sources and definitions of variables: see Table 9.3

Table 9.3 *Sources and definition of variables in Table 9.2*

Sources

2, 4, 5, 9, 11 *1981 World Population Data Sheet, PRB Inc.*

3, 6, 7, 8 numbers of *FAOPY* and *World Statistics in Brief*

10 *World's Children Data Sheet*, PRB Inc. 1979.

Definition of variables

1 Total area in thousands of square kilometres
2 Population in millions, 1981
3, 4 Percentage of total area under arable or permanent crops *3* and forest *4*, 1978
5 Persons per square kilometre of arable plus pasture land, 1978
6 Percentage rate of annual natural change of population around 1980
7, 8 Economically active population *not* engaged in agriculture as a percentage of total economically active population in 1960 and 1979
9 Consumption of energy in millions of tonnes of coal equivalent per inhabitant in 1979
10 Annual number of deaths to infants under one year of age per 1000 live births in late 1970s
11 Number of school-age children per teacher, 1975
12 Gross national product in US dollars per inhabitant, 1979

industrial processes has been eroded by the development particularly of the USA, Russia and Japan as major industrial powers. Virtually every colony outside Europe has now gained or been given independence. The coalfields and other mineral deposits in West Europe that formed the basis for industrialization in the 19th century have been depleted and in some areas largely used up. Militarily West Europe is in a precarious position, with its population and industries highly concentrated, and a limited nuclear capacity in France and the UK. In an all-out nuclear war involving Europe little would be left at all in West Europe. In the rest of the section it is assumed that there will not be such a conflict and that the problems of West Europe will be political, economic and social.

Natural resources and production

West Europe contains a variety of natural environments, from the cold wastes of parts of Iceland and the north of Scandinavia to the warm temperate conditions of the Mediterranean. Mountain areas, hill country and plain alternate over the region, giving a wide variety of physical conditions. Climatically almost everywhere in West Europe can be described as humid, though summers in the southern part tend to be very dry. Where soil and other conditions are suitable the growing season is adequate for cultivation except in the extreme north of Scandinavia. Most of the land not under cultivation is either under forest or is used for pasture.

The main uses of land in West Europe are about 33 percent forest, 25 percent arable and permanent crops

like France and Spain are numerous, weakening to some extent the cohesion and defence capability of countries.

Relatively, West Europe, as a whole, has declined dramatically as a force in the world since the 19th century. Its early near monopoly of modern

and 19 percent pasture. Since the 1950s the proportion under field crops has diminished somewhat while the proportion under forest has increased. There are more than twice as many persons per square kilometre of cultivated land in West Europe as in the world as a whole, but yields are much higher than the world average. The removal of land from crops in marginal areas of cultivation as, for example, where land is terraced or is steeply sloping has therefore been more than offset by higher yields. West Europe as a whole probably has enough land to feed its population but it cannot produce many of the plant and animal raw materials it needs, nor, of course, beverages grown only in tropical conditions. The third of the area under forest, mainly softwood, is still inadequate to satisfy the timber needs of the region. The extensive fisheries have suffered setbacks due to overfishing and have been the cause of political disputes.

Although West Europe has only about 8 percent of the population of the world it consumed in 1980 about 18–19 percent of all energy (Japan consumed about five percent). It only produces about 40 percent of its total energy requirements. The German FR is credited with several percent of the world's coal reserves and it also has lignite. The UK also has adequate coal reserves, but elsewhere in West Europe reserves are very limited. Of the world's 'published proved' oil reserves, West Europe had only 3.6 percent in 1980, 3100 million tonnes, largely shared by the UK and Norway. In the same year it had 6.1 percent of the world's natural gas reserves, most of them located in offshore areas belonging to the Netherlands, the UK and Norway. West Europe has a considerable hydro-electric potential, now largely harnessed, but this is concentrated mainly in the regions of high mountains. There are also modest reserves of uranium. The prospects for discovering large new deposits of oil and natural gas depend on techniques for exploring in deeper off-shore areas and in the harsh conditions off northern Norway.

The industries of West Europe now derive about 90 percent of their non-fuel mineral requirements from outside the region. The countries indicated have the following percentages of the world's reserves of the minerals listed:

Iron ore   Sweden 2%
Bauxite   Greece 3%
Lead   German FR 3%, Spain 2%
Tin   UK 3%
Zinc   Ireland 5%
Potash   German FR 2%
Mercury   Spain 29%, Italy 8%
Antimony   Italy 3%

The following important minerals are not found in commercial quantities in West Europe: copper, gold, nickel, sulphur, manganese, silver, industrial diamonds, phosphate, tungsten, molybdenum, chromium, asbestos and cobalt. The chances of finding more than modest new reserves of non-fuel minerals seem remote. West Europe faces a bleak future in which it will depend increasingly on recycling materials, being less lavish in their use, substituting scarce ones with more abundant materials and keeping on good terms with other parts of the world. The USSR and South Africa have large deposits of many of the minerals needed by West Europe.

Production

The relative importance of the agricultural sector in West Europe has been declining (*see Table 9.2 columns 7 and 8*). In the most highly industrialized countries of West Europe (the UK, Belgium, the German FR and Sweden) less than five percent of the economically active population is employed in agriculture. Greece and Portugal, in contrast, still have a relatively large agricultural sector. The reduction of the labour force in agriculture has been made possible by the rapid development of the mechanization of many processes, coupled with an increase in farm and field size. Higher yields have been achieved through an increase in the use of fertilizers and the introduction of new strains of plant. Policy in the EEC has been both to help farmers by guaranteeing prices in some sectors and by encouraging them to leave less productive land or to change its use.

While it is difficult to generalize about West European agriculture a broad contrast may be noted between the lands to the north and to the south of the Alps. In the northern and central parts of West Europe, mixed farming predominates. Wheat, barley and oats are usually the main cereals grown though maize has been increasing in popularity. Sugar beet is the main 'industrial' crop. Roots, grasses and other fodder crops are widely grown and together with extensive natural pastures support the large livestock sector. In the southern part of Europe, relief and climatic conditions produce a different agricultural mix. Greater emphasis is placed on permanent crops (the vine, olive, fruit trees) while maize and even some rice are produced. Natural pastures and forests tend to be less productive than further north and sheep are relatively more important than cattle in some areas.

The situation of agriculture in West Europe as described above is one in which drastic changes seem unlikely. The area under crops cannot be greatly extended in the near future and increases in yields cannot go on indefinitely. Changes may come through new irrigation projects and a further reduction in the labour force in some countries and there is a possibility that in the future West Europe could even become a net exporter of temperate agricultural products. Agriculture now only accounts for a few percent of the gross national product of EEC yet takes about 70 percent of the budget, although since the early 1970s mining and industry have become the main problem sector of EEC.

New industrial techniques using water power and steam power, coke instead of charcoal for producing iron and the mechanization of many factory processes were first applied mainly in Britain in the 18th century. The increase in the use of coal before the rapid growth of railway systems in the middle of the 19th century meant that early development took place mainly on or near coalfields. Many of the problem areas of West Europe are precisely the older coalfield areas with their associated industries. *Figure 9.6* gives the location of the coalfields while *Figure 9.7* shows trends in energy consumption in the 1970s.

The development of rail transport and steamships allowing the movement of coal over longer distances

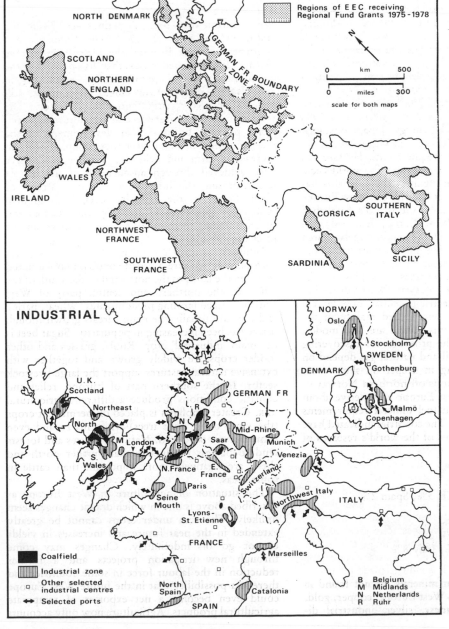

Figure 9.6(a)   *Regions of EEC receiving Regional Fund Grants*

Figure 9.6(b)   *Coalfield and industrial regions of West Europe*

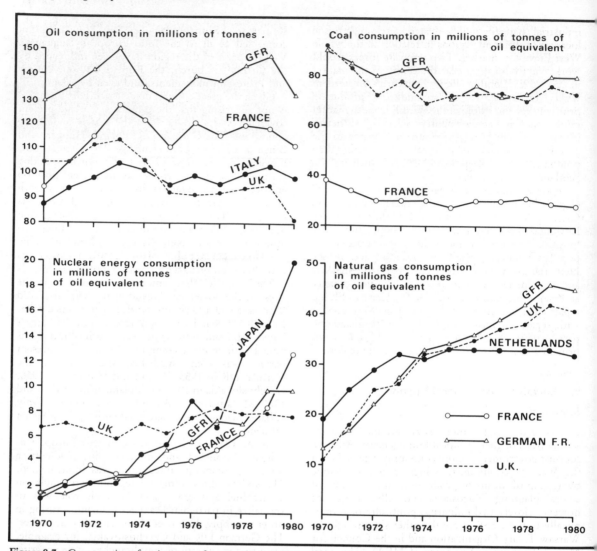

Figure 9.7 *Consumption of major types of energy in selected West European countries in the 1970s*

and later the development of hydro-electric power attracted industry to many additional areas in West Europe, as in northern Italy, Switzerland, Sweden and northeast Spain. It became attractive to locate some industries near to large markets (London, Paris, for example) away from sources of fuel.

Since the rapid increase in the use of imported crude oil after the Second World War, and also of coking coal and high grade iron ore, coastal sites have become favoured places for the establishment of new industries, as along the Seine and Rhine estuaries. In the Mediterranean the Marseilles area was a centre for new industries while the less industrialized peninsula and islands of Italy were the scene of the creation of new industrial 'poles'. Even these new developments with their oil refineries, petrochemicals and iron and

steel works are threatened as dependence on oil declines. Their oil refineries have worked well below capacity since the early 1970s, and comparatively few jobs have actually been directly created in the areas where it was hoped their influence would stimulate development. Two further sources of energy in West Europe may also be noted, nuclear power and natural gas. With the construction of natural gas pipelines and an increasingly complete electricity grid these sources of energy allow industries to be located virtually anywhere.

Availability of energy has not been the only influence in the rise and decline of different industrial regions in Europe. Raw materials and large markets have played their part. Ultimately if energy is readily available anywhere in West Europe then future industrial changes could be related increasingly to the location of different regions in relation to the whole West European market. Two opposite forces would work. Firstly location near the centre of the market in an area London–Paris–Frankfurt–Amsterdam would be attractive to an industry producing primarily for the whole of the West European market rather than for a single country. Secondly, sufficient incentives might be given to attract industry to the periphery for social (and political) reasons rather than economic ones, to keep the industries in such areas as Scotland or Sicily going.

Figure 9.6 shows the 'negative' regions of the EEC eligible in the period 1975–78 for Regional Fund Grants. They cover more than half of the total area of EEC and have about half of the population. The 'positive' regions include the concentrations of population around Paris and London and in the Netherlands–Belgium–Ruhr area with large markets and immediate access to research facilities, finance and major international airports. The German FR has a particular advantage over the rest of West Europe with regard to East Europe through language and past experience of trade. The North Sea is the principal internal source of energy for the region.

9.3 Socialist East Central Europe

Introduction

In this section eight countries are grouped together on the basis of their political homogeneity. All have socialist governments, more commonly referred to in the West as communist, a high degree of public ownership of means of production, and a system of central planning. Yugoslavia and Albania are not however closely tied militarily, politically or economically to the USSR as are the other six through the Warsaw Treaty Organization and in the Council for Mutual Economic Assistance (CMEA). The total population of the eight countries in 1981 was about 135 million or about 3 percent of the total population of the world. The total area is about 1 280 000 square kilometres.

The eight countries under consideration, referred to henceforth simply as East Europe, form a buffer zone between West Europe and the USSR (*see Figure 9.8*). The present territories of the eight countries date either from changes following the First World War or from changes agreed during the Second World War and applied in 1945.

East Europe lies between countries that have been powerful militarily in the last few centuries: Germany and Austria to the west, Russia (now the USSR) to the east and Turkey (the former Ottoman Empire) to the south. Linguistically the region contains speakers of Slavonic languages (Poland, Czechoslovakia, Yugoslavia, Bulgaria), German, Romanian (a Romance language) and Hungarian (not connected at all to the Indo-European languages). Various parts of the region have been influenced by the Orthodox church, the Roman Catholic church and Protestant movements, and even by Islam.

Given the great cultural diversity of East Europe it is not surprising that the boundaries of the present sovereign states do not exactly match underlying conditions. The German DR, Poland, Hungary and Bulgaria are fairly homogeneous. The German DR, however, is only a small part of German speaking Europe. Poland was forced into its present position and national homogeneity by displacement of both land and people. In contrast, Czechoslovakia consists mainly of two cultural groups, the Czechs and Slovenes, while Romania contains a large Hungarian minority and also claims (in theory) Moldavia, now the Moldavian Republic of the USSR. Yugoslavia is extremely complex, with several cultural groups.

The countries that came to be referred to in the West as the Soviet satellites after the Second World War have had a fairly uneventful history since 1945. In 1948 Marshal Tito took Yugoslavia out of Stalin's Soviet orbit and in the same year Czechoslovakia was pulled into it by a communist coup. The present pattern was then established and unrest in the German DR in 1953, Poland and Hungary in 1956, Czechoslovakia in 1968 and Poland again in 1970 and 1980–81 failed to shake off Soviet control. Only Albania managed to isolate itself entirely from Soviet influence.

In the 1950s Soviet influence in East Europe was primarily political and military, but gradually a policy of economic integration was implemented. The six 'satellites' came to depend increasingly on Soviet fuel and raw materials. their role was to specialize in particular branches of manufacturing in order to supply the Soviet Union and also each other. The German DR and Czechoslovakia, for example, were to concentrate on key branches of engineering while Poland was to produce a surplus of steel, coal and food. In the 1970s the USSR sold oil to its partners at prices below those on the world market. The USSR was a 'captive' market for the manufactured products of its East European partners.

The economies of the CMEA countries have become increasingly sophisticated since the 1950s. Capital goods, raw materials and food have been obtained outside the CMEA bloc but East Europe has not competed wholeheartedly in the tough world of the market and mixed economies outside. By 1980 the CMEA bloc had accumulated debts of some 70 000 million US dollars which, given no political or military change or catastrophe, could rise to 100 000–130 000 million by the mid-1980s. Thus western economic influence on CMEA has actually grown enormously in the 1970s, through an increase in both trade and loans.

160

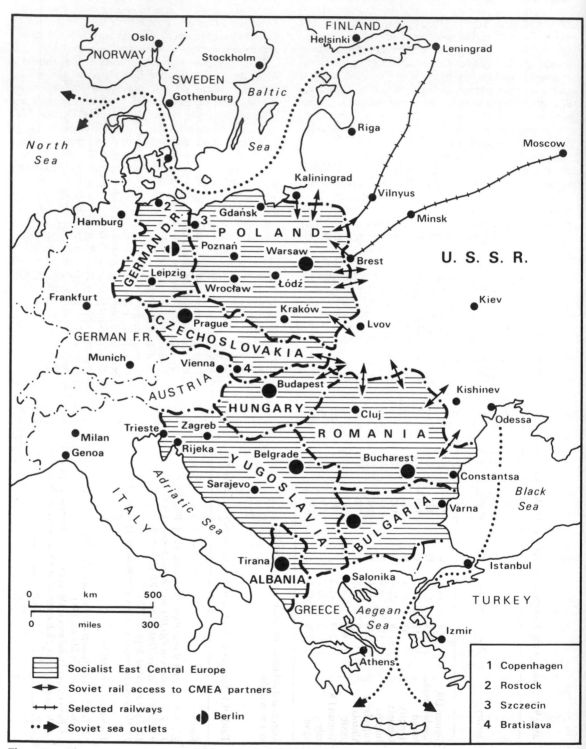

Figure 9.8 *The countries and main towns of Socialist East Central Europe*

Table 9.4 Data for the European members of CMEA and for other selected countries

Country	Area 1	Popn 2	% crops 3	Popn/ arable 4	Popn/ change 5	Non- agric 1960 6	Non- agric 1979 7	Energy cons 1979 8	Infant mort 9	Pupils/ teacher 10	GNP 11	Catholic 12	Imports from USSR 13	Soviet trade 14
German DR	108	16.7	47	267	0.0	82	90	7120	13	25	6430	8	30	10
Czechoslovakia	128	15.4	41	222	0.6	74	89	7530	19	32	5290	72	35	8
Poland	313	36.0	48	189	1.0	52	69	5600	22	36	3830	94	31	9
Hungary	93	10.7	58	160	0.2	61	83	3450	24	29	3850	61	28	6
Romania	238	22.4	44	150	0.9	35	52	4040	30	33	1900	1	16	3
Bulgaria	111	8.9	39	144	0.5	43	66	5020	22	34	3690	1	57	8
Yugoslavia	256	22.5	31	158	0.9	36	61	2040	32	38	2430	32	18	3
Albania	29	2.8	23	226	2.2	29	39	1000	na	34	840	10	na	–
USSR	22 402	268.0	10	44	-0.2	58	83	5500	36	28	4110	1	–	–
German FR	249	61.3	32	465	-0.2	86	96	6020	15	30	11 730	45	2	5
Italy	301	57.2	41	326	0.2	69	88	3230	15	28	5 240	99	3	3
Japan	372	117.8	13	2145	0.8	67	88	3830	8	38	8 800	0	2	3

Sources and definition of variables see Table 9.5
na not available

Table 9.5 Sources and definition of variables in Table 9.4

Sources

2, 4, 5, 9, 11 1981 *World Population Data Sheet*, PRB Inc.
3, 6, 7, 8 numbers of *FAOPY* and *World Statistics in Brief*
9 *World's Children Data Sheet*, PRB Inc. 1979
12 *The Economist*, Sept. 5, 1981
14 *Vneshnyaya torgovlya SSSR v 1979 godu*, Moscow, 1980

Definition of variables

1 Total area in thousands of square kilometres
2 Population in millions, 1981
3 Percentage of total area under arable or permanent crops, 1978
4 Persons per square kilometre of arable plus pasture land, 1978

5 Percentage rate of annual natural change of population around 1980
6, 7 Economically active population *not* engaged in agriculture as a percentage of total economically active population in 1960 and 1979
8 Consumption of energy in millions of tonnes of coal equivalent per inhabitant in 1979
9 Annual number of deaths to infants under one year of age per 1000 live births in late 1970s
10 Number of school-age children per teacher, 1975
11 Gross national product in US dollars per inhabitant, 1979
12 Roman Catholics as a percentage of total population
13 Percentage of value of imports supplied by the USSR
14 Percentage of total value of Soviet foreign trade with country in question, 1979

Natural resources and production

East Europe is a region of considerable physical diversity. Nevertheless the mountain areas, mainly in the central and southern countries, are not excessively high. Winter temperatures are harsh but not extreme and rainfall is everywhere adequate for some kind of cultivation though it may in places be supplemented by irrigation. About 45 percent of the total area of the eight countries is arable land or under permanent crops (*see* column 3 in *Table 9.4*). Most of the remaining land is used for pasture or forest. There is adequate cropland for the region as a whole to be roughly self-sufficient in food.

The fossil fuel reserves of East Europe are limited in both extent and type. Only Poland has very large known recoverable reserves of coal, enough to last about 100 years at current production rates, but Czechoslovakia and Hungary also have some. The German DR has about one-sixth of the world's

known recoverable reserves of lignite and all the other countries of the region except Yugoslavia also have some. Oil and natural gas reserves are however very limited. Romania's oilfields have been exploited for many decades and in the late 1970s had only enough oil for about ten years of production. Proved oil reserves elsewhere are negligible and natural gas reserves are very limited. A modest hydro-electric potential has now largely been brought into use in the various mountainous areas.

Various non-fuel mineral deposits occur in East Europe but there are no very large scale reserves. In the lists of the top ten countries for reserves of over 20 major non-fuel minerals only Poland figures for sulphur and copper, the German DR for potash and Yugoslavia for lead, zinc and mercury.

The rapid industrialization of East Europe since 1945 could not have taken place without the importation of fuel and raw materials from outside. With its vast reserves of fuels and mineral raw

Figure 9.9   *The main industrial regions of the German DR, Poland and Czechoslovakia*

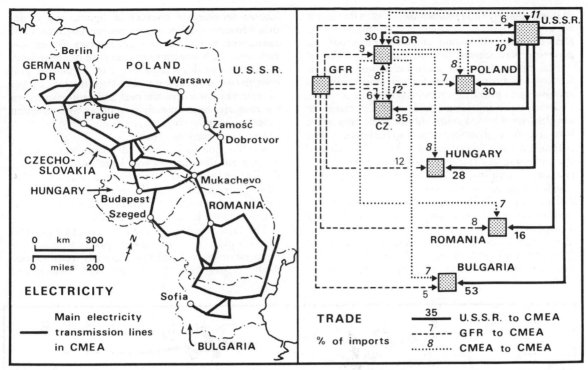

Figure 9.10(a)   *The electricity transmission system of CMEA countries*

Figure 9.10(b)   *The largest sources of imports of the CMEA countries of East Europe*

materials, its forests and subtropical agricultural products the USSR has been the main supplier of the region. Coking coal, oil, natural gas and electricity are all exported to the region from the USSR, together with such plant and animal materials as timber, cotton and wool. East Europe therefore broadly resembles Japan and West Europe in its economic structure and trade pattern with the rest of the world but it has very little trade with developing countries.

Agriculture in East Europe is mostly of a mixed kind with roughly equal weight given to crops and to livestock. Cereals are widely cultivated as well as sugar beet and vegetables, particularly potatoes. Permanent crops include the vine in the southern part of the region. In the late 1970s perhaps for both climatic and organizational reasons, cereal production has faltered, imports have been necessary and sources other than the USSR have been sought. Food consumption levels are generally high, with a large intake of meat and dairy products.

Energy consumption per inhabitant is at a very high level in Czechoslovakia and the German DR while even in Poland it is as high as in the USSR. Energy consumption has increased greatly since 1945, continuing into the late 1970s even though the production of coal in Czechoslovakia and of lignite in

the German DR stagnated throughout the decade, as did oil production in the region as a whole (18 million tonnes in 1970, 19 million in 1980). Only Polish coal production continued to rise in the 1970s but exports of coal also increased.

The energy prospects for East Europe are not promising, but there is apparently scope for improving the efficiency of use. Increased imports are expected to come from the USSR though in the form of natural gas rather than oil, since Soviet oil production targets are not being achieved. It would be difficult under present circumstances for the region to plan to obtain much oil from OPEC sources, given the high price.

Industrialization has been given high priority in all Soviet influenced economies. High priority was given in the 1950s to the establishment or expansion of heavy industry, particularly the iron and steel industry and heavy engineering. Parts of the German DR and Czechoslovakia already had an industrial base before the Second World War. In 1945 the former German industrial region of Silesia was ceded to the new Poland.

Heavy industry was expanded or newly established in the regions noted above and also on a smaller scale in the other countries. The main industrial region of East Europe is indicated in *Figure 9.9*, the coalfield of

southern Poland and northern Czechoslovakia forming the core. Existing food and textile industries were
expanded after 1945 and the development of engineering was encouraged in a large number of
towns. Other branches of manufacturing such as
chemicals and the working of wood and its products
are also widespread.

East Europe is well provided with rail transport
but road transport is relatively much less important
than in West Europe. Although Hungary and

Czechoslovakia are landlocked, nowhere is more
than a few hundred kilometres from the sea coast and
ports, either on the Baltic, the Black Sea or the
Adriatic. Oil and natural gas enter by pipeline from
the USSR and there is an integrated electricity grid
system (*see Figure 9.10*). On the other hand the
movement of workers, tourists and visitors between
the countries seems greatly restricted and integration
is only partial.

Developed Regions Rich in Natural Resources

10.1 The USA

Compared with most of the larger and more powerful countries of the world the USA is comparatively young as a nation. It has only been a sovereign state for about two centuries. Its rise to its present prominence has been rapid and most impressive.

In the 15th century the present area of the USA was inhabited by an indigenous population of perhaps several million, thought by some to have filtered into the Americas from Asia many thousands of years previously. Modern USA is very generously provided with natural resources, but the technology of the American 'Indians' at the time Columbus discovered the New World was very limited and the bioclimatic resources were the only ones used to any extent. Maize and other New World crops were cultivated and hunting and fishing were practised in the grasslands, forests and inland waters.

European colonial powers put less effort into exploring and conquering North America than they did in the rest of the Americas or in parts of Asia. In the 18th century much of what is now the USA was still the domain of the 'Red Indians' who however had adopted some European ways, including the use of the horse and of firearms. In the 18th century the region might have been claimed by the French, converging via the St Lawrence (Quebec) and the Mississippi (Louisiana), by the Spaniards (from Mexico and Florida) or even by the Russians (via Alaska). The 13 British colonies were not the obvious eventual conquerors. They exported less in the 18th century than the small French colony of Haiti in the Caribbean. Their expansion westwards was blocked by mountains and Indians.

After independence from Britain was achieved in 1783 the new USA took some decades to consolidate its position. With the advent of the steamship and the railway and with the impulse from numerous new migrants from Europe expansion westwards began. Settlers were attracted by the prospect of both good farm land across the Appalachians and minerals in various places, especially the gold of California. The outcome of the Civil war (1861–65) ensured the integration of the USA, and the search for new techniques and strategies by the north to win the conflict boosted industrial growth and hastened the westward penetration of settlement. In the middle of the 19th century, however, the USA was still heavily dependent on agriculture. Abundant reserves of coal, oil, natural gas, many non-fuel minerals, and agricultural and forest raw materials gave it an excellent base for industrialization. The 'young' USA had a much higher ratio of natural resources to population than that found in contemporary industrializing countries of Europe.

Until the 20th century the USA largely confined its growing political and military influence to the Americas. President Monroe stated in 1823 (The 'Monroe doctrine') that

'the American continents, by the free and independent condition which they have assumed and maintained, are henceforth not to be considered as subjects for future colonization by any European powers'.

Although the USA was unwillingly involved in the First World War it withdrew from European affairs shortly after by declining to join the League of Nations.

After the Second World War the USA adopted its current view that the world is indivisible and that to maintain influence it had to be prepared to act militarily outside the Americas. Its ideological and political preoccupation has been to counter Soviet influence and to deter Soviet military expansion. Economically it has come to depend heavily on the importation of many different products from both developed and developing countries. The aim of the rest of this section is to describe some key internal features of the USA and to relate them to world affairs in general.

In 1980 the USA had a population of 226 500 000, much of it concentrated in the urban areas of the northeastern quarter. *Figure 10.1* shows the largest urban centres. Over 30 million people live in and around the five between Boston and Washington. The concentration of population is of concern to those who believe that a nuclear war might be necessary to achieve the aims of the USA.

Over the last two hundred years the population of the USA has spread westwards and the centre of gravity of population of the country has shifted accordingly. The inset map in *Figure 10.1* shows how the trend has continued even in the 1970s. Birthrate tends to be higher in the south and the west of the USA than elsewhere but the greater increase in these areas than in the northeast is also due to migration in the direction of greater perceived opportunities and of the sun (*see Figure 10.2*). The mobility of US citizens might be envied by Soviet planners for whom a major problem is the difficulty of moving more

people towards the natural resources in the thinly settled inhospitable north and east.

It would be unfair to attempt to define the original or 'true' citizens of the USA. Newcomers are rapidly absorbed or assimilated in certain respects, being compelled for example to learn English, though left to practise their chosen religions. It is not so easy to suppress in them all allegiance to the part of the world from which they came originally and it is impossible to change their physiological appearance quickly. Generations of mixing still left 11.5 percent of the population in 1976 defined as 'non-white' or even 'black' (though in practice of all possible shades of colour).

Even in the 1980s settlers of different origins in the USA can organize themselves into influential pressure groups perhaps to sway the fortunes of a political party in a particular election or to infiltrate individuals into positions of power. Such a situation would have only modest influence on world affairs in

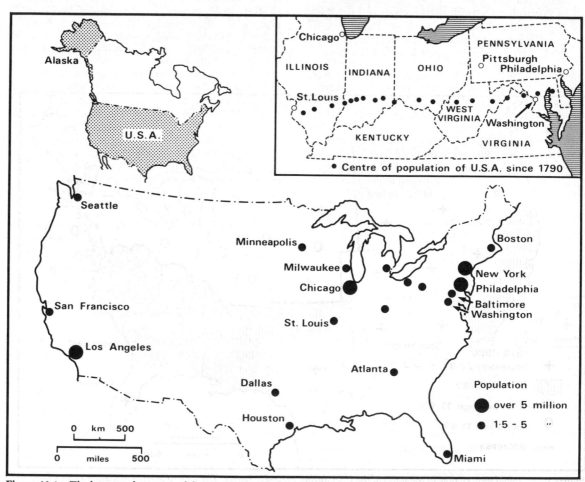

Figure 10.1   *The largest urban areas of the USA and the westward shift of the centre of population since 1790*

any country other than the USA or possibly the USSR. It is of interest to note briefly some of the many influences:

(1) Red Indians might one day head a movement in the Americas to claim some of their original 'rights' to the resources of the continent.
(2) The 'black' descendants of slaves imported from Africa must be seen to get reasonable treatment in the world's leading democracy if the USA is to have a good image in Africa. The distribution of black population by states is shown in *Figure 10.3*.
(3) The Chinese and Japanese settlers in the USA, mainly in the west, have only a minor influence or 'use' politically.
(4) The Latin Americans are taking over or taking back parts of the USA, at least numerically. In 1976 5.3 percent of the total population of the USA was classed as 'of Spanish origin', hardly the most appropriate term. From the US possession of Puerto Rico Spanish speakers have migrated in large numbers especially to New York. From Cuba around 1960 some 300 000 entered the USA, particularly through Florida. Even larger

numbers live legally or illegally in the southwest, ironically mainly in areas that were once part of the Viceroyalty of Mexico. Their distribution is indicated in *Figure 10.3*. The USA may soon be glad to import some of Mexico's rapidly growing production of oil and natural gas.

(5) The bulk of the immigrants to the USA from the middle of the 19th century to the middle of the 20th came from Europe. One may note in particular the connection of Irish immigrants and the independence of Ireland (except the North) from the UK after the First World War, the concern of the USA over the fate of Israel after it achieved sovereign status in 1948, and even the influence of the Poles and other East European settlers including Ukrainians and Russians. There are still strong cultural links between the USA and Italy, Scandinavia and the German FR.

It can be seen that the USA is a very varied society, involved through different groups of its own citizens with all the major countries and regions of the world. Equally it is linked commercially to virtually every country in the world through trade or investment (or both). It was shown in chapter 7 that the value of US

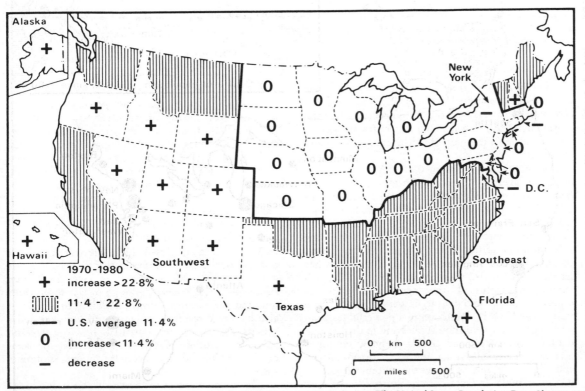

Figure 10.2 *Population change in the states of the USA during 1970–80. Source: The United States Population Data Sheet, Population Reference Bureau Inc, Washington, 1981*

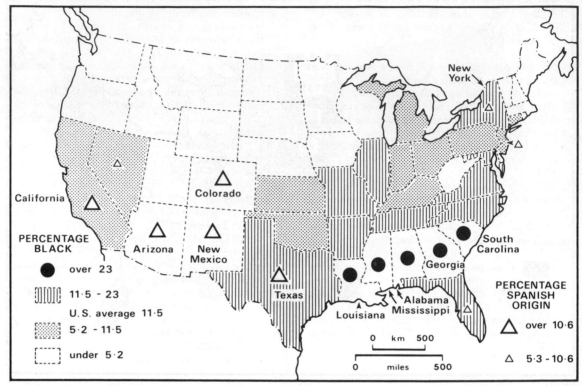

Figure 10.3   *The distribution of 'black' and 'Spanish origin' peoples in the USA. Source: The United States Population Data Sheet, Population Reference Bureau Inc, Washington, 1981*

imports (or exports) is only a few percent of the value of total gross national product. Even so, the US economy is so large (about 25 percent of the total world economy) that it is the largest trading nation and in the 1970s its share of world trade was actually growing.

The area under crops in the USA has not been extended greatly in recent decades (180 million hectares in 1961–65, 192 million in 1978 or 20 percent of total land area). There have, however, been other marked changes in US farming. 'Agro-industry' means ever increasing mechanization and the consolidation of traditional family farms into larger units of organization. The increased use of fertilizers has led to high yields. Output has risen and the US exportable surplus of some agricultural products, especially maize and wheat, is very large in some years. The USA is, therefore, in a powerful position with regard to world food suplies and in the last two decades it has exported cereals to many countries including the USSR, India, the Sahel areas of Africa, and China. *Figure 10.4* shows that within the USA it is the triangle Indiana–Texas–North Dakota that holds the key to success in agriculture, having fertile soil, a good rainfall and good thermal conditions.

The USA is well endowed only with some non-fuel minerals, but has virtually no prospect of finding large reserves of many others, a feature clearly shown in *Table 4.8*. It is a net importer of nearly every non-fuel mineral and depends especially on Canada and certain Latin American countries to supply its needs. Even though substitution, recycling and saving can cut the use of some minerals, if the USA is to continue its current levels of production it will have to rely on countries outside the Americas as well as in them for minerals and its interest in South Africa and Australia is considerable.

With regard to fossil fuels and uranium the position of the USA is complex. On the positive side it has about 35–40 percent of the world's known recoverable coal reserves, enough to last about three hundred years at rates of production in the 1970s. It also has large reserves of lignite and uranium and a generous hydro-electric potential. Its large size and fairly low latitude give a large potential for solar power. In Colorado it has very large reserves of oil in shales.

Like most countries of the world, both developed and developing, the USA has come to rely very heavily on oil and natural gas. In 1980 oil (43 percent)

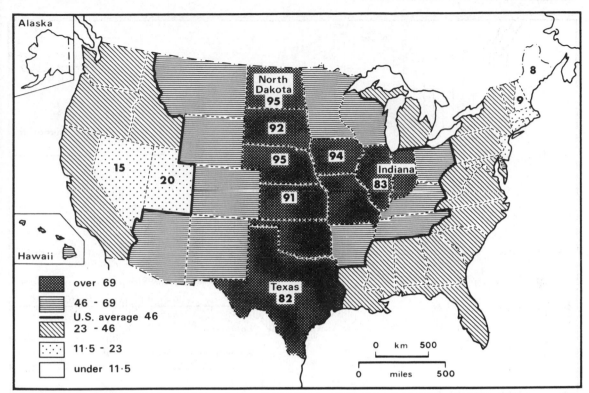

Figure 10.4   *Percentage of total area in farms in 1978. Source: The United States Population Data Sheet, Population Reference Bureau Inc, Washington, 1981*

and natural gas (27 percent) accounted for 70 percent of all US primary energy consumption (their share was 76 percent in 1970). The proved reserves of oil in the USA in 1980 were only equal in quantity to one-quarter of the amount extracted since 1859 and smaller than the amount extracted in the period 1971–80. Proved natural gas reserves around 1980 also had a life of barely a decade at current production rates. The energy situation in the USA is changing dramatically and is having a great impact on the fortunes of different regions within the USA and on the importance the USA attaches to particular regions elsewhere in the world.

Figure 10.5 compares the energy situation of the USA in the 1970s with that of the USSR (*see* section 10.3). Values are all in millions of tonnes of oil equivalent.

(1) During 1970–80 US energy consumption increased by nine percent but actually dropped slightly per inhabitant because population grew by over 11 percent.
(2) US oil production declined between 1970 and 1977 but picked up slightly then (*see Figure 10.5*); natural gas liquids are not included in the total.

Oil imports, however, roughly doubled between 1970 and 1980, changing oil from being a small to a very prominent item of import. In 1980 the USA imported 337 million tonnes of oil (but exported 29 million tonnes), more than 20 percent of world oil imports, more than Japan imported and more than half as much as West Europe took. Its suppliers included (millions of tonnes): Latin America 105, the Middle East 77, North Africa 51 and West Africa 47.

(3) US production and consumption of natural gas were declining in the 1970s and new home and foreign sources could not be expected to raise present levels of use.
(4) Hydro-electric capacity has mostly been harnessed by now and the meteoric rise of nuclear generated electricity until 1978 may be halted at least for some years. Less 'conventional' sources of energy would need a long lead period and much investment before they could make a major contribution.
(5) The USA seems, therefore, to be forced to fall back on its coal industry. Production rose considerably after 1975 but is still below the highest levels of some decades ago. *Figure 10.6*

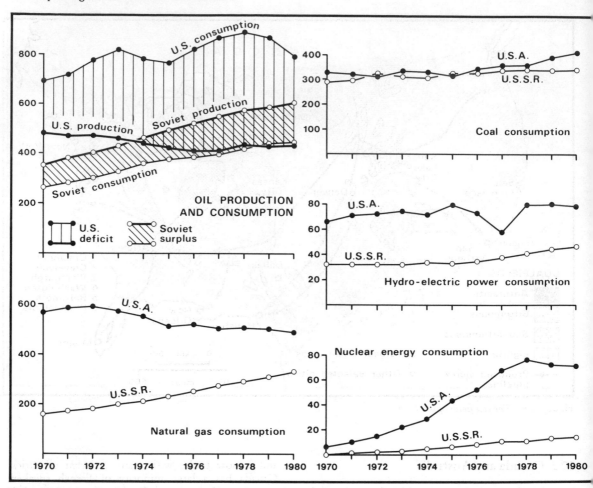

Figure 10.5 *A comparison of energy production and consumption in the USA and USSR during the 1970s. All units are millions of tonnes of oil equivalent. Source: BP Statistical review of the world oil industry 1980*

shows the distribution of the coal deposits of the USA. Those in the eastern part of the country have been most used so far but many of the western reserves are in areas distant from the major clusters of population. Coal is usually either moved by rail or by waterway, or is converted locally to electricity for transmission. In the future it may be more economical to move the coal by slurry pipelines, and some possible courses of future lines are shown in *Figure 10.6.* The rejuvenated coal industry could bring jobs and prosperity but also pollution to new areas in the west of the USA. There would be an accompanying relative decline in such traditional oil and gas states as Oklahoma, Texas and California and eventually Alaska.

Oil and natural gas must retain a major role in the USA for at least two or three decades. US concern about keeping supplies of oil open from the Middle East and other parts of Asia and Africa is not likely to diminish for a long time, especially in view also of the dependence of its industrialized allies on these supplies. Fossil fuels and non-fuel minerals are both likely to continue to figure prominently among US imports, together with such items as timber, wool and coffee. These cannot be paid for by food exports alone. The USA is also both a large exporter and importer of manufactured goods. If it reduced its imports of manufactured goods, Japan and West Europe would suffer, as would industrializing countries such as Mexico and the Caribbean area, the Korean Republic and Taiwan, to mention a few.

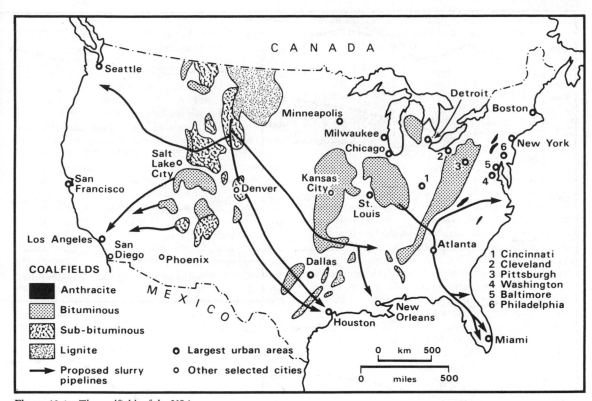

Figure 10.6   *The coalfields of the USA*

10.2 Canada and Australia

Canada and Australia are considered together here because they have many features in common, though in other respects they are very different. In common they have:

(1) Large area in relation to population size.
(2) Extensive bioclimatic, fossil fuel and non-fuel mineral resources in relation to population.
(3) They are highly developed and industrialized, with largely urban populations and high living standards.
(4) Their exports are mainly primary products and their imports mainly manufactures.
(5) Both form part of the British Commonwealth.

In certain respects Canada and Australia differ. Canada is jammed against the USA, sharing a long land (and lake) boundary with its neighbour. Australia has no land neighbours. Canada is situated between the USA and the USSR while Australia is far from both the superpowers. About 75 percent of Canada's trade is with the USA whereas Australia's 'captive' market in the UK has largely disappeared

and it now trades with many different countries. Canada has a large minority of French speakers concentrated in one part of the country. Australia's British tradition has only been substantially modified by immigrants from other parts of the world, particularly Europe, since the Second World War, but the newcomers are not concentrated in any one area. Before the Europeans came both Canada and Australia were inhabited by populations with very simple technology and small in number. Neither country could be considered densely populated even after two centuries of immigration by Europeans, but the reasons for the emptiness of most of the territory of each country are different. In Canada low temperatures preclude cultivation and make living conditions difficult over most of the country while in Australia low and unreliable rainfall and high temperatures have the same effect.

Canada and Australia must be regarded as world powers of considerable importance but on account of their sheer size and of their bioclimatic and mineral reserves, not on account of their population sizes. Together Canada and Australia have less than one percent of the world's population. Their combined percentage of the world total of most natural resources is several times that percentage, as shown by the following data for bioclimatic, fossil fuels and major types of non-fuel minerals:

Zinc	27%	Permanent pasture	15%
Uranium	26%	Forest	11%
Bauxite	24%	Lignite	7%
Nickel	24%	Copper	6%
Iron ore	23%	Area under arable	
Lead	23%	and tree crops	6%
Total land area	16%	Natural gas	5%
		Coal	5%
		Oil	1.5%

Reservations can be made about the quality of the bioclimatic resources in both countries and home requirements of the various products must be allowed for. Even so, it is clear that the two countries possess sufficient natural resources to give them a very strong position in the future as and when non-renewable resources dwindle elsewhere in the world and pressure grows on arable land. Put another way, if Australia and Canada are increasingly concerned at the prospect of running out of minerals (eg Canada of oil, Australia of uranium) then nearly everywhere else in the world is at crisis point already. Only Siberia, the Amazon and Congo basins and the privileged small oil states of the Gulf have perhaps a comparable natural resource/population ratio.

Canada

After the War of American Independence the British colony of Canada did not join the 13 states of the new USA. Modern Canada began with French and British colonies in the St. Lawrence and Great Lakes area. After the 18th century, however, very few French immigrants came to the continent. In the 19th century the British settlers spread westwards, always north of the 49th parallel boundary with the USA.

In the 19th century Canada was regarded as a land of apparently unlimited resources with comparatively few people to settle it. The forest yielded timber and furs and the rivers and lakes provided fish. In the Lakes peninsula and later in the prairies grazing and then grain cultivation provided new opportunities. The discovery of gold attracted settlers to the northern regions.

The British North America Act of 1867 united Ontario, Quebec, New Brunswick and Nova Scotia in the Dominion of Canada. The newly independent country continued to receive immigrants from Europe. Integration was helped by the completion of two transcontinental railways. The Canadian Pacific Railway across Canada was formally opened in 1887. Under a federal system considerable power was left in the hands of the provinces, a situation that ensured that the mainly French speaking province of Quebec retained some autonomy.

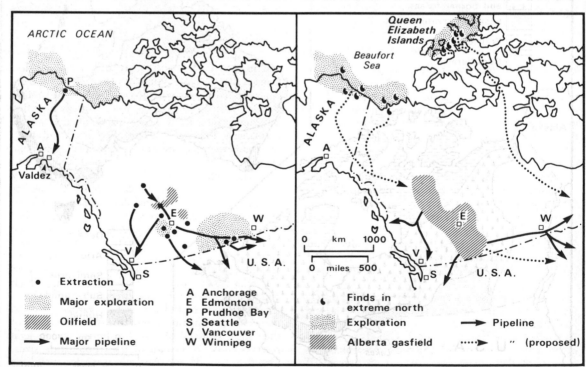

Figure 10.7(a) *Deposits of oil in western Canada and Alaska*

Figure 10.7(b) *Deposits of natural gas in western Canada and Alaska*

In the 20th century coal, hydro-electric sources, oil, natural gas and uranium have been discovered and exploited in many parts of Canada. The country also has large reserves of non-ferrous metals, ferro-alloys, asbestos and potash. The main deposits of oil and natural gas in western Canada and Alaska are shown in *Figure 10.7*.

There are two main areas of cultivation in Canada (*see Figure 10.8*), the Great Lakes Peninsula, where hot summers allow the cultivation of warm, temperate crops such as maize, and the prairies of the west, where the growing season is limited and wheat and barley are the preferred cereals. The area under crops in Canada has grown by more than ten percent between the early 1960s and 1980, but is unlikely to be extended much more in the future. The land defined as permanent pasture is considerable but of limited use. Canada is able both to support itself in food and to export agricultural products. Any tropical foods, beverages and agricultural raw materials that it needs must, however, be imported.

Canada has coal reserves both in the east and in Alberta and British Columbia. Most of the oil and natural gas deposits are in Alberta (*Figure 10.7*) and

in the 1970s this province was accounting for about half of the value of Canadian mineral production. Canada has a large hydro-electric potential, much of which has now been brought into production. Hydro-electric energy is worth about 80 million tonnes of coal equivalent and provides more than one-quarter of Canada's total energy consumption.

Canada's known reserves of conventional oil were expected to last only 10–15 years in 1980 and its natural gas 20–30 years. Great importance is, therefore, being given to exploration for oil in islands off the north coast of Canada and in waters in the Arctic Ocean itself. Frozen ground and freezing of the sea for most of the year require special methods of exploration and extraction while transportation of the oil to markets, either by pipeline or by sea is also a great challenge. In addition Canada has large possible reserves of oil in the form of bituminous tar sands, from which small quantities of oil are already being extracted.

Metallic minerals account for about 30 percent of the value of Canada's mining production and in order of value are iron ore, nickel (Sudbury area), copper, zinc, gold, silver and lead. Canada is also a large

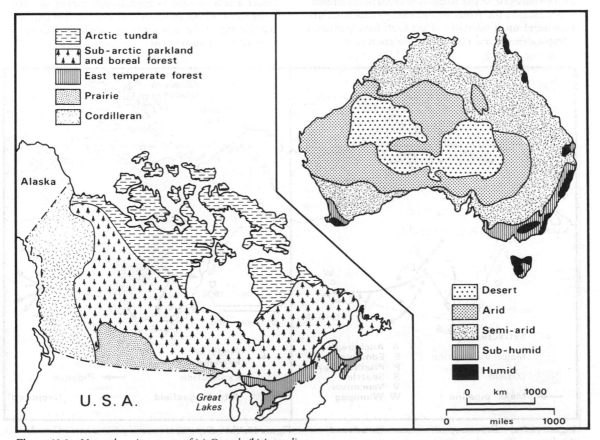

Figure 10.8 *Natural environments of* (a) *Canada* (b)*Australia*

producer of the non-metallic minerals potash and asbestos.

About 22–23 percent of the economically active population of Canada is in manufacturing and the country has both heavy and light industry. Steel production is considerable and still growing (1970–78 from 11 to 15 million tonnes) and various branches of engineering are represented. The scale of the Canadian home market is, however, overshadowed by that of neighbouring USA. Much of the industrial capacity is actually owned by US corporations.

Compared with the problems of most other regions of the world those of Canada are of modest proportions. Among them may be noted political, economic and environmental problems:

(1) Canada's role as a world power is limited partly on account of the divided nature of its population. The province of Quebec contains most of the French speaking population of Canada and is culturally apart from the rest of the provinces. Complete independence from the rest of Canada has been hoped for by extreme groups but the possibility has not been pursued since a referendum in which the moderate opinion majority chose to remain in Canada. The Canadian constitution was still ultimately the responsibility of the UK until the early 1980s. *Figure 10.9* shows how a completely sovereign Quebec would weaken Canada greatly and would be a nightmare for the USA, controlling not only access from the interior of Canada to the Atlantic

via the St Lawrence Seaway but also the outlet of US ports on the Great Lakes such as Chicago and Cleveland.

(2) Canada was described by one economist as a long footnote on the edge of the USA. Such a description seems unkind but the truth is that more than 90 percent of the population is within about 300 kilometres of the USA and each distinct cluster of population in Canada is close to some larger cluster of population in the USA. What is more, although the USA is well provided with natural resources of its own, its 5 percent of the world's population consumes about 25 percent of the world's energy and raw materials and produces about 25 percent of the manufactures. Canada happens to have many of the primary products needed in the US economy. It is comparatively easy for Canadian oil, gas, timber and pulp and non-fuel minerals, either raw or processed, to reach some nearby part of the USA. The USA adds the value and Canada imports manufactured goods back. Machinery and transport equipment account for about half of all its imports.

Canada is the largest single source of US imports but it supplies less than 20 percent of these. The USA is also the largest single source of Canada's imports but it supplies more than 70 percent. The 'dependency' relationship is very uneven. It is not surprising, therefore, that US investment in Canada is very high. In 1979 the USA had abou 40 000 million US dollars of direct investment in Canada, over 20 percent of its total

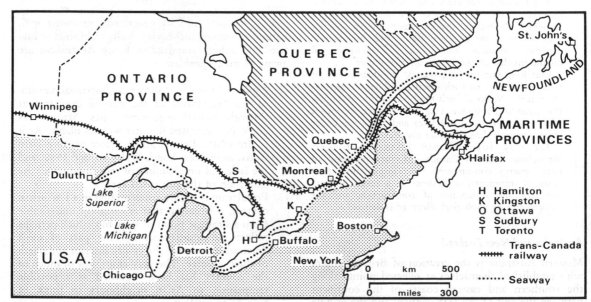

Figure 10.9 *The position of the province of Quebec in relation to eastern Canada*

direct investment abroad (190 000 million dol-lars). Nearly half was in manufacturing and nearly a quarter in oil.

Canadians are naturally sensitive about poten-tial US influence on the economy and increasing-ly see their natural resources being used up to feed the industries and sustain the high consump-tion levels of their neighbour. Canadian oil production reached a peak in 1973 and dimi-nished in the following years. Oil consumption has continued to rise and in the late 1970s Canada was a net importer of oil, obtaining some from Venezuela.

(3) Canada is increasingly concerned about the depletion of its natural resources, the concentra-tion of population in large urban centres, and problems of conservation. These and related preoccupations are clearly expressed in the following passage from the official *Canada Handbook* (1979, pp 1–2):

'Both Canadians and people abroad tend to assume that Canada's population can continue to increase and that we can still export large quantities of food. Unfortu-nately, relatively little of Canada's area is easily habitable for those living average southern Canadian lifestyles and only about 7 percent of the land is economically viable for farming at present. Recent years have seen increased concern about how we use or abuse this environment and how we can better adapt to our demanding climate while making much more efficient use of finite resources.

Canada's rapid population growth has been concen-trated mainly in cities. Growing cities increasingly threaten surrounding agricultural lands with demands for more land . . .

The Canadian economy and society were founded upon cheap, abundant resources and limitless horizons. However, the realities that must now govern policies, lifestyles and designs are finite and increasingly expensive resources and social and environmental conditions that can no longer tolerate abuses . . .

Our long non-growing season, the distribution of water resources and such specific problems as perma-frost in the North all limit food production and mean that renewable resources take longer to renew, wastes take longer to decay and flora and fauna are under greater stresses than in milder climates . . .

The very large distances between settlements throughout most of Canada introduce problems of heavy energy consumption for transportation, high costs of transportation systems as a result of both distances and difficulties of construction and the cultural and psychological effects of isolation.'

Australia and New Zealand

Modern Australia is the creation of British settlers who established themselves at selected points along the southern and eastern coasts of the continent between the end of the 18th century and the middle

of the 19th. They were administered as separate colonies by Britain until 1901 when the Common-wealth of Australia came into being and developed separately, even building rail systems with different gauges. In the 19th century minerals (especially gold), arable land and pastures formed the basis of the economy of the colonies. Since settlers all arrived by sea, whatever their origins and jobs in Britain, those who went inland first passed through an urban sieve and many actually stayed close to the coasts. The new settlers had to learn or relearn in new conditions how to make use of the abundant land available in the hinterlands of the ports.

The Australian economy depended on various products at different periods but in the first half of the 20th century was based especially on the cultivation of wheat, the raising of sheep and the production of non-ferrous metals. In 1915 the first Australian iron and steel works was opened in Newcastle and from that decade heavy and light manufacturing were developed alongside existing processing industries.

After the Second World War economic growth in Australia was impressive. Between the 1930s and 1980 population roughly doubled. Improvements in arable and pasture land, increased exploitation of coal and lignite resources, the discovery of oil and natural gas and schemes such as the Snowy River project all contributed. The data in *Table 4.8* give Australia's estimated share of the world total of selected minerals and *Figure 10.10* shows the location of some of the major reserves. Many deposits have been discovered or their size reassessed upwards in the 1970s, a decade, however, when world demand has been fluctuating rather than rising fast as it did in the 1960s.

In spite of being so generously provided with natural resources and having well established indus-tries and a high standard of living Australians are aware of many problems:

(1) *Figure 10.8* map (*b*) shows an Australian version of the limitations of the natural environment through natural vegetation types. Many areas could be cultivated if water were available. As it is droughts often seriously affect existing farm lands, as in New South Wales around 1980, and the lack and unreliability of rain discourages the extension of the cultivated area.

(2) Population is highly concentrated in a few urban centres, a situation that is certainly undesirable from a strategic point of view but which perhaps reflects an indifference on the part of most Australians towards the empty interior and north of their country. The very emptiness of much of the country has been regarded as a possible temptation to Asian neighbours to think of helping to settle it.

(3) Until the 1960s Australia had been concerned about preserving the 'whiteness' of its population. Apart from the indigenous population of some 300 000, now either mixed and absorbed or relegated to the outback or the edge of the towns, non-Europeans have been prevented or discouraged from settling permanently in Australia. In the late 1970s restrictions were even put on traditional flows of immigrants. Whether Australia could take in several million Indonesians or Indians is of academic interest, firstly, because such an event would not be contemplated by

Australia and, secondly, because it would do little to change the demographic situation in the Asian countries. India's population grows by 14–15 million people *a year*, equal to the *whole* current population of Australia.

(4) Integration of the national territory of Australia has been a preoccupation since the Commonwealth was established. A modest advance was the establishment of a 'neutral' capital, Canberra, in a federal territory. More important, psychologically, perhaps was the extension of standard gauge railway track on major lines into all the

Figure 10.10   *The natural resources of Australia*

states, adopting the gauge of New South Wales. Local lines continue to use their own gauges. A great dependence on air transport also helps to keep Australians in touch with each other and with Europe.

New Zealand

New Zealand is overshadowed in size by Australia but its influence in world affairs should not be overlooked. Although it has only about 3 million inhabitants, compared with nearly 56 million in the UK it is larger in area than the UK. Its mineral reserves are limited in size and variety and only a small proportion of the total land area is cultivated. Even so its extensive pastures provide about one-eighth of the wool produced in the world and it is also an exporter of various other livestock products. With growing interest in the resources of the oceans New Zealand is well placed to control and form a base for the exploitation of a large area of the southwest Pacific.

10.3 The USSR

Introduction

The USSR is by far the largest country in the world territorially, covering 22 400 000 square kilometres, or about 17 percent of the world's land area. In 1981 it had about 268 million inhabitants, six percent of the total population of the world. According to the Soviet census of 1979 only 52 percent were Russians. Ukrainians made up another 16 percent and there were nearly 100 recognized national groups, each with its own language.

When the Communist Party gained power in Russia in 1917–18 it inherited a large empire, built up over some five centuries (*see* chapter 2). In the years after 1917 the status of Soviet Socialist Republic was given to 14 non-Russian peoples (*see Figure 6.2*) but by far the largest Republic in the Soviet federal system is the RSFSR (Russian Soviet Federal Socialist Republic), covering about three-quarters of the national area.

After internal conflict and external intervention during 1917–22 the Communist Party became firmly established in power first under Lenin, then Stalin. Soon after the Revolution most means of production were taken over by the state, though for a time under the New Economic Policy most farmers were left to run agriculture themselves. In the First Five-Year Plan (1928–32) there was a big drive to expand heavy industry and also to collectivize agriculture. Although much of European USSR was occupied by the Germans for a time in the Second World War (1941–5), the USSR had by then considerable industries to the east of the area of conflict.

Apart from the war period and a lapse during the 1956–60 plan the economic life of the USSR has been run according to national plans. Under such a system the state can control the amount of consumption by the population of the country and determine how much capital investment takes place. It is possible under state planning in the USSR to allocate investment among sectors of the economy and also by regions.

Since 1928 priority has been given to the expansion of branches of the economy producing capital rather than consumer goods. Sectors that have generally been favoured include the energy industry, particularly electricity, the production of metals, the engineering industry and, in transport, the railways. Many resource rich but thinly populated central and eastern areas have received heavy investment in mining, hydro-electricity and heavy industry. Health and educational standards have also been improved enormously since the 1920s. On the other hand, agriculture has been neglected, the chemicals industry lagged behind until the early 1960s, housing is very limited in space and quality, and consumer goods are often criticized as being short in supply and of poor quality.

Natural resources

For its population size the USSR is well endowed with bioclimatic resources. With ten percent of its total area under arable or permanent crops it has only 44 inhabitants per 100 square kilometres of arable plus pasture land (USA 53 but Australia only 3). 17 percent of the total area is defined as permanent pasture and over 40 percent as forest (*see Figure 10.11*).

In the 1950s extensive new areas were brought under the plough in the USSR but since then the sown area has changed little. The short growing season, with severe winters, a short frost free period and the presence of permafrost make cultivation impossible in much of the north and east of the country. In the south, low rainfall, high summer temperatures and drying winds reduce yields drastically in some years. The main belt of cultivated land extends eastwards from the western boundary of the USSR to about half way across. The areas of fertile black earth soils are almost all under cultivation but soils are of poorer quality north of the belt in the coniferous forest and south in the semi-desert fringe. In Soviet Middle Asia limited but very productive irrigated lands allow the cultivation of sub-tropical crops, especially cotton.

The USSR has abundant reserves of timber and much of the forest is still difficult to reach and far from concentrations of population. In western USSR large areas have been cut but replanting has not been as extensive or thorough as it should have been. The natural pastures of the USSR are mostly in the region of semi-desert, where lack of water and unreliable fodder supply reduce their usefulness. Most of the livestock in the USSR is raised within the main zone of crop farming.

The USSR has abundant reserves of fossil fuels. *Figure 10.12* shows the location of the major deposits, which include the following:

Coal 83 000 million tonnes of known recoverable reserves (160 years of life at present rates of extraction, and 20 percent of the world total)
4 000 000 million tonnes of possible reserves (50 percent of the world total)
Lignite 54 000 million tonnes of known recoverable reserves (a third of the world total)
1 720 000 million tonnes of possible reserves (65 percent of the world total)
Oil 8 600 million tonnes (nearly 10 percent of the world total but less than 15 years of life at current rates of extraction)
Natural gas 26×10^{12} cubic metres (35 percent of the world total and about 60 years of life at current rates of extraction).

Additional sources of energy include oil shales, peat, hydro-electric power (about 65 million tonnes of coal

equivalent in 1980), geo-thermal power and nuclear power as well as the traditional source, firewood, now used only locally.

Oil is the only source of energy in the USSR threatened with depletion in the near future. It is expected that new reserves will be discovered, but they may be located in the extremely difficult conditions of the frozen north or offshore in the Arctic Ocean.

Virtually every economic mineral is found somewhere in the USSR. The data in *Table 10.1* show how well provided the country is with most of the major non-fuel minerals. The locations of the deposits of these minerals are shown in Soviet atlases, but the size of any particular reserve is difficult to gauge. Areas that have proved particularly rich in minerals include the Ural mountains, the plateau of Kazakhstan and mountain ranges in the south and east of the country. Large areas have yet to be properly explored and even in European USSR new deposits of minerals are still being found.

At a rough estimate the USSR has about 20 percent of the non-fuel minerals of the world and if proved reserves of coal, oil, and natural gas are arbitrarily given equal weight, also about 20 percent of the

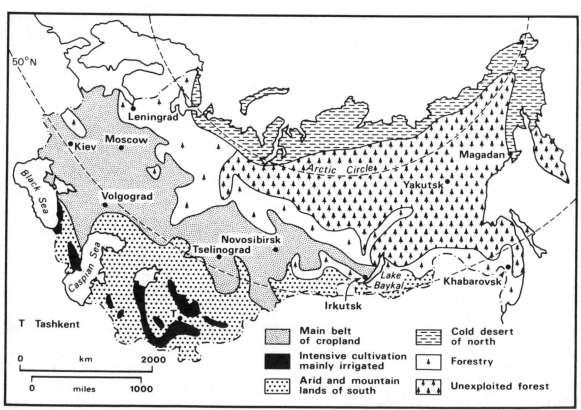

Figure 10.11   *Natural environments of the USSR*

Table 10.1 *Percentage of total world reserves of selected non-fuel minerals attributed to the USSR*

Mineral	Percentage	Mineral	Percentage
Platinum	47	Nickel	10
Manganese	38	Molybdenum	9
Iron ore	30	Tungsten	8
Asbestos	30	Copper	7
Silver	20	Zinc	7
Gold	20	Antimony	6
Potash	15	Tin	6
Sulphur	13	Phosphates	5
Lead	13	Bauxite	5
Mercury	10	Industrial diamonds	4

world's fossil fuel reserves. It is, however, committed to supplying CMEA partners with fuel and raw materials and the whole bloc is claimed to have about 20 percent of the world's industrial production.

Soviet agricultural production has grown faster since the Second World War than population, but farming is still regarded as the weakest sector of the economy. Cereals account for about 60 percent of

total sown area but yields are low by comparison with those in West Europe or the USA on account of lack of care with cultivation, the misuse of fertilizer resources and droughts. Grain yields vary greatly from region to region and are generally low in the newer areas of cultivation in central USSR. Since a particularly bad harvest in 1963 the USSR has in most years imported grain, mainly to support its livestock industry.

About 30 percent of the sown area of the USSR is under animal fodder crops. Meat and dairy products should be abundant in the country given the efforts to develop the livestock sector and the size of the livestock population. A vital 10 percent of the crop land is dedicated to other crops including sugar beet, the sunflower (for its oil), vegetables, tobacco, flax, hemp and cotton, or to permanent crops including tea, wines and fruits. The USSR is not able to grow tropical crops and must import such commodities as rubber (especially from Malaysia), coffee, cocoa beans (eg from Ghana) and tropical fruits, but consumption per inhabitant of these products is very low.

The importance of the fuel and power to the development of industry was appreciated in the early

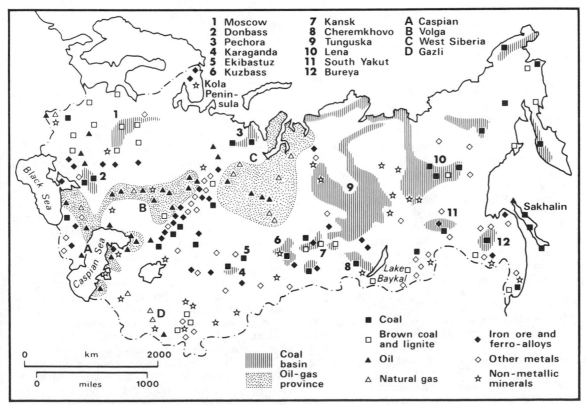

Figure 10.12 *Mineral deposits of the USSR*

years of Soviet rule and Lenin frequently emphasized hydro-electricity and the need to ensure a supply of electricity throughout the country. Until the 1960s, however, coal was actually the main source of energy in the USSR and in the early 1950s it still accounted for two-thirds of all energy produced. The oldest coalfield in the country, the Donbass, still produces large amounts of coal but many of the more accessible seams have already been worked and production costs are high. The Kuzbass and Karaganda fields were developed in the 1930s and many newer fields, some for regional use, some of national importance, now also contribute. *Figure 10.13* shows, however, that expansion of coal production slowed down after 1957 and in the late 1970s was actually declining.

A new fuel policy in the mid-1950s led to increased emphasis on the more cheaply produced sources, notably oil and natural gas. In the 1950s and 1960s the many oilfields of the Volga region dominated production, eclipsing the oldest Caucasus–Caspian–Baku oilfields. Pipelines take Volga oil westwards to energy deficient regions of European USSR, to CMEA partners and to ports for export. In the 1970s production in the newer West Siberian oilfields increased rapidly and these, though less accessible than the Volga fields, supply places both to the west and to the east.

Natural gas deposits occur in many regions of the USSR. Some gas is used locally but the many pipelines built in the last 30 years distribute it also to energy deficient regions and into other European countries. The very large Yamal deposits near the Arctic coast will supply gas by a long pipeline to the German FR and to other West European countries.

With abundant natural resources at their disposal Soviet planners have been able to develop industry in the last 50 years without relying to any great extent on foreign sources of fuel or raw materials. In the 1930s, however, equipment had to be imported and it is ironical in retrospect to find that most of this was obtained from Germany, Britain or the USA. Since the Second World War the USSR has manufactured most of its own producer goods or has obtained them from CMEA partners, especially from the German DR and Czechoslovakia. In the 1970s it seems to have come to depend again on some kinds of capital equipment produced only in sophisticated Western industrial countries, especially Japan, the German FR and the USA.

In the 1930s the production of steel and engineering goods increased rapidly in the first two Five-Year Plan periods (1928–37) with older industrial centres such as the Donbass, Moscow and Leningrad being expanded and new industries in the Urals and Kuzbass being established. Investment was allocated also to industry in more backward regions such as Transcaucasia and Central Asia and to remote regions

such as the Far East. *Figure 10.13* gives a rough idea of the distribution of industrial activity in the USSR at present.

Internal problems

While the USSR carries on comparatively little foreign trade, the flow of goods within the country between regions is on a massive scale and it has been estimated that the Soviet rail system carried about 40 percent of all ton-kilometres of goods carried by rail in the world. Inland waterways, coastal shipping, roads, pipelines and the electricity grid contribute increasingly to the internal movement of goods in the country. *Figure 10.13* shows the main features of the transportation system.

In a year around 1980 on average one tonne of goods was transported some 22 000 kilomeres *within* the USSR for *each* citizen of the country. Such quantities of goods are carried by international shipping on behalf of Western industrial countries. Only in the USA, Canada and Australia are comparable large quantities carried internally. The USSR is one of the few countries in the world that is still extending its rail system. Until about 1960 Soviet railways carried over 80 percent of internal traffic, but their share was less than 60 percent in the early 1980s with sea, pipeline and road increasing their contribution.

An adequate transportation system is particularly important in the USSR. While most of the population lives in European USSR, on about one-quarter of the area, most of the natural resources are in central and eastern USSR. Attempts to move large numbers of people permanently into Siberia have been of limited success in spite of various measures available to persuade if not compel people to move to new jobs in the east. The eastern regions of the USSR contain most of the fuel, non-fuel mineral, forest, water and hydro-electric resources and potential of the country as well as an extensive agricultural zone.

Soviet planners, then, have to take into account the great size of their country, additional long hauls of goods often required by new projects and the harsh environmental conditions of the north, the east and the deserts of Middle Asia. The way they allocate investment by sectors of production and by regions and the production targets set are affected by several distinct, often conflicting considerations: ideological, strategic, economic, social, international.

Ideologically the Communist Party of the USSR has committed itself to building communism in the USSR. Differences in living conditions between regions and between types of employment should be eliminated or at least minimized. The industrialization of recently colonized, underdeveloped, non-Russian regions has been impressive.

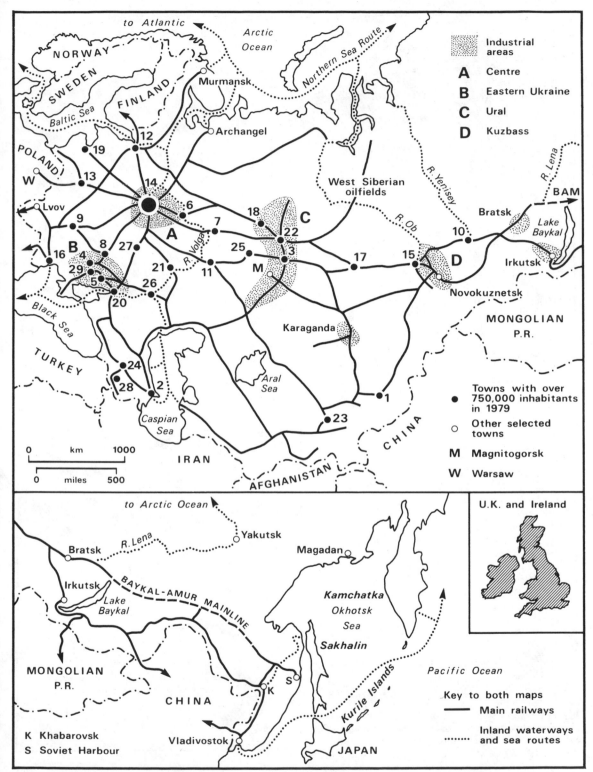

Figure 10.13 *Industries and the transportation system of the USSR. The lower map shows only the southern part of eastern USSR. The scale is the same for all three maps*

182

Figure 10.14(a) *Is a theme recurring regularly since the mid 1950s:*

Caption: In view of the adverse weather conditions of this last summer it is particularly important to take care to protect the harvest. Unfortunately, in a series of places the managers did not give sufficient attention to the gathering and storage of the grain.
Source: Krokodil, no 26, 1981

Figure 10.14(b) *Caption: So you have put up the new chimney? But why isn't it made of brick? We did not fulfil our brick production plan. Source: Krokodil, no 33, 1977*

Strategically concern was felt in the 1930s about the possibility of a new invasion by Western capitalist countries. It was regarded as essential to set up new industries east of the Volga, especially in the Ural region but also even further east where, however, regional needs were limited at the time. Even in the age of intercontinental missiles with a range adequate to hit any part of the USSR a determination to develop areas as far from the borders of the country as possible seems to linger.

Economically it is just as important in the USSR as in western market economies to place new productive capacity where production costs are as low as possible (bearing in mind also costs of movement to markets). For the production of fuel, hydro-electricity, some minerals, grain and some industrial products the cheapest areas of production are roughly between the Volga and Lake Baykal. Development here conforms with strategic, but not with ideological goals because heavy investment creates well paid industrial employment and deprives other regions of industrial growth. It is also argued by some in the USSR that the low production costs are illusory because they do not take into account the advantages of the infrastructure, skills and large market in European USSR.

Social considerations may also influence Soviet planners in the way they influence politicians and planners in market economy countries. If the high cost coal mines of the Donbass were closed overnight or if the cotton mills of the Moscow area were run down to relocate cotton manufacturing in Middle Asia where the cotton is grown then there would be social problems in the old mining and industrial areas.

The growing international ties of the USSR through increased trade and aid mean a more serious approach to the production of goods for export and their delivery to neighbouring countries by land or to seaports. Some of the main outlets for Soviet foreign trade, Murmansk in the northwest, and Vladivostok and other ports in the even more remote Far East, are at great distances from many areas of production.

For simplicity the USSR may be divided into three distinct regions in order to see the broad geographical problems affecting internal planning strategy:

(1) *European USSR* including the Ural region but excluding Transcaucasia. Although this part of the country has some predominantly rural areas in the western Ukraine and Blackearth Centre regions it has a long established industrial tradition, sophisticated industries, skilled labour and research institutes. The bulk of the popula-tion of the USSR is here but not many of the mineral resources. The population, which is mainly Russian, Belorussian or Ukrainian is growing only very slowly.

(2) *Siberia and the Far East* have new agricultural lands but depend mainly on extractive and processing industries. These regions have little more than ten percent of the population of the USSR, but most of the natural resources apart from the agricultural ones. The original population was non-Russian (Yakuts, Chukhots) but most of the present population consists of Russian and Ukrainian settlers originally from European USSR. Population is growing but more through natural increase than in-migration.

(3) *Transcaucasia, Middle Asia and southern Kazakhstan* are still heavily dependent on agri-culture, but have some mining and industrial projects. Agricultural production could be in-creased by the transfer of water, at present flowing into the Arctic Ocean, south into the region. Health and educational services and purchasing power are appreciably lower in this region than in most parts of the other two regions (*see* chapter 6, section 1). The population is essentially non-Slav, though there are consider-able Russian minorities in the larger towns. Population is growing fast and its age structure resembles that in developing countries. In two or three decades the population could grow from some 50 million in 1980, less than 20 percent of the population of the USSR, to 90 million in 2010, perhaps 30 percent of the population of the USSR.

Planning strategy in the USSR, then, is concerned with the relationship of industrial capacity (European USSR), natural resources (Siberia) and growing population (southern regions). Should population be transferred to the natural resources of Siberia? If so, then should it come from the non-Russian southern regions even though there would be a drastic environmental and cultural shock for people from here settling in Siberia? Should it come from European USSR? Will the USSR continue (like Canada and Australia) to export primary products from Siberia without adding great value to them?

Planning itself is facing many problems. The more sophisticated and complex the Soviet economy becomes, the more difficult it is to formulate and implement the directives. The Soviet press has frequent references to problems in the economy. Two cartoons from the satirical magazine *Krokodil* are shown in *Figure 10.14.*

CHAPTER ELEVEN

Developing Regions Poor in Natural Resources

South Asia

Introduction

In 1981 the region defined here as South Asia had just over 900 million inhabitants. The largest country, India, alone had nearly 690 million, Bangladesh 93 million and Pakistan 89 million. Sri Lanka, Nepal, Bhutan and the Maldive Islands are the remaining countries. The whole region has 20 percent of the population of the world, but its area of 4.4 million square kilometres is only 3.3 percent of the total land area. Some data are given in *Table 11.1* for the main countries of South Asia.

South Asia gradually came under British control during the 17th to 19th centuries and from early in the 19th century to 1947 formed one of the major elements of the British Empire. From the very complex administrative framework of British South Asia three modern independent states came into being: India, Pakistan and Ceylon (now Sri Lanka). In the northern mountains Kashmir remained in dispute between India and Pakistan, while Nepal and Bhutan now form 'buffer' states between northern India and China (Tibet).

Of the various cultural and physical criteria that might have led to a fragmentation of British India into many smaller units after independence in 1947 only the religious division between Hindu and Muslim played a major part. As the two religious groups were territorially mixed it was possible only with great difficulty and ensuing bloodshed to separate two predominantly Muslim areas to form together two distinct regions of Pakistan, one area being in the Ganges-Brahmaputra delta in the east, the other mainly the Indus basin in the west. In 1972 East Pakistan, which had hitherto been dominated by West Pakistan, broke away to become the independent state of Bangladesh.

British influence in South Asia was very great though it varied in intensity from place to place. The economy of India was developed to be complementary to that of Britain. India exported cotton, jute, tea, spices and numerous other items, and factory made manufactures were sent back. Railways and ports made various parts of the region accessible to trade. Only in the 20th century did much modern industrial development take place in India.

In 1947 the newly independent countries of South Asia already had impressive irrigation projects, a reasonably complete rail system, a civil service and the basis for a Western style multiparty political system. It can be argued that improvements in hygiene, food supply and medicine during British rule led to or at least accelerated a population explosion with which the governments of the region have been trying to cope since the 1950s. The population of South Asia is increasing by about 2.2 percent per year and could double in three decades at that rate.

Natural resources and production

After Africa, South Asia is the most rural and agricultural major region of the world. In 1979 the following percentages of the economically active population were engaged in agriculture: Nepal 93 percent, Bangladesh 84 percent, India 64 percent, Pakistan and Sri Lanka each 54 percent. Any future development of the region must be accompanied therefore by progress in the agricultural sector. The region has about 200 million hectares of land under arable or permanent crops. This land is farmed by about 210 million people. There are therefore on average about 105 persons working each 100 hectares compared with 270 in China, but less than one in Australia.

About half of the total area of India is defined as cropland and the percentage is over 60 percent in Bangladesh, though lower in the other three countries. Only about 20 percent of the region is forested and much of this vegetation is of poor quality commercially, as is the limited area defined as permanent pasture. There are large areas of desert in the northwest and rugged mountain areas in the

Table 11.1 *Data for South and Southeast Asia and other selected countries*

Country	Area 1	Popn 2	Popn change 3	Popn/ arable 4	Non- agric 1960 5	Non- agric 1979 6	Energy cons 1979 7	Infant mort 8	Pupils/ teacher 9	GNP 10
Burma	677	35	2.4	340	36	47	60	140	111	160
Thailand	514	49	2.0	273	16	24	330	68	64	590
Malaysia	330	14	2.3	220	37	51	720	44	52	1320
Indonesia	1904	149	2.0	524	25	40	290	91	91	380
Philippines	300	49	2.4	538	26	52	340	65	52	600
Viet Nam	330	55	2.8	512	20	29	130	115	54	na
Kampuchea	181	6	1.8	152	18	26	4	150	124	na
Laos	237	4	2.4	217	17	26	60	175	130	na
Taiwan	36	18	2.0	na	na	na	na	25	58	na
Korean R	98	39	1.7	1719	33	60	1000[b]	37	71	1500
Korean DPR	121	18	2.4	808	38	53	3000[b]	70	82	1130
Mongolia	1565	2	2.9	1	30	50	1240	70	49	780
Singapore	1	2	1.2	[a]	91	98	2460	13	47	3820
Hong Kong	1	5	1.2	[a]	95	97	1300[a]	13	na	4000
India	3288	689	2.1	381	25	36	180	134	80	190
Pakistan	804	89	2.8	356	24	46	170	142	115	270
Bangladesh	144	93	2.6	954	24	16	40	139	120	100
Sri Lanka	66	15	2.2	592	44	46	110	42	46	230
Nepal	141	14	2.4	358	6	7	10	133	157	130
China	9597	985	1.2	309	33	39	840	56	46	230
Japan	372	118	0.8	2145	67	88	3830	8	38	8800
Australia	7687	15	0.9	30	89	94	6620	12	25	9100

Definition of variables see Table 11.2

[a] not meaningful
[b] author's estimate
na valid data not readily available

Table 11.2 *Definition of variables in Table 11.1*

1 Total area in thousands of square kilometres
2 Population in millions, 1981
3 Percentage rate of annual natural change of population around 1980
4 Persons per square kilometre of arable plus pasture land, 1978
5, 6 Economically active population *not* engaged in agriculture as a percentage of total economically active population in 1960 and 1979
7 Energy consumption in millions of tonnes of coal equivalent per inhabitant in 1979
8 Annual number of deaths to infants under one year of age per 1000 live births in late 1970s
9 Number of school-age children per teacher, 1975
10 Gross national product in US dollars per inhabitant, 1979

north and also in places in the peninsula. The area under cultivation has been extended considerably since the 1940s and, with new irrigation works, could still be extended and improved. Even so, the amount of cropland per inhabitant is much lower than it was a few decades ago and only the increase in yields has kept food supply adequate.

There are three main types of soil in South Asia. Much of the cropland in northern India, Pakistan and Bangladesh is fertile irrigated alluvial land. In peninsular India there is an extensive area of fertile black soil formed on lava. In the east and south of India the red and yellow soils, formed mainly on old rocks, are generally less fertile, but along river valleys and in the coastal areas alluvial soils occur more locally.

The success of Indian agriculture is related closely both from year to year and in the long term to the supply of water. Most of the rain comes during a rather short summer monsoon period. *Figure 11.1* shows great regional contrasts in the average amount falling, with more than 4000 millimetres in exposed mountain slopes in the north (the Himalayas) and northeast and along the western coast of the peninsula (the Ghats), but less than 100 millimetres in northwest India (Thar Desert) and in much of Pakistan. There are however also great variations around the average from year to year.

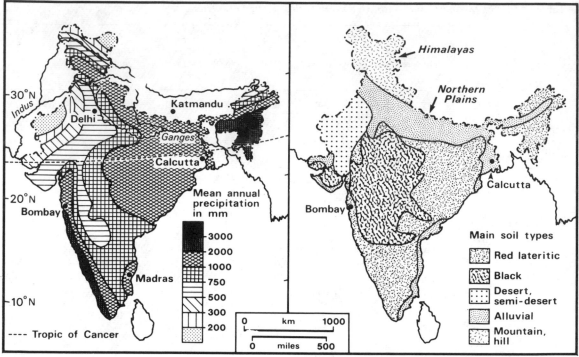

Figure 11.1 *India* (a) *mean annual precipitation. Simplified from Johnson (1979)*

Figure 11.1 *India* (b) *Main soil types. Simplified from Johnson (1979)*

The big rivers from the Himalayas (Indus, Ganges and their tributaries) ensure a reasonably abundant and reliable supply of water, distributed by canals to the plains of the north and northwest. The eastern half of India is on the whole wetter than the western part. In the west of the peninsula the dark lava soils retain moisture well, while various east flowing rivers provide water locally stored in small reservoirs (tanks) and numerous wells provide a further supply. Even so, much of the cropland of South Asia is at risk if the monsoon rains are late, much below average or even fail to come at all, and cultivation without irrigation is impossible in places in the northwest. Less than 25 percent of the total area of Pakistan is under crops, but much of the cultivated area is irrigated.

South Asia has 20 percent of the total population of the world but its percentage of the world total of most mineral reserves is far below that proportion. It has about 2.5 percent of the known recoverable coal reserves of the world (mainly in India), but less than one percent of the oil and natural gas reserves (mostly in Pakistan). In the 1970s only about one-third of all the oil consumed in the region was produced there. The hydro-electric capacity is about three percent of the world total.

Reserves of many different non-fuel minerals are known in South Asia, but by world standards nearly all of them are small. India has about six percent of the world's iron ore reserves, three percent of the zinc ore and some manganese ore, but fertilizer minerals and most metallic ores are inadequate for the needs of the region.

About two-thirds of the cropland in South Asia, apart from that in Sri Lanka, is under cereals. Where rainfall exceeds about 1000 millimetres and local conditions are suitable rice is the preferred cereal and it occupies about 25 percent of all cropland. In the north of India and in northern Pakistan wheat is widely grown, while in the drier and less favoured, parts of the peninsula millets are most common. Maize and barley are comparatively unimportant. Many varieties of each main type of cereal are cultivated and these are adapted to local conditions and preferences. Cereal yields in India are usually below the world average and there is much room for improvement in farming methods.

About a quarter of the cropland in India is devoted to pulses (eg chickpeas) and oilseeds (eg groundnut). Such plants are the main sources of protein and cooking oil in a diet where little meat and animal fat is produced and where dairy products are almost unknown. It is indeed more efficient to make use of plant foods than to derive animal foods through an expensive food chain.

A small but important proportion of the cropland

of South Asia is devoted to sugar cane, cotton and jute, tea and condiments and spices. Cotton is cultivated in the drier western and northwestern regions of India and in Pakistan. Jute cultivation is restricted mainly to the lowlands of northeast India and adjoining Bangladesh. Sri Lanka specializes in tree crops, producing coconuts and tea.

Many different ways have been tried to increase agricultural production in South Asia. Soil conservation, irrigation and reclamation help to prevent erosion and extend cultivation. Double cropping, most prevalent in irrigated areas, considerably increases output. The development of hybrid rice and other cereals has greatly increased grain yields in some areas, but such sophisticated practices require the use of fertilizers and pesticides and a reliable supply of water. Farms tend to be very small and capital for new ventures limited. Attempts at land reform have led both to improvements and conflicts. Machinery is often too expensive, uses valuable energy supplies and reduces the number of jobs in agriculture. Fields are small especially in hilly, terraced areas, making mechanization difficult.

A much publicized matter of controversy in India is the large number of livestock. Oxen are the main work animals but they consume large quantities of fodder, produced on land otherwise grow food for hu not usually slaughtered in perhaps sentimental reasons.

Industry

South Asia is regarded as a de a very low gross national about 200 US dollars around world average of about 2 surprising, therefore, to fin highly industrialized and that manufctured goods provide a major proportion of exports. Household industry is very widespread in South Asia and this sector is encouraged though in practice it is in competition with factory industry.

The development of modern industry in South Asia came with the processing and manufacture of such raw materials of the region as cotton and jute. Coal and iron ore west of Calcutta formed the base for the first large iron and steel plant in the 1930s. After independence in 1947 industrialization was a high priority.

The energy used in South Asia is derived mainly from the coalfields of east central India (see Figure

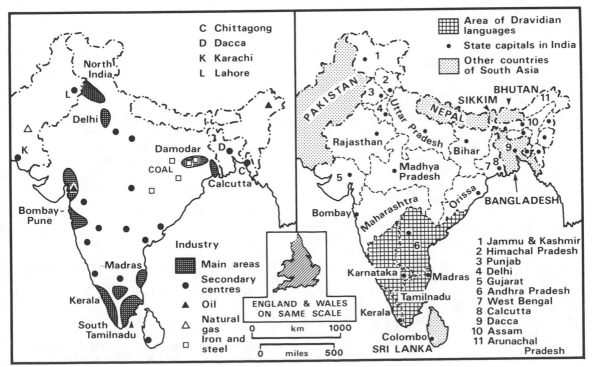

Figure 11.2 *India* (a) *Industry. Simplified from Johnson (1979)* (b) *Political divisions*

dro-electric power, home produced oil and
gas in Pakistan, western India and the
eme northeast of India, together with imported
il, processed in many refineries. *Figure 11.2* shows
the main concentrations of industry. The Bombay
(cotton textiles) and Calcutta (Hooghlyside) areas
each have about ten percent. The Damodar valley and
adjoining regions west of Calcutta have five iron and
steel works. The extreme south of India and the
north around Delhi are also comparatively highly
industrialized. Central India has few modern industries and the Himalayas, including Nepal and
extreme northeast of India virtually none. Pakistan,
Bangladesh and Sri Lanka are less industrialized than
India.

Internal features and problems

Far from being a country that is uniformly poor and
'underdeveloped' India is a land of great contrasts.
Some of its internal problems will be noted. India is
subdivided administratively into states, which broadly coincide with linguistic groups. When India
became independent the Indo-European language
Hindi was chosen as the most appropriate for the
whole country. It is, however, the language of
education only in northern India. Several powerful
states including West Bengal and Maharashtra do not
use Hindi and the four states of southern India have
languages from a completely different language
family (*see Figure 11.2*). The linguistic situation is
complicated by the wide use of English throughout
India. The result is that some people in India may use
two or even three languages and attempts to enforce
the use of a particular language can be a source of
conflict or friction.

Religion has a large influence on life in South Asia.
The Hindu religion of India carries with it cultural
practices and customs including the existence of
castes. India still has the underprivileged caste of
untouchables (Harijans). Such people are traditionally isolated residentially from the rest of the
community. With urbanization they tend to mix
more and 'positive' discrimination may be used to
help them in education and employment.

Great contrasts exist in India between rich and
poor and between town and country. Altogether 40
percent of the population of India is defined as being
below the poverty level (of India). Over 90 percent
would be by US standards. Maharashtra and Punjab,
which include Bombay and Delhi, are the richest
states, while eastern and southern India include the
poorest (*see Figure 6.3a*). Even in the most prosperous towns enormous numbers of people live in very
poor housing, often in dwellings of their own
making, barely integrated into the urban life. Much
of the private investment in urban housing goes into
high income dwellings.

The dual economy of India is seen in many ways.
Air services and fast express trains cater for a top
layer of society while bullock carts serve most local
needs in the countryside. It has been estimated that
the bullock cart sector of the Indian transport system
employs about 20 million people, while modern rail
and road transport employ about 25 million but carry
more than 30 times the traffic. Nuclear research
(delayed by safety and other problems) contrasts
with the widespread use of manual labour.

India has a mixed economy with planning through
Five-Year Plans since 1947. The public sector covers
about 20 percent of the economy and corresponds to
the more organized and planned part, including
government services, some health and education,
many large-scale industries, gas, electricity and the
railways. Railway technology and electrical engineering have done well whereas iron and steel and the
sugar industry have been disappointing and in India's
space, defence and nuclear programmes achievements
have been varied. The private, largely unplanned
sector covers most of the agriculture as well as
internal trade and small household industries.

With only 100 US dollars per inhabitant, Bangladesh (*see Figure 11.3*) is poorer than most of India.
It resembles closely the adjoining Indian state of
West Bengal, but is less urbanized and industrialized.
It is smaller in area than England and Wales, yet in
1981 already had about 90 million inhabitants and by
the year 2000 could have 150–160 million. Agriculture is difficult to organize and improve as there are
some 70 000 villages, dispersed in a country crossed
by numerous rivers and prone in many places to
frequent flooding. Economic improvements in the
future depend on an increase in agricultural yields
and the discovery of minerals. Unless population
growth is curtailed, however, the future is dismal.
Family planning is contrary to traditional Muslim
thinking and rural commonsense, since children help
on the land and are eventually expected to support
their parents. Around 1980 a serious campaign was
started to achieve the voluntary sterilization of
something like one million young people per month.
Such a procedure is necessary where early marriage is
common and three-quarters of the women have their
first child by the age of 17.

International

The influence of South Asia in world affairs does not
match population size. With some 20 percent of the
population of the world the region only has two to
three percent of the natural resources and accounts
for two to three percent of gross national product.
The hope of the founders of modern India that the
country might become a major world power have not
materialized. Economically India has little to offer
the rest of the world and militarily its limited strength

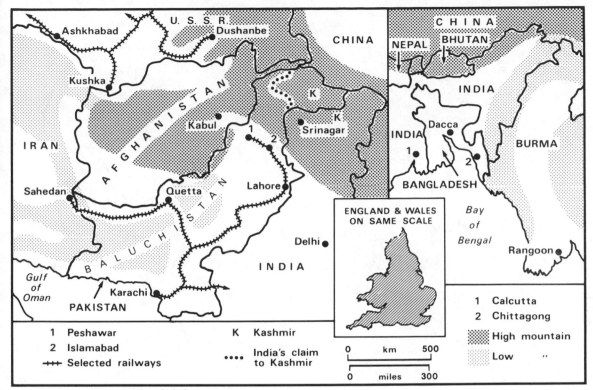

Figure 11.3 *Pakistan and Bangladesh*

has been deployed without distinction in conflicts against its immediate neighbours, China and Pakistan. It tends to maintain a neutral role in east–west ideological conflicts.

India trades with the major industrial countries and has received aid from West Europe, the USA and the USSR both to relieve famines and for long term development. Bangladesh is of very limited strategic value and even Sri Lanka's commanding position in the Indian Ocean is no longer particularly vital when smaller islands can be more easily used as bases. Pakistan, however, is in a more sensitive position, sharing boundaries with China, its rival India and since 1979 virtually with the USSR through Afghanistan. Like India, modern Pakistan is a mixture of linguistic groups and associated cultures. *Figure 11.3* shows that the eastern part includes the densely populated irrigated lands of the Indus valley and the major towns of the country while the western side is inhabited by peoples more closely associated with Iran and Afghanistan. In southern Afghanistan, Soviet forces are only 400 kilometres from the coast of the Indian Ocean, a dream location for a port both ice free and unimpeded in its access to the Ocean by narrow straits.

South Asia is both poor and largely self-contained. Around 1960 India provided about 1.2 percent of world exports but in 1980 only 0.5 percent. Is South Asia on an endless downhill path? Two centuries ago Adam Smith wrote about the area that is now eastern India and Bangladesh. In Bengal and in some of the other of the English settlements in the East Indies he saw a state of decline:

'In a fertile country which had before been much depopulated, where subsistence, consequently, should not be very difficult, and where, notwithstanding three or four hundred thousand people die in one year, we may be assured that the funds destined for the maintenance of the labouring poor are fast decaying.'

His view of China was less pessimistic:

'China has been long one of the richest, that is one of the most fertile, best cultivated, most industrious and most populous countries of the world. It seems, however, to have been long stationary . . . The poverty of the lower ranks of people in China far surpasses that of the most beggarly nations in Europe.'

He concluded, however, that China 'does not seem to go backwards.'

Figure 11.4 *Southeast Asia*

11.2 Southeast Asia

Introduction

In this section an 'extended' Southeast Asia (*see Figure 11.4*) is covered in order to bring in the smaller countries of East Asia, leaving Japan (chapter 9) and China (next section) for separate consideration (*see Figure 11.5*). In 1981 'tropical' Southeast Asia had 363 million inhabitants, to which may be added 81 million in the Koreas, Taiwan and Hong Kong. The whole region has almost exactly ten percent of the population of the world. *Table 11.1* contains a selection of data for Southeast Asia as well as for South Asia. China, Japan and Australia are included for comparative purposes.

Although Indonesia is the fifth largest country in the world in population size, Southeast Asia consists of countries that are modest in terms of total gross national product. The whole region is broadly regarded as 'developing' though in places some striking industrial growth has taken place since the Second World War. It is a negative area in the sense that powers outside the region had, and still continue to have, considerable influence, whereas the region has made little impact on the rest of the world apart from its contribution of primary products to industrial countries.

Southeast Asia has long been influenced by China, India and Arabia. The Muslim religion came to the islands from the west in the distant past. In the 16th century, on account of its tropical agricultural products, especially spices, Southeast Asia attracted the attention of European explorers. Spain colonized the Philippines while Portugal took various footholds, including Macau and Timor, as trading posts. Later the Netherlands gained control of what is now Indonesia. Britain occupied Burma, Malaya and a number of other territories while France took Indo-China, now the three countries of Viet Nam, Kampuchea and Laos. Japan held Taiwan (formerly Formosa) and Korea for some decades. Even the USA and USSR have been involved in affairs in Southeast Asia with the cession of the Philippines by Spain to the USA in 1898 (independent 1945) and the presence of US forces and Soviet advisers and arms in the Viet Nam conflict, in the 1960s and early 1970s.

Indeed only Thailand escaped direct colonial rule by an outside power but the instability in Viet Nam and Kampuchea has forced it to rely on the USA for military aid.

The presence of China is felt throughout the region. China shares a boundary with Burma, Laos, Viet Nam, Hong Kong and the Korean DPR. It regards Taiwan as its own territory. Many Chinese have settled in Viet Nam, Hong Kong, Singapore and Indonesia. The most recent positive political move in Southeast Asia has been the formation of ASEAN to counter the military strength and potential threat of Viet Nam, now a full member of CMEA and a Soviet sphere of influence in the region.

Natural resources

Southeast Asia (*see Figure 11.5*) is characterized by the large proportion of its area that is rugged. Flat and gently sloping land are rarely extensive. Thus although there is adequate rainfall for cultivation almost everywhere, agriculture is difficult on account of relief and also on account of poor soils. The equatorial climate brings hazards. Viet Nam, for example, was afflicted seriously in 1978 and again in 1980 when typhoons wiped out freshly planted rice and destroyed many buildings. The data in *Table 11.3* show that only a small part of the total area of most of the countries is defined as arable or under permanent crops. Much of the cultivated land is on alluvial soils or, especially in Java (Indonesia), on volcanic soils. In many areas hillsides have been meticulously terraced. Irrigation is very widespread and is especially critical for the cultivation of rice. Double cropping is possible in most of the region.

Forests cover about 60 percent of the total region and consist mostly of tropical rain forest, which contains a large number of tree species mostly hardwood, some like teak and mahogany being highly valued. There is little permanent pasture in the region.

The two Koreas (*see Figure 11.5*) are the only countries in Southeast Asia with coal reserves of more than very local importance. Indonesia is the only country with large oil reserves (about two percent of the world total) and like Malaysia it also

Table 11.3 *Cropland and forest in Southeast Asia*

Country	Percentage cropland	Percentage forest	Country	Percentage cropland	Percentage forest
Burma	15	48	Malaysia	19	66
Thailand	34	41	Indonesia	9	64
Laos	4	63	Philippines	27	43
Kampuchea	17	76	Korean R	22	67
Viet Nam	18	38	Korean DPR	18	75

has some natural gas. With ten percent of the population of the world, Southeast Asia is very poorly provided with fossil fuel reserves, since it has less than one percent of the world total. The hydro-electric potential is not great.

A number of non-fuel minerals are found in Southeast Asia and the data below show the main reserves as percentages of the world total. Iron ore, bauxite and fertilizer minerals also occur.

Malaysia	tin	8%	antimony	3%
Philippines	copper	4%	cobalt	13%
Indonesia	nickel	8%	tin	25%
Thailand	tin	12%		
Korean DPR	tungsten	6%		

Southeast Asia is broadly self-sufficient in food though Hong Kong and Singapore depend on nearby mainland areas for supplies. Agricultural yields in the region have improved gradually in recent decades, but population has roughly doubled since the 1940s.

Spectacular increases have been achieved in experimental conditions and the Korean R has high yields. New strains of rice, experimented with in the Philippines, tend to be more susceptible to pests than traditional ones.

The colonies of Europe and Japan in Southeast Asia provided agricultural products such as sugar, coprah and spices for export. They were also the main producing region of natural rubber extracted from the Brazilian tree *Hevea brasiliensis*, grown in plantations. Tea and coffee are also products of Southeast Asia.

Tropical Southeast Asia is one of the least industrialized regions of the developing world. Indonesia exports about two-thirds of its oil production, and raw materials like tin, rubber and timber are hardly used at all. Tropical Southeast Asia has very little heavy industry and produces virtually no steel or engineering goods. Processing industries are widely found but only the national capitals and

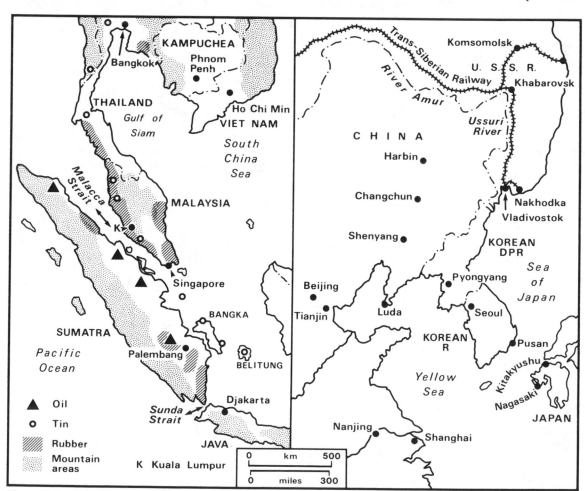

Figure 11.5 *Malaysia and part of Indonesia*

ports have much manufacturing and the politically sovereign island of Singapore alone stands out as an industrial city.

The situation is now different in the two Koreas, Taiwan and Hong Kong. In both Koreas coal production has grown impressively since the Second World War, and is about 35 million tonnes a year in the Korean DPR and 20 million in the Korean R. Energy consumption (*see Table 11.1*) here and in Taiwan, Hong Kong and Singapore is much higher than in the more agricultural tropical countries of the region. With considerable Soviet aid in the Korean DPR, and US and Japanese aid and investment in the Korean R and Taiwan, various branches of modern industry have developed. The Korean R, for example, now produces motor vehicles, ships and other items which Japan has so successfully produced and exported since the Second World War.

Present features and future prospects

Southeast Asia is a region of great contrasts, both physical and man-made. Early in the 1970s a small group of people, the Tasadays, were discovered deep in the forests of southern Mindanao in the Philippines living in 'stone age' conditions. In many areas of tropical Southeast Asia people live in very simple conditions. In contrast, Hong Kong and Seoul have virtually all the features of a modern industrial urban environment: banks, an underground rail system, high rise flats, skilled industrial workers. Most people in Southeast Asia are still engaged in agriculture, but pressure on cultivated land of both total population and actual agricultural labour is very great. It can be seen in *Table 11.1* that in Australia there are only about 30 inhabitants per square kilometre of cropland compared with several hundred in most of Southeast Asia and over 1700 in the Korean R.

In most parts of Southeast Asia population is growing by more than two percent per year. The total population of 440 million in 1980 could reach 640 million in the year 2000. Such a growth would put pressure on land resources, leading probably to attempts to extend cultivation into areas that are rugged and with soils that could easily be eroded. Forest is already being cut for timber and the whole physical environment is at risk where the tropical rain forest is cleared, while the existence of delicately balanced communities of forest dwellers is also threatened.

Southeast Asia has three special features in world affairs:

(1) Malaysia and the western part of Indonesia are still a source of oil, and mineral and plant raw materials for industrial countries of the developed world, especially Japan (*see Figure 11.7*).

(2) From small beginnings, the Korean R, Taiwan and Hong Kong have grown into industrialized countries capable of qualifying as members of the developed world by outdoing Japan in some of its lines of production.

(3) Malaysia, Singapore and Indonesia control key straits linking the Indian and Pacific Oceans and are, therefore, of interest to the leading world powers for the passage of both commercial and naval vessels.

Soviet influence in Southeast Asia has in a way diminished with the end of conflict in Viet Nam and the unification of the country. Economically Viet Nam could be a liability to the USSR now that it is in CMEA because it has little that the USSR needs, yet if it is to be transformed into a model socialist state with the expected industrial development then its 55 million people, rising to nearly 80 million in the year 2000, could absorb a considerable amount of Soviet and other East European machinery and equipment. The South China Sea could turn out to be a major oil province, but China is in a stronger position than Viet Nam and the USSR to uphold claims to offshore areas for oil exploration.

11.3 China

Introduction

The Chinese People's Republic is the largest country in the world in population (approaching 1000 million in 1981) and the third largest in area after the USSR and Canada. Its present limits and political organization were established in 1949 when the Communist Party under Mao defeated the Nationalist Chinese under Chiang and a one party system with selective party membership was established. Taiwan (formerly called Formosa) and ultimately also the British and Portuguese colonies qf Hong Kong and Macau are regarded as part of China by the Communist Party.

China has been a recognizable cultural unit for several thousand years. The origins of Chinese civilization are found in the area of fertile, easily tilled loess soil of the middle Huanghe basin. More than 2000 years ago Chinese influence spread south taking in and developing the lowlands of the Huanghe, Hua, Changjiang and other valleys, where rice could be grown on irrigated land. Some 2000 years ago China of the Han Dynasty controlled an area roughly comparable in extent to that held by the contemporary Roman Empire in Europe and the Mediterranean and with roughly the same number of inhabitants (50–60 million).

The exact limits of the Chinese Empire have varied over the centuries. Control of the Chinese Empire has at times been strong and widespread, but at other times it has become weak and fragmented, with semi-independent provinces. Nation states characteristic of Europe over the last thousand years did not, however, develop in China. The Chinese language has been held together by its symbols and has helped to give cohesion to the Chinese culture area.

The Chinese name for China, Tsung guo, means middle country. China has been the centre of the world in its own eyes, and other peoples have been regarded as outsiders or barbarians. Even so, invaders have frequently attacked China, infiltrated and settled there. In the 17th century (1644) the last dynasty, the Qing (Manchu), was controlled by non-Chinese peoples from Manchuria, now north-east China.

Like Japan, China only became seriously involved with European powers and the USA around the middle of the 19th century. In the 17th century Russia had already taken lands on the northern fringe of the Chinese Empire but Russian aspirations were checked by the Treaty of Nerchinsk in 1689. In the 19th century foreign powers gained concessions in various Chinese ports and China was forced to trade. Hong Kong was one of the main footholds and it became a British colony. China did not adopt modern industrial methods as readily as Japan did and apart from a few centres it remained deeply traditional and agricultural. In the Revolution of 1911 the Manchu dynasty was overthrown and a Republic established. In the turmoil throughout the first half of this century the modernization of China, though considered desirable by various Chinese leaders, was limited. Textile and other industries were established in some places along the coast and some railways were built. Manchuria was occupied during 1931–32 by the Japanese and a heavy industry was developed, though for Japanese, not Chinese needs.

When it was clear that the Communist Party was in control in China, the first countries to establish diplomatic relations with the new regime were the socialist countries of Europe and Asia, all in October 1949: USSR, Bulgaria, Romania, Hungary, Czechoslovakia, the Korean DPR, the Mongolian PR, and the German DR. The following year a few non-socialist countries also established diplomatic relations with China: India, Burma, Sweden, Denmark, Switzerland, Finland. Though the UK recognized the *de facto* existence of the Communist government in China it did not have diplomatic links until 1972, while the USA waited until 1979.

For about ten years the USSR strongly influenced the form of development in China. The overall strategy of the Soviet model meant public ownership of means of production, central planning, emphasis on coal, heavy industry, rail transport and hydro-electric power, and the collectivization of agriculture. Existing industrial capacity, damaged in the war with Japan, was restored and expanded and new heavy industrial centres were created (eg in Wuhan and Baotou). Soviet and East European equipment and technicians were used to start China's modernization. It was perhaps ambitious to follow the Soviet experience too closely since conditions in China in 1949 differed considerably from those in the Soviet Union immediately after the 1917 Revolution. Per inhabitant China had and still has far fewer natural resources than the USSR and it was less industrialized in 1949 than the Russian Empire had been in 1917.

In 1958 Chinese leaders decided they did not need the Soviet model and embarked on the Great Leap Forward. Ideologically their aim was to move quickly from 'socialism' to 'communism'. Organizationally the period is marked by the establishment of communes formed of groups of agricultural settlements, and by the introduction of industries into rural areas (eg backyard blast furnaces). Chaos, worsened by bad harvests, soon put an end to the new line of development.

In 1966 Mao sponsored a new development, the Cultural Revolution (1966–76, off and on). Emphasis on agriculture was maintained, urban growth was discouraged and educational advances halted. In the early 1970s China was very isolated, having virtually cut its links with the socialist countries around 1960 and with few formal links with the western industrial countries. It had a few 'friends' in Africa. China's admission into the United Nations in 1972, replacing Taiwan, prepared the way for the termination of its isolation, and Mao's death in 1976 finally gave China's leaders a chance to break with the agricultural socialism of the Mao 'dynasty', discrediting some of Mao's closest associates including the 'Gang of Four'. In 1981 about three-quarters of the population of China was under the age of 40. They would barely remember the conditions of pre-1949 China.

Some of the outstanding achievements of China since 1949 may be noted. Between 1949 and the early 1980s the population of China grew by about 400 million people, an increase of about 70 percent. Concern over population growth in the early 1970s prompted the introduction of the drastic methods now being used to limit the birthrate. China could well have done without its increase, allowed by the Marxist, anti-Malthusian view that 'overpopulation' is a problem only in capitalist economies.

Production in many sectors of the economy grew by more than 70 percent between 1949 and the early 1980s and on average each person in China was better fed, clothed, educated and treated medically than 30 years earlier. There is little conspicuous affluence, or one might say that poverty has been fairly shared. One of a series of general policy statements is the aim to raise production per inhabitant from 300–400 dollars in 1980 to about 1000 in the year 2000. Such an aim implies roughly a threefold increase in the production of goods and services to cater for some

1200 million people at the hoped for higher level. The forthcoming transformation of China is to be through 'four modernizations': agriculture, industry, science and technology, and defence.

Natural resources and production

China is roughly comparable in area to Europe (including European USSR) but it actually has about 1000 million people compared with 700 million in Europe. It is much more rugged than Europe (*see* *Figure 11.6*). Much of its population lives on the ten percent of its total area that is flat or gently sloping lowland. China is located further south than Europe and the eastern part has very hot summers, the period when most of the monsoon rain comes. In the winter northeast China is extremely cold. Much of the interior is high, rugged and dry.

The limitations of the land resources of China are evident from the following data:

	Area in million hectares		Percentage	
	1961–65	1978	1961–65	1978
Total	960	960	100	100
Arable and permanent crops	105	100	11	10
Permanent pasture	220	220	23	23
Forest	104	115	11	12

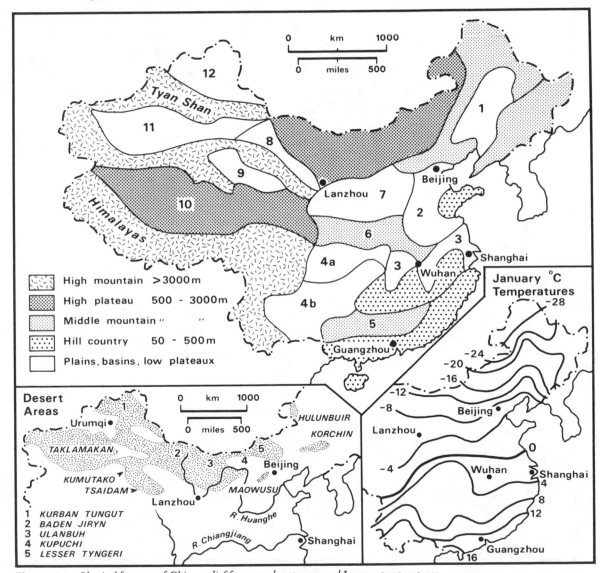

Figure 11.6   *Physical features of China: relief features, desert areas and January temperatures*

196

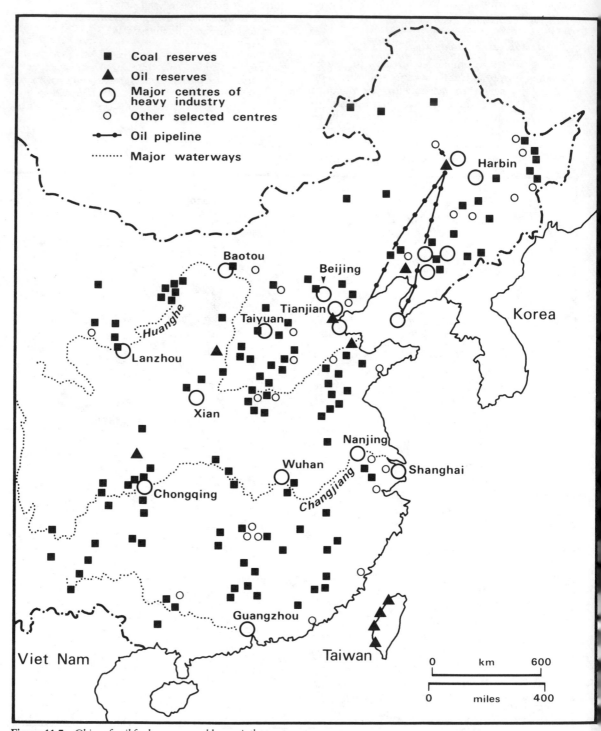

Figure 11.7 *China: fossil fuel reserves and heavy industry*

Only about one-tenth of the total area of China is used to produce crops but as a substantial part of this area is cropped twice and in the extrme south even three times, the area actually harvested is greater. The data show that the area of cropland has been diminishing in spite of efforts to extend cultivation by drainage schemes, the terracing of hillsides and irrigation projects in arid areas. Concern has been expressed over the felling of forests and there has been a campaign to replant both for conservation purposes and to provide future timber. The permanent pastures of China, mostly in the north and west of the country, are of limited quality. Any substantial increase in agricultural production in China must come essentially from the areas already under cultivation. These are mostly inherently fertile, with deep soil and the constant addition of new materials carried by irrigation water. There is no lack of labour in Chinese agriculture. With 267 million persons working in agriculture there are on average about 270 persons working each 100 hectares of cropland compared with less than one in Australia.

China is apparently fairly well endowed with energy resources. In particular it is credited with nearly 20 percent of known recoverable reserves of coal, enough to last several hundred years at current rates of production. Much of this is concentrated in one part of China (*see Figure 11.7*), but most other provinces of eastern China have at least some coal reserves. At the end of 1980 it was estimated to have 2800 million tonnes of oil (3.1 percent of the world total), enough to last about 25 years at current production rates, and less than one percent of the world's natural gas reserves. Much of the oil is in deposits in the West of China, in northeast China (Daching) and, it is hoped, in offshore areas. China also has a large hydro-electric potential and some of this has been developed in multi-purpose projects on the Huanghe and other rivers to control flooding and supply irrigation water as well. There is a very large potential in the southwest of China but this is inaccessible at present.

Information about the non-fuel minerals of China is difficult to obtain and much of the country awaits thorough exploration. The following are the percentages of the world's totals of the minerals indicated credited to China:

Antimony	50	Silver	10
Tungsten	47	Iron ore	3
Mercury	19	Molybdenum	2
Tin	15	Lead	2

Most of the minerals listed are of secondary importance and there are apparently only very limited deposits of bauxite, copper, zinc and fertilizer minerals.

Chinese agriculture is to some extent planned centrally but it is also organized locally at several levels. The basic unit of organization since about 1960 has been the commune, a group of settlements (villages), averaging 12 000 inhabitants. There are 50–60 000 communes altogether in China. As well as providing crop and livestock products both for local needs and for delivery to urban centres, communes are expected to provide many of their other needs, producing building materials, repairing machinery and also processing their own farm products. Although almost all the land belongs to the commune (or the state), farmers are allowed very small plots of land for their own use. There were moves in 1982 to reduce the role of the communes and to give more freedom and flexibility to management at a more local level.

In spite of double cropping in many areas, cropland is very limited in China. The livestock sector is therefore comparatively small. Horses, water buffalo and oxen are used in different parts of China as work animals but they are increasingly being replaced by small tractors. Pigs and poultry are the main forms of livestock in many areas. Most of the crops produced are for direct human consumption or are raw materials (cotton). The production of fodder crops for livestock and the fodder from permanent pastures are very restricted.

Chinese agriculture concentrates on the production of cereals (especially rice, millet and wheat), tubers (potatoes), root crops (sweet potatoes), soybeans, groundnuts, and cotton. In the drier northeast wheat, millet, maize and barley are widely cultivated without irrigation. Yields fluctuate greatly from year to year. Where land can be irrigated, however, rice is the preferred cereal as it gives larger amounts of grain per unit of area. Yields are more reliable with the controlled water supply.

Yields in Chinese agriculture have increased in the last 30 years through great efforts to control flooding, provide more water for irrigation and increase the use of chemical fertilizers (in addition to the traditional use of 'night soil' from humans). Yields of some crops in China are higher than those for the world as a whole whereas in others they are lower. China lags behind developed countries like Japan, the USA and France in yields, but does better on the whole than developing countries. It is now difficult to feed the population adequately and at the same time produce enough plant and animal raw materials for industry. Considerable quantities of cereals have been imported especially since the early 1970s, the suppliers including Canada, Australia and the USA.

China still has to concentrate on the improvement of agriculture. Progress may be made at both national and local levels. A giant-scale project, for example, is the diversion of water from upstream in the Changjiang drainage basin into the Huanghe, along which it could be transferred for irrigation in the drier northers parts of eastern China. At more local level an 8-point improvement scheme proposed in China in the late 1970s would be applicable in many developing countries: soil (deeper ploughing, levelling), fertilizer (organic and inorganic), water conservation, seed selection, close planting, plant protection (tree belts, pesticides), tools, field management.

Although energy production and consumption in China have grown enormously since 1949, the 1980 level of consumption of roughly 800 kilograms of coal equivalent per inhabitant was only about one-third of the world average. Coal has been the main source of energy throughout the last 30 years but oil production has increased rapidly since the late 1960s. In 1980 coal accounted for 78 percent of primary energy consumption, oil for 17 percent and hydro-electricity and natural gas for almost all the rest. In 1980 China still had no nuclear power production.

Coal has been of key importance in the economy of China (*see Figure 11.7*), providing fuel for electricity generation, steam power, coking, transport and domestic needs. It is mined in many parts of China and though consumed mainly in the industrial and urban areas of the country is widely distributed

regionally in dust form by rail and then locally by road to many small towns and villages, where often it is compressed into briquettes for easy use. Electrification of the rural settlements seems to be widespread also. With the increased consumption of oil products it has been possible to increase the motorization of the towns and countryside, but for commercial traffic, not private motoring. Since the mid-1970s China has been an exporter of oil and it is hoped that new offshore finds will enable it to increase oil production even further. The help of western industrial countries is needed and in 1982, for example, bids were made by many transnational companies for exploration concessions (*see Figure 11.8*) off the coast of China.

In the 1950s high priority was given in China to the production of iron and steel (*see Figure 11.9*), and many provinces now have works using local coal and iron ore. Much of the capacity is still found in the northeast of China and in a few other large works. Much of the equipment dates from the 1950s and a large labour force is required to run the industry. The engineering industry of China now produces a wide range of producer and consumer goods. Mines, railways, road transport and factories are largely run on equipment produced in China, and bicycles are

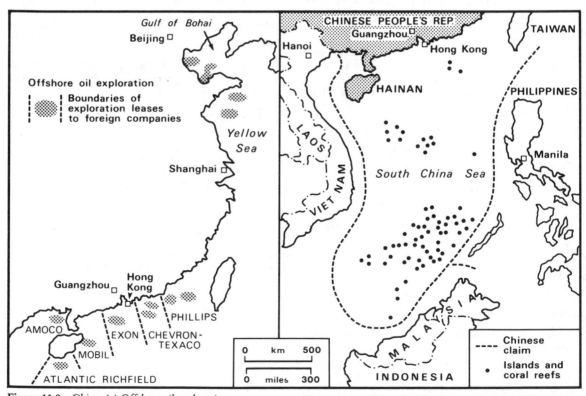

Figure 11.8 *China: (a) Offshore oil exploration*   **Figure 11.8** *China: (b) The claim to the South China Sea*

mass produced. In more sophisticated branches of engineering China is very backward, as in the aircraft and electronic industries. A wide variety of other industries have been developed in China in a mixture of large factories and small workshops. The concentration of industry in certain towns and regions of China has already been noted in chapter 6.

Internal features

While China is often described by western observers as a developing country in which many problems have been solved, conditions are far from being uniform over the national area. Contrasts between regions result from differences in ethnic and cultural background, natural resource base, economic specialization and location in relation to the coastal regions and to railways. The lifestyle of a top urban administrator, manager or skilled worker differs enormously from that of a cultivator in a deeply rural commune or a herder in the interior of the country.

Over 90 percent of the population of China consists of Han Chinese, a fairly homogeneous 'national' group using the Chinese script though speaking Chinese in many different dialects. In the interior of China, however, Mongols, Uighurs and Tibetans are the original inhabitants of a very large extent of the national area and they are administered in special autonomous regions distinct from the traditional Chinese provinces (*see Figure 11.10*). Since 1949 road, rail and air links between eastern and western China have been strengthened and several million Han Chinese have moved west to settle among the non-Chinese. In the extreme south of China other non-Chinese peoples form regional majorities (*see Figure 11.11*).

Contrasts occur in China on account of the impossibility of achieving uniform development everywhere. Where coal, oil and other minerals have been extracted industrialization has often followed and new towns have been established. In eastern China the older urban centres, particularly the ports with their infrastructure as well as some industrial experience and skills, have been the obvious places to put larger factories. In the late 1970s special coastal industrial 'estates' were created (eg near Hong Kong) to attract foreign investors.

Much of the longer distance movement of goods in China is carried by sea, inland waterway, especially the Changjiang river, or rail. For a country with 1000 million people and an area as large as Europe, China has a very limited rail system (*see Figure 11.12*). The route length is about equal to that of the French system, but France has only about one-twentieth the area and one-twentieth as many people (*see Figure 11.12*). Much of the line in China is single track. In the southwest of the country three provinces with a total population of 120 million had almost no railway

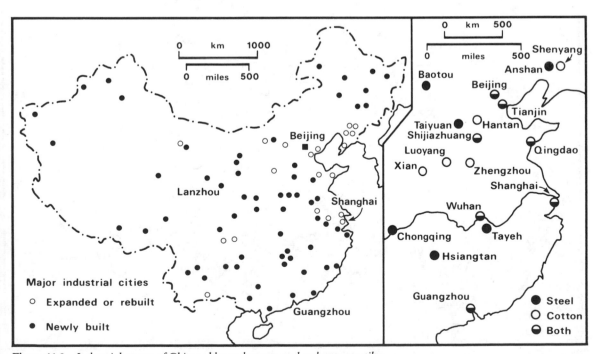

Figure 11.9 *Industrial centres of China: older and recent, steel and cotton textiles*

at all before 1949. It is impossible to move large quantities of goods between most major regions of China on such a system and impossible for interior provinces to carry on more than minimal transactions with the outside world via the coastal ports. The sheer difficulty of moving imported equipment far inland into China alone hinders the modernization of many places.

The limited capability of the transportation system of China prevents the large-scale mobility of people between regions. Differences in spoken Chinese further alienate people from different parts of the country. As in many developing countries great contrasts remain between town and country. The larger Chinese cities are indeed experiencing many of the problems found in other countries: pollution,

Figure 11.10   *The provinces and associated politico-administrative divisions of China*

traffic congestion, housing shortages. Even in socialist China a dual economy exists, with an upper level of fairly sophisticated towns and industrial centres, linked moderately well by air services, and a lower level with a vast number of largely self-contained rural settlements. Future modernization and industrialization in China is likely to further differentiate the coastal region of China from the rest for at least two or three decades, especially if foreign trade continues to increase.

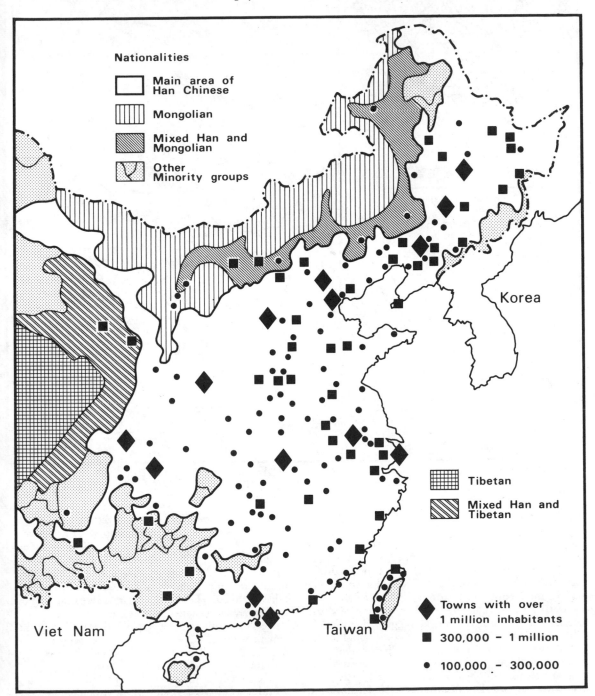

Figure 11.11 *Nationalities of China and largest cities*

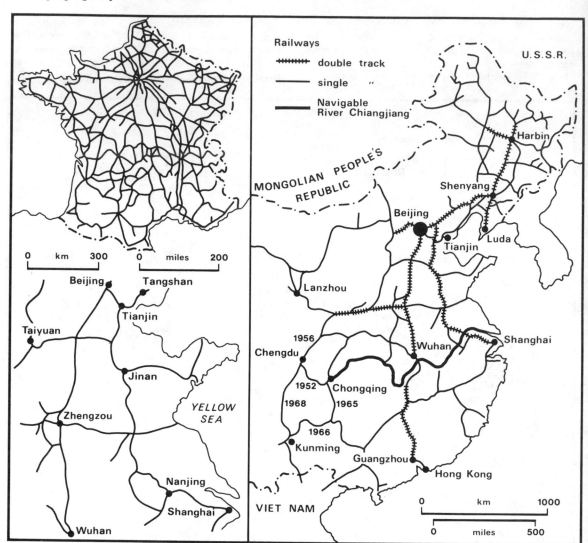

Figure 11.12   *The railway system of China*

Foreign relations

In view of China's great size its direct impact on world affairs has been modest since 1949 in spite of the hoped for role of leader of the Third World. Militarily it has not overrun any adjoining countries and its intervention in Korea in 1950, and invasions of India in 1962 and of Viet Nam did not lead to escalation. Common sense alone suggests that even if it were capable of doing so, China would gain little if it occupied areas to the south which, if anything, are more densely populated and less developed than China itself. Any aspirations in the north and northwest are blocked by a massive Soviet military superiority. Almost everywhere the boundary of China runs through rugged or desert areas. China's nuclear capability may be seen simply as a deterrent.

Can China become another Japan economically? In the 1960s and 1970s it was largely inward looking. Its foreign trade is only a few percent of its total production and China is, therefore, almost entirely self-contained. The reasons for this isolation are not too clearly appreciated by western governments and writers on business prospects, who are misled by the apparent size of China. *Table 11.4* compares the size of China's area and population with its share of resources and production.

If China is to make an impact on the rest of the world it will have to do this through trade, but there is very little that it can produce that it does not need itself. Raw materials in great abundance are confined to a few minerals such as tungsten and antimony. Any form of modernization and economic growth would preclude China from exporting much oil. Its engineering goods are also needed at home and they can hardly match up to those of Japan for quality. Chinese exports of consumer goods tend to be items using few raw materials and even include luxury goods laboriously worked by craftsmen. Given such constraints on exports China is not in a position to obtain large quantities of either raw materials or capital goods from foreign sources.

As a result of the limited scale of its land and mineral resources China already has too many people to achieve and sustain a high level of material consumption even if it acquires the technology and equipment to produce the goods needed.

One way of alleviating the problem of lack of natural resources is to slow down and stop the growth of population. One proposal is to control the number of children born in such a way that after peaking at 1200 million in two to three decades time the population should be reduced to an eventual

700–800 million by the decade 2070–2080. A long term plan of this kind would require strict measures and a consistent policy of applying them.

Late marriage and birth control are being encouraged. One child families get special rewards, two-child families are barely tolerated and families with more than two are penalized. In the 1970s the average number of children per woman decreased, but for the next two decades the actual number of women of child-bearing age will increase. The population of China could rise to somewhere between 1200 and 1400 million. If, however, cuts in the number of births are too drastic for too long, a very unbalanced age structure would be produced and would last as such for many decades. The eventual reduction in the total size of the population would require control of births since increases in mortality would not be desirable or acceptable.

Between about 1700 and 1850 the populations of China and Europe both grew from about 150 million to 450 million. The figures are now about 1000 million and 700 million respectively. About 150 years ago, when growing population and widespread poverty were coped with and gradually reduced in Europe population growth slowed down. Europe was also able to benefit from mass emigration, industrialization, imported food and raw materials and its own advanced technology. China's position now resembles that in Europe around 1850. Its options now are much more restricted and a European type of cure and future is unlikely.

Socialist governments of centrally planned economies carry more direct responsibility for the success of their economic and social programmes than the governments of market economies. It seems then a devastating admission of failure to say that the ten years following the Soviet model and later the ten years of the Cultural Revolution were for China the wrong path. That is 20 out of 28 years between 1949 and Mao's death in 1977. But we have it from the 'horse's mouth' in *China, a general survey*, Foreign Languages Press, Beijing, 1979, introduction:

'The backward and poverty-stricken China of old has been transformed into a new socialist state with initial *prosperity*. . . To build socialism in a country as vast as China is an undertaking never before attempted and is not a simple task. It was inevitable that we made unnecessary detours on our way forward. In the past 30 years we reached many targets but also suffered setbacks; we achieved a great deal, but still have many shortcomings. In particular, during the 10 years from 1966 to 1976, China suffered immense disruption and damage due to domestic reasons. After reviewing the positive and negative experiences of the past and drawing useful lessons from them, China will be able to manage her affairs more competently in future. Now the Chinese people are embarking on a new Long March – that of building the nation into a strong socialist state with modernized agriculture, industry, national defence and science and technology by the end of the century.'

Table 11.4 *China's percentage share of the world total of selected items*

Item	%	Item	%
Sweet potatoes	79	Timber (broadleaf)	7
Tung oil	71	Energy production	7
Pigs	39	Energy consumption	7
Rice	35	Iron ore	7
Millet and sorghum	34	Cultivated land	6
Silk	33	Pig iron	6
Agric population	28	Newsprint	5
Coal reserves	28	Cattle	5
Tungsten ore	22	Phosphate fertilizers	5
Population	22	Cement	5
Coal production	20	Steel	4
Tea	19	Sugar	4
Salt	17	Forest area	4
Tobacco	17	Wool	4
Cotton seed	17	Asbestos	4
Antimony ore	17	Potatoes	4
Cotton lint	16	Lead ore	3
Soybeans	16	Manganese ore	3
Meat	16	Phosphate rock	3
Groundnuts	15	Oil reserves	3
Horses	11	Rayon etc	3
Wheat	10	Oil production	3
Barley	10	Aluminium	2
Maize	9	Bauxite	2
Fish catch	9	Zinc ore	2
Nitrogen fertilizers	9	Natural gas reserves	1
Coke	8	Copper ore	1
Timber (coniferous)	8	Potash fertilizers	1
Sheep	8	Tractors in use	1
Land area	7	World trade	1
Pasture	7	Potash	1

CHAPTER TWELVE

Developing Regions Rich in Natural Resources

12.1 Latin America

Introduction

The region of Latin America refers here to all of the Americas south of the USA. It therefore includes Mexico, Central America and the islands of the Caribbean (*see Figure 12.1*) as well as South America (*see Figure 12.2*). In area it is roughly equal to the USA and Canada combined, but in the early 1950s it passed them in population and in 1981 had a total of 366 million inhabitants (USA plus Canada only 254 million), rather more than eight percent of the world total. Twenty-three countries of Latin America each have over one million inhabitants, seven of them in Central America including Mexico, six in the Caribbean and ten in South America. There are many other smaller political units in Latin America, mostly in the Caribbean, some of them still dependencies of European countries or of the USA.

Latin American countries are for the most part more 'developed' than the developing countries in

Africa and Asia and some, especially Puerto Rico in the Caribbean, Venezuela, and the southern part of South America, might in some respects be thought of as developed countries. Latin America is both the oldest area of colonization by Europe and the first area after the USA to gain independence from Europe.

During the 16th to 18th centuries virtually the whole of what is now Latin America plus Florida and the southwest of the USA was in the empires of Spain or Portugal. Metals, precious stones, sugar, coffee, cotton and other products were sent to Europe in exchange for manufactured goods.

During the colonial period, Spaniards and Portuguese settled in the region in comparatively small numbers and formed a class of administrators and landowners. Much of the labour in the extractive and agricultural sectors was provided where possible by American Indians. Large numbers of African slaves were brought into the parts of Latin America where Indians were few. In Mexico, Central America and

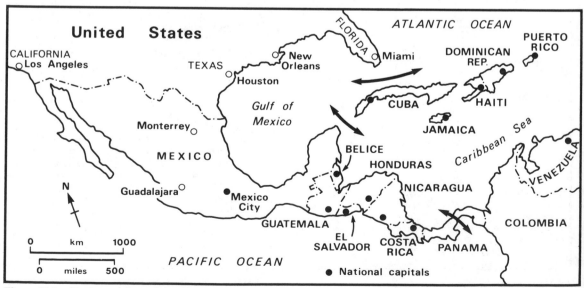

Figure 12.1 *Mexico and Central America*

Figure 12.2 *South America: general features and mineral deposits. Florida and the British Isles are shown on the same scale as the main map*

the Andean countries, the American Indian element was particularly strong and the African element very restricted. A mestizo population, Spanish mixed with American Indian, had become a major element. In the Caribbean and northeast Brazil (as in southeast USA), the negro element was strong. After independence further migration took place, especially into southern South America, mainly of Europeans but also of Chinese, Japanese and Indians. Modern Latin America is therefore a region of great ethnic diversity, each country having a particular mix of population, from almost exclusively African in Haiti to largely American Indian in Bolivia and European in Argentina.

When independence was achieved over most of Latin America in the 19th century the Portuguese colony of Brazil remained intact as a single country. The larger and less compact Spanish Empire, extending from California in the north to Tierra del Fuego in the south, broke up into many countries. In the 19th century some small areas remained as colonies.

Natural resources

Physically Latin America is very varied. Some of the driest areas in the world are found along the Pacific coast of South America in Peru and Chile. Mexico and the Andean countries contain some of the highest mountain ranges in the world and have large numbers of people living at high altitudes, mainly at 2000–4000 metres. Elsewhere dense forest has been cleared in places for the cultivation of tropical crops. In Argentina and Uruguay natural grasslands with fertile soil provide arable land and pastures. The interior of South America contains the largest area of tropical rain forest in the world, shared by Brazil and several smaller countries. In Mexico, Chile and southern Brazil there are extensive areas of pine forest.

Only about six percent of the total area of South America is defined as cropland compared with about 20 percent in the USA. However, 25 percent is permanent pasture and over 50 percent forest. It seems that the trend to increase the area of cropland will continue. In Mexico and Central America there is also the potential for bringing new areas under cultivation, but in the islands of the Caribbean the limits may have been reached.

Already much of the better quality land in Latin America is under cultivation. Outside the pampa of Argentina and the fertile lava soils of São Paulo state in Brazil, areas of high fertility are disappointingly few. Oases in the deserts of Peru and Mexico, certain

valleys in the Andes and special soils along the coasts of tropical South America are very productive locally. New areas of cultivation will tend to be on poorer soils or will require costly reclamation or irrigation and will mostly be remote from existing concentrations of population.

The lack of good quality land is to some extent compensated for in Latin America by very large areas of natural pastures which, with better control of herds through the help of fencing, and with improvements to the pasture and the quality of livestock, could yield more than they do. The vast area of tropical forest in the Amazon region is a resource in which interest has increased in the 1970s, especially in Brazil. In places the forest has been cleared for new agricultural settlements, cattle ranches, plantations of special fast growing trees or simply for the plant material itself to serve as a raw material.

Although new discoveries of coal are being made Latin America has only a very small share of known recoverable reserves, about 0.5 percent of the world total. These are mostly in Brazil, Mexico and Colombia. Peru, Chile and Venezuela have small reserves of coal and also some lignite. Latin America's share of world oil and natural gas reserves is larger: about ten percent of world published proved reserves of oil in 1980 and six percent of the natural gas. In 1978 Mexico was estimated to have about 50 percent of Latin America's oil reserves, Venezuela 30 percent and Argentina about six percent. Other countries with modest reserves include Ecuador, Brazil, Colombia, Peru, Chile and Trinidad and Tobago. Venezuela, Mexico and Argentina also together have most of the natural gas. Since large areas of Latin America remain to be explored for fossil fuels and assuming that both private and state companies understate reserves, actual reserves are probably greater than those given above. Mexico, for example, is thought to have several times its official reserves of oil and Venezuela has very large reserves of oil in the tar sands of the lower Orinoco. The hydro-electric potential of Latin America is large and Brazil already has some very large power stations in the basin of the Parana River.

Latin America covers about 16 percent of the world's land area and includes areas of both old and very recent folding. The highlands of Brazil and Guyana (iron ore, bauxite, manganese ore), the Andes of South America and the mountains of Mexico (copper, lead, zinc, silver ores) have some of the largest deposits of non-fuel minerals in the world. The following data show the percentage of the world total of reserves of selected minerals attributed to Latin American countries. Most metallic ores are found in commercial quantities in the region, but non-metallic minerals are less abundant:

Chile	copper 20% molybdenum 31%
Peru	copper 6% lead 3% zinc 5%
	silver 10% molybdenum 3%
Mexico	copper 4% lead 4% zinc 3%
	silver 14% mercury 5%
	sulphur 11% antimony 5%
Bolivia	tin 10% antimony 8%
Brazil	iron ore 18% tin 6% zinc 2%
	manganese 2%
Venezuela	iron ore 2%
Guyana	bauxite 4%
Surinam	bauxite 2%
Jamaica	bauxite 8%
Cuba	nickel 6%

Production

In the 1950s more than half of the economically active population of Latin America was in the agricultural sector. By 1980 the proportion was only a third. In spite of this relative decline, the labour force has not changed greatly because total economically active population has grown. Since the cultivated area of Latin America has continued to expand and yields have increased total production has also grown. Relief, climatic and soil conditions vary greatly over Latin America and virtually every major crop and form of livestock can be found somewhere in the region.

During the colonial period, plantations specialized in the production of sugar, cotton, coffee and other tropical crops, especially in the Caribbean islands and the coastal regions of Brazil. After independence Latin American countries continued to produce for export to the industrial countries of Europe and the USA. Exports still include coffee from Brazil, Colombia and Central America and meat and grain from Argentina and Uruguay. With the increasing cultivation of sugar beet in Europe, sugar cane became an unreliable product to export but it is still the mainstay of the export sector in some Caribbean countries. Cocoa beans, tropical fruits and cotton are additional agricultural exports of Latin America.

Maize, cassava and more locally wheat and potatoes are among crops grown mainly for internal needs. With increasing population Latin America has gradually been using more of its output of many plant and animal products formerly exported. Setbacks such as the frosts that have damaged Brazilian coffee trees and the use of synthetic instead of plant fibres have contributed to the relative decline. Some Latin American countries are now net importers of food, and trade in agricultural products between Latin American countries themselves has increased.

There is much scope for increasing agricultural production in Latin America. Very high yields are obtained, for example, in the irrigated lands of northern Mexico, where high yielding wheat has been developed. In the state of São Paulo soya beans have been successfully introduced and yields of most other crops are also high. The irrigated lands of the coast of Peru have very high yields of sugar cane and produce excellent quality cotton. In contrast agricultural yields are very low in southern Mexico, northeast Brazil and the Andes of Peru. In Argentina the inherently fertile soil of the pampas produces good crops of maize, wheat and fodder for livestock but there is still room for improvement.

Some decades ago mineral production in Latin America was almost entirely destined for export, a tradition carried through from colonial times into the 19th century as copper, guano, nitrates, lead and zinc were added to silver, gold and precious stones. When large oil reserves were found in Mexico and later in Venezuela these also were mainly exported to industrial countries.

Although Latin America continues to be one of the main sources of oil, metallic minerals and tropical agricultural products, most countries have tried to industrialize and to use more primary products at home. In the 19th century the processing of raw materials was already widespread in Latin America and, stimulated by the two World Wars, when manufactures became difficult to obtain, import substitution was encouraged. By the 1950s, when Latin American countries began to consider collaboration, Mexico, Brazil and Argentina in particular already had some industrial capacity. In the Alliance for Progress sponsored by President Kennedy in the early 1960s plans were outlined for the massive development of heavy industry and the growth of engineering. The progress hoped for at the time has not been achieved, but Brazil is emerging as an industrial power of considerable importance, serving the needs not only of its own large home market but also of markets elsewhere in Latin America and in Africa. With abundant oil, Mexico is also becoming industrialized.

Latin America in world affairs

Latin American economic and eventual political integration seems a distant though not impossible prospect. Spanish is the language of all the main countries except Brazil (Portuguese), Haiti (French) and Guyana and some Caribbean islands (English). To some extent all the countries feel a cultural connection with Iberia and France. The official religion is Roman Catholic. Air travel and television have made it possible for the wealthier, more influential Latin Americans to visit and learn about each other's countries and the migration of professional people is common. More locally, workers cross interstate boundaries in many parts of Latin America, more often causing friction than greater contact. Colombians move into Venezuela, Chileans into Argentina and migrants and refugees move between the small Central American countries.

A number of other factors are against unification in Latin America. Latin American countries tend to produce similar goods. Trade has grown between some pairs of countries, but most countries have stronger trade links with partners outside the region. Domination by the larger countries could be seen as a possible consequence of integration by smaller, weaker countries such as Paraguay, Uruguay, Bolivia, Guatemala. Political independence was won more than 150 years ago but is still valued. A number of interstate boundaries are not finally defined or accepted (eg Bolivia with Chile, Argentina with Chile, Venezuela with Guyana). It must also be

appreciated that enormous distances separate Mexico from Chile or the Caribbean from Argentina.

Individual countries in Latin America have strong connections with particular countries or blocs outside the region. Cuba is now economically and politically tied to the Soviet bloc. Venezuela's economic bargaining position is strengthened by membership of OPEC. Millions of Mexicans live and work in the USA, legally or otherwise, and Puerto Rico is virtually a state of the USA though technically termed a Commonwealth.

What prospects lie ahead for Latin America? Demographic, economic and strategic prospects may be distinguished. Column 3 in *Table 12.1* shows that the rate of natural increase is still high in many countries of Latin America though the 1970s showed a downward trend. The attitude of the Roman Catholic church to family planning is only one reason

Table 12.1 *Data for the largest countries of Latin America in population*

Country	Area 1	Popn 2	Popn change 3	Popn/ arable 4	Crops/ total area 5	Forest/ total area 6	Non- agric 1960 7	Non- agric 1979 8	Energy cons 9	Infant mort 10	pupils/ teacher 11	GNP 12
Mexico	1973	69.3	2.5	71	12	36	45	63	1380	70	58	530
Guatemala	109	7.5	3.1	282	17	53	33	44	260	69	97	1020
Cuba	115	9.8	0.9	185	28	18	61	76	1170	19	28	1410
Haiti	28	6.0	2.6	425	32	7	17	33	60	130	186	260
Dominican R	49	5.6	2.8	205	25	23	33	43	460	96	85	990
Venezuela	912	15.5	3.0	70	6	54	65	81	2990	45	59	3130
Colombia	1139	27.8	2.1	120	5	68	48	72	700	77	62	1010
Ecuador	284	8.2	3.1	159	9	52	42	55	510	70	54	1050
Peru	1285	18.1	2.7	59	3	57	47	62	650	92	54	730
Bolivia	1099	5.5	2.5	18	3	51	39	49	370	168	45	550
Chile	757	11.2	1.5	63	8	27	70	81	1000	38	42	1690
Argentina	2767	28.2	1.6	16	13	22	80	87	1870	41	27	2280
Uruguay	176	2.9	1.0	19	11	3	79	88	1050	48	32	2090
Paraguay	407	3.3	2.6	20	3	50	44	51	650	58	43	1060
Brazil	8512	121.4	2.4	58	5	60	48	61	790	84	43	1690

Definition of variables see Table 12.2

Table 12.2 *Sources and definition of variables in Table 12.1*

Sources

2, 3, 4, 10, 12 1981 World Population Data Sheet, PRB inc
5, 6, 7, 8, 9 numbers of *FAOPY* and *World Statistics in Brief*
11 World's Children Data Sheet, PRB Inc 1979

Definitions

1 Total area in thousands of square kilometres
2 Population in millions, 1981
3 Percentage rate of annual natural change of population around 1980
4 Persons per square kilometre of arable plus pasture land, 1978

5 Percentage of total area under arable or permanent crops, 1978
6 Percentage of total area defined as forest 1978
7, 8 Economically active population *not* engaged in agriculture as a percentage of total economically active population in 1960 and 1979
9 Consumption of energy in millions of tonnes of coal equivalent per inhabitant in 1979
10 Annual number of deaths to infants under one year of age per 1000 live births in late 1970s
11 Number of school-age children per teacher, 1975
12 Gross national product in US dollars per inhabitant, 1979

for the high fertility rate in much of Latin America. The lack of information and facilities among the rural and poorer urban sectors of the population should also be noted. The 1980 population of about 360 million could rise to some 560 million by the year 2000. Given the tendency of Latin Americans to move from crowded rural areas into towns, rather than out into pioneer areas, the urban nightmare ahead for Latin American planners is daunting. In and around Mexico City there could be 25 million of Mexico's 100 million population. The city of São Paulo in Brazil could have 20 million.

The increase in energy consumption and industrial development have been impressive in Latin America in the 1970s (*see Figure 12.3*). Frustration with the way fuel and raw materials have been exported to industrial countries with little value added has been an incentive to develop industry. Many foreign enterprises have been nationalized and the control of most natural resources is firmly in the hands of host countries.

Latin American governments would do well to put greater effort into improving the agricultural sector, though there are dangers. The search for and exploitation of new resources in the tropical forest region could lead to the destruction of a system of natural vegetation evolved over perhaps tens of millions of years, and with it many simple American Indian societies and plant and animal species. On previous experience the Indians could be wiped out,

put into artificial reservations or assimilated into the rest of the population.

Strategically the most sensitive part of Latin America is the Caribbean area and Central America. The many small countries here are particularly susceptible to pressure by the great powers. The USA has been involved in the internal affairs of several countries during the 20th century. Having a piece in Cuba in the world chess game is evidently worth a heavy financial subsidy from the USSR. While the USA could easily occupy Cubs militarily it is hard to know what it would do with it. If 'free' elections were held, the Communist Party might get voted back into power. Panama is another sensitive area in view of the importance, especially to the USA, of the Panama Canal. The spread of Soviet backed socialism and communism into Central America is a nightmare of successive US administrations and by the late 1970s Nicaragua had become highly suspect. Its neighbours include Costa Rica, which has no credible military forces, and the notoriously unstable Republic of El Salvador.

The Argentinian occupation of the Falkland Islands in 1982 was a reminder that many territorial disputes remain in the Americas from colonial days. Many other areas in Latin America have been occupied since 1833, the date Britain claimed the islands, so on similar grounds Mexico might set about occupying Texas or California and Peru and Bolivia could retake northern Chile.

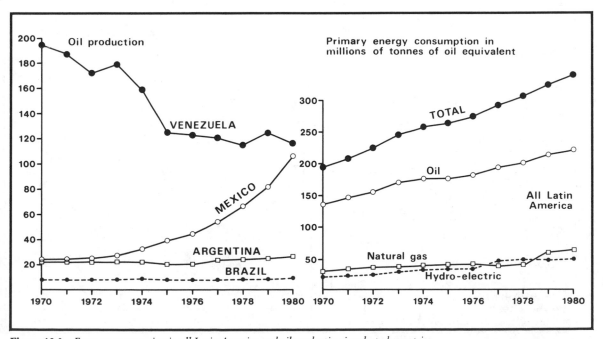

Figure 12.3   *Energy consumption in all Latin America and oil production in selected countries*

12.2 Africa south of the Sahara

Introduction

Africa south of the Sahara had about 390 million inhabitants in 1981, nearly nine percent of the total population of the world and rather more than Latin America had. It is nearly 25 million square kilometres in area, rather less than one-fifth of the earth's land area. There are almost 50 countries, counting small island units in the Atlantic and Indian oceans.

Most of Africa has been much less affected by European colonization than Latin America or South Asia. Before the 1880s European influence was largely restricted to the coasts where forts and trading posts supplied the Portuguese, Dutch, British and French naval and commercial vessels on the route to Southeast Asia. Trade in slaves and in exotic products

such as ivory flourished. Even after 1880 when much of Africa was divided among various European powers (*see Figure 2.5*) European influence was superficial and sporadic in many areas.

After about 1880 most of Africa was arbitrarily divided into a number of colonial possessions of European powers. Portugal, the first European power in Africa, consolidated its three main areas of influence, Angola, Mozambique and Guinea. The British aimed for continuous influence from Egypt to southern Africa. In the sharing out of the continent the European powers in the last century paid little attention either to the nebulous cultural areas and old kingdoms of Africa or to the more immediately meaningful tribal areas. Although there was some attempt to attract Europeans to settle in such climatically attractive areas as the highlands of eastern

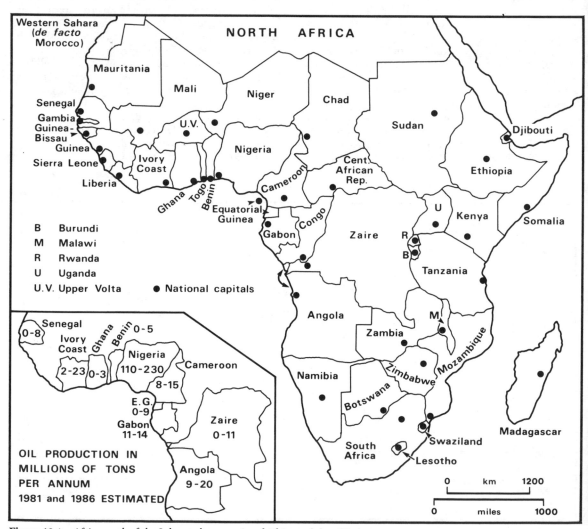

Figure 12.4   *Africa south of the Sahara: the countries of Africa and the oil industry*

Africa and the plateaux of Rhodesia, the numbers settling were small except in South Africa.

From the 1880s the colonies of Africa were organized to provide tropical agricultural products such as cocoa beans and palm oil (West Africa), coffee, cotton, sisal and various other crops, and also minerals, for the industrial countries of Europe. Rail systems linked areas producing for export to ports.

By the time that Italy occupied Ethiopia in the mid-1930s virtually all Africa was under European control or protection. Liberia was an exception. South Africa had, however, by then become independent within the British Empire. A few countries became independent in 1945 as a result of the Second World War. Around 1960 France, the UK and Belgium gave independence to most of their colonies and only Portugal attempted to retain its possessions. By 1980 there were no European colonies on the continent of Africa, the Southern Rhodesia dispute between the UK and the white minority being settled in 1979. The future of Spanish (Western) Sahara remains in dispute but in practice the whole area is controlled by Morocco. *Figure 12.4* shows the location of the countries of Africa south of the Sahara.

Natural resources

In Africa south of the Sahara about 145 million hectares is under arable or permanent crops, only six percent of the total area. About 25 percent is defined as forest or woodland and the same proportion as permanent pasture. In some countries the actual area under field crops is considerably less than that defined as arable because areas may only be cultivated periodically. Soils are generally low in plant nutrients in tropical Africa except locally where alluvial or volcanic material is found, as in the lava plateau of Ethiopia and around the volcanoes of East Africa. Although there is an extensive area of tropical rain forest in central Africa much of the area defined as forest is of limited value commercially, is increasingly being cut for firewood, yet is a protection in many places against soil erosion.

In spite of the gloomy description of African bioclimatic resources the area under crops in many African countries has been extended in the last two decades. Temperature and slope constraints to cultivation are only a local inconvenience. To the north and south of the tropical forests of central Africa, lack and unreliability of rain is the main problem (*see Figure 12.5*). Droughts cause devastating reductions in yields. In the 1970s a wide belt of semi-arid land between the Sahara desert and the humid tropics received virtually no rain at all in some years. Insect pests, such as the tsetse fly, attack livestock in many areas and the locust, before it was controlled, reduced the fodder supply.

Africa south of the Sahara is not distinguished by extensive reserves of coal and the geological conditions over most of the continent make large discoveries unlikely. South Africa is exceptional in having two to three percent of the world's known recoverable reserves of coal, enough to last over 100 years at production rates in the late 1970s. Neighbouring Zimbabwe and Botswana have most of the rest of Africa's reserves of coal. Until the 1960s Africa south of the Sahara was not considered to have major oil reserves, but discoveries in Nigeria now give that country more than two percent of the world's oil. Moderate reserves of oil have been found in Gabon, Cameroon, Angola, several other West African states and Zaire (*see Figure 12.4*, inset map). Nigeria also has more than two percent of the world's natural gas reserves. The hydro-electric potential in certain localities in Africa is considerable, but not much has yet been used.

Southern Africa is one of the world's major regions of non-fuel minerals. Other parts of Africa have not been so fully explored and could in the future prove to have large reserves. The following data show the percentage of total world reserves of selected minerals attributed to the countries listed. South Africa is extremely well placed with regard to many minerals vital to the industries of developed countries and it even shares a virtual monopoly of certain minerals with the USSR:

Guinea	bauxite 31%
Ghana	industrial diamonds 4%
Cameroon	bauxite 3%
Gabon	manganese 5%
Zaire	copper 5% industrial diamonds 74% cobalt 30%
Zambia	copper 7% cobalt 8%
Zimbabwe	chrome 30%
Botswana	industrial diamonds 7%
South Africa	iron 1% lead 4% gold 49% nickel 3% manganese 43% industrial diamonds 7% phosphate 11% chrome 30% asbestos 6% antimony 7% platinum 34%

Production

Data for agricultural production in much of Africa are only approximate, but one feature comes out clearly: in many areas the crops grown for export, whether in large plantations or on small farms, are well cultivated, achieve good yields and are of good quality. In contrast the cultivation of crops for local food consumption is less well organized and yields tend to be low. West Africa produces cocoa beans, groundnuts and palm oil for export while eastern Africa specializes in coffee production and South Africa in fruits. Other agricultural exports of tropical

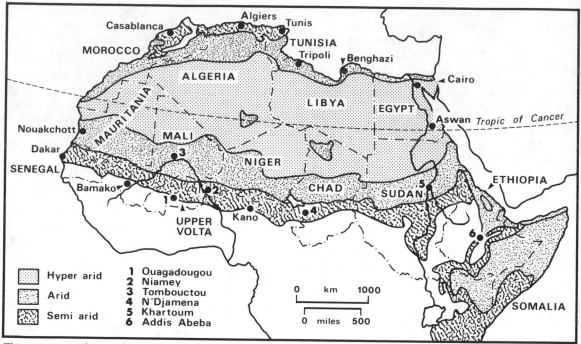

Figure 12.5 *Arid areas of northern and northeastern Africa*

Africa include cotton, sisal, tobacco and meat. Food crops widely grown include the cereals, millets, sorghum and maize and various roots and tubers. Virtually every aspect of agriculture in tropical Africa could be improved. New crops could be tried in many areas and existing strains improved. Rice could be more widely grown in appropriate areas. Chemical fertilizers are hardly used at all.

In 1980 the whole continent of Africa consumed about 230 million tonnes of coal equivalent of energy, but of this about 120 million was consumed in South Africa alone and about 60 million in the five northern countries. The rest of Africa, with 360 million people therefore consumed only about 50 million tonnes of coal equivalent, about as much as Sweden (with 8 million people), and less than 200 kilograms per inhabitant. The data in column *8* of *Table 12.3* show great contrasts in the level of consumption of energy per inhabitant even excluding South Africa (3180 kilograms). In southern Africa the comparatively high level in Zambia and Zimbabwe is to some extent due to mining and refining activities. In contrast the average level in Nigeria is dismally low and around 1980 the country was using at home only about one-tenth of the oil it was producing. There is little need for domestic heating in tropical Africa, though cooking does require some fuel, often in the form of firewood. When transportation and the processing of materials for export are allowed for, very little energy is available in rural Africa and the

widely found level of 10–30 kilograms per inhabitant is found elsewhere in the world only in the poorest countries of Asia.

Lack of local fuel in most countries and the increasing price of oil have been one reason for failure to develop heavy industry. In Africa south of the Sahara only South Africa is a major producer of steel and Zimbabwe is the only other producer. Steel consumption is very low in many countries. *Table 12.4* shows a recent decline in consumption per inhabitant in some. More than 20 countries produce cement, a product that can be made in comparatively small plants without being at a disadvantage. In the late 1970s Africa south of the Sahara, excluding South Africa and Zimbabwe, produced seven to eight million tonnes of cement, about one percent of the world total. Light industries such as the manufacture of cotton textiles, clothing and other consumer goods are factory produced in some African countries.

Problems and prospects

Most countries of Africa are very small. As a result they are very weak militarily and their home markets are minute. Together all the countries of Africa south of the Sahara have a total gross national product roughly equal only to that of Belgium or to the turnover of one of the largest transnational corporations. There is little opportunity to benefit from economies of scale.

Table 12.3 *Selected countries of Africa south of the Sahara*

Country	A	Area 1	Popn 2	Popn change 3	Popn/ arable 4	Total area/ crop 5	Non- agric 1960 6	Non- agric 1979 7	Energy cons 8	Infant mort 9	Pupils/ teacher 10	GNP 11
West												
Senegal	F	196	5.8	2.6	62	12	16	25	180	160	196	430
Guinea	F	246	5.1	2.5	72	17	12	19	90	220	112	270
Ivory Coast	F	322	8.5	3.1	72	12	11	20	360	138	91	1060
Ghana	U	239	12.0	3.1	90	11	38	48	170	115	63	400
Nigeria	U	924	79.7	3.2	178	26	29	46	110	157	140	670
Mali	F	1240	6.8	2.8	21	2	6	13	30	210	203	140
Upper Volta	F	274	7.1	2.6	37	21	8	18	30	182	534	180
Niger	F	1267	5.7	2.9	46	2	5	11	40	200	420	270
Eastern												
Sudan	U	2506	19.6	3.1	62	3	14	23	170	141	155	370
Ethiopia	I	1222	33.5	2.5	43	11	12	20	20	178	342	130
Uganda	U	236	14.1	3.0	132	24	11	19	50	120	133	290
Kenya	U	583	16.5	3.9	273	4	14	22	140	83	52	380
Rwanda	B	26	5.3	3.0	361	36	4	10	20	127	168	210
Burundi	B	28	4.2	2.7	245	46	10	16	10	140	265	180
Tanzania	U	945	19.2	3.0	38	5	11	18	70	125	182	270
Zambia	U	753	6.0	3.2	17	7	21	33	470	144	89	510
Malawi	U	118	6.2	3.2	150	19	7	16	50	142	163	200
Zimbabwe	U	391	7.6	3.4	104	6	31	41	580	129	98	470
Mozambique	P	783	10.7	2.6	23	4	22	35	130	148	272	250
Madagascar	F	587	8.8	2.6	24	16	7	16	80	102	121	290
Central and southern												
Chad	F	1284	4.6	2.3	10	2	5	16	20	190	468	110
Cameroon	F	475	8.7	2.3	55	16	12	19	120	157	90	560
Zaire	B	2345	30.1	2.8	97	3	16	25	70	171	88	260
Angola	P	1247	6.7	2.4	22	15	30	42	190	192	111	440
South Africa	U	1221	29.0	2.4	30	12	68	71	3180	97	68	1720

Sources and definitions of variables: *see* Tabel 12.2 (excluding 6 in that table)

A = Colonial power at time of independence: B Belgium, F France, I Italy, P Portugal, U UK.

Table 12.4 *Steel consumption in kilograms per inhabitant in selected years*

Country	1970	1975	1979	Country	1970	1975	1979
Ivory Coast	28	23	30	Ethiopia	3	1	1
Ghana	7	9	6	Tanzania	6	5	8
Nigeria	12	22	19	Mozambique	8	1	5
Kenya	17	9	18	Uganda	3	1	0
Zaire	9	4	2	Sierra Leone	4	4	3
Zambia	14	14	6	Guinea	15	4	5
South Africa	209	263	169	Gabon	66	208	74

In Africa south of the Sahara only South Africa can be regarded as more than superficially industrialized. The desire to be self-sufficient as far as possible has guided South African policy and the iron and steel industry, engineering, chemicals and light industry are all represented. Elsewhere the hoped for industrialization following independence in many countries around 1960 has not materialized.

The failure of African countries to collaborate economically, as groups of countries have been doing elsewhere in the world, is accompanied in the continent by political instability. As there are so many countries the number of different pairs of countries sharing a boundary is very large and the potential for local disputes great. Existing international boundaries are not everywhere accepted either by the governments of states or by local people whose tribal territories may have been divided. Fourteen countries have no coast, compared with only two in Latin America, and are dependent on good relations with neighbours to ensure outlets to the ocean. Distances between countries are on average much greater than between European countries yet interstate links are very poor or non-existent. The transportation network of each country tends to deteriorate in quality towards the interior.

The existing contrasts between different countries and regions of Africa may continue to exist and even increase. The expected growth of oil production in many West African countries brings problems as well as benefits. The oilfield areas, ports and capital cities attract excessive numbers of people from the countryside. Agricultural production tends to be neglected and food is even imported. In 1980 over 90 percent of Nigeria's foreign exchange came from the exports of oil. As occurred in Venezuela when oil exports grew, money from oil that could have been used to develop other sectors including agriculture and industry tends to be spent on prestige projects and the importation of consumer goods.

Each African country has its own individual problems. Two 'disappointments' will be noted here. After independence, Tanzania was regarded with interest as a very poor and backward country in which a new socialist experiment would produce a reasonable standard of living for a predominantly rural, agricultural population. Participation of the rural masses was to be achieved through the creation of special villages and the dissemination of information and appropriate means of production. After 20 years it was admitted (December 1981) that the economy was in crisis. Oil price increases had cut the amount of oil that could be imported. The world recession and a lack of demand for Tanzanian products such as coffee, cotton and sisal, one drought after another, the military assistance given to neighbouring Uganda and inefficient government were all blamed.

In South Africa the problem is different. In spite of many forecasts to the contrary the white minority has managed to achieve an expanding economy and at the same time to keep much of the non-white population from enjoying many of the benefits of development. *Figure 3.3* shows areas in South Africa designated as homelands for the various Bantu peoples of the country. Eventually most non-whites would live in such territories, some of them sovereign states in the eyes of South Africa, and would be permitted to enter white South Africa only as guest workers. Most homelands have little in the way of natural resources.

Like tropical Southeast Asia (*see* chapter 11.2), tropical Africa is a negative or vacuum area which has had little positive influence on the rest of the world except in the slaves it supplied to the Americas and its various tropical primary products. Firmly held in the grip of European colonial administrators for some decades, Africa suddenly became independent and vulnerable. The following influences now seem to threaten Africa:

(1) Disputes within Africa involving both fragmentation of existing countries (eg Biafra in Nigeria in the 1960s, Eritrea from Ethiopia), Shaba (Katanga) in Zaire, and disputed areas (eg the Ogaden area of Ethiopia claimed by Somalia).

(2) Continuing influence of former colonial powers, mainly now via commercial and financial links with the European Economic Community. European companies play a major part in the development of oil in West Africa. The French government is interested in Gabon and Niger because these have reserves of uranium ore.

(3) South Africa is inevitably involved in the future of southern Africa. Workers from Lesotho, Swaziland and even Malawi find jobs in South Africa. Namibia (South West Africa) is *de facto* controlled by South Africa although its mandate from the League of Nations to administer the country was terminated by the United Nations General Assembly in 1966. Politically the Marxist governments of Angola (with a Cuban presence) and Mozambique are seen as a grave threat to South Africa in an area where it is prepared to take military action if necessary. Economically the whole of southern Africa could be an outlet for South African manufactures.

(4) Northern Africa has long influenced tropical Africa. The Muslim religion spread far south from the Red Sea and Mediterranean coastlands. Libya, small in population but rich from oil, has become involved in the affairs of Chad and Niger to its south.

(5) As the European colonial powers moved out of Africa the Soviet Union attempted to move in. In the 1960s its influence was felt in Ghana (ex-British), Guinea (ex-French), and the Congo,

now Zaire (ex-Belgian). Perhaps the Soviet Union did not have the right things to offer in order to create socialist allies at that time. When Portugal finally withdrew from Mozambique and Angola it was better prepared and with the help of Cuban forces established its influence in Angola. For reasons that were more opportune than consistent the USSR also switched support from Somalia to Ethiopia. There is very little in Marx to support the view that any African country was suitable for imminent communism. With many commitments already outside its own boundaries, the USSR can hardly afford to offer large amounts of aid to developing countries.

Some indication as to the African countries with which the Soviet Union is most involved is given by the value of Soviet foreign trade (imports plus exports in millions of roubles in 1979). The countries of northern Africa are included for comparison:

Northern Africa		West Africa		Other	
Libya	438	Ghana	152	Angola	64
Egypt	325	Ivory		Ethiopia	63
Morocco	126	Coast	75	Sudan	29
Algeria	116	Guinea	50	Mozambique	21
		Nigeria	50		
		Liberia	21		

Trade with Africa south of the Sahara accounted for only 0.7 percent of all Soviet foreign trade and there was virtually none with southern Africa at all.

(6) Ideologically China, with Mao's emphasis on rural development, might have been a more suitable ally for African countries and a model for their development but its modest influence in some countries (eg Sudan, Tanzania) in the 1970s had almost disappeared by the early 1980s.

(7) The United States has been much less of an influence in Africa south of the Sahara than in Latin America. Even so it has found itself increasingly involved both economically through its need to import oil, and militarily to counter Soviet influence.

Africa south of the Sahara has some of the poorest regions of the world. Famines, deadly tribal disputes and battles fought on behalf of powers outside the continent leave it very unstable in the 1980s. Total population could increase from about 390 million in 1980 to about 680 million by the year 2000 if the present annual rate of growth of three percent continues. Southern Africa is a vital reserve of key minerals for the Western industrial countries and the prospect of a new source of oil in West Africa, if modest compared with that in the Middle East, is welcome, especially for Europe. None of the industrial countries can afford to lose influence or sympathy in Africa.

In the late 1940s the British government itself sponsored the groundnuts scheme in Tanzania. Even to a socialist government of Britain, Africa was still seen as a part of the world that should be developed to supply raw materials to Europe. In an article in Unilever's quarterly journal *Progress* for Spring 1954 Lord Milverton wrote:

'. . . the development of modern Africa owes everything 'to business' and its manifold operations. The European has supplied the enterprise, capital, skill, brains and experience without which the miraculous transformation of the past fifty years could never have taken place. Everywhere it has been the activities of 'business' that stimulated even the control and conscience of Government.'

A book by Rodney (1974) in *How Europe Under-developed Africa* opens with a quotation from Che Guevara (1964) which could hardly be more different from that quoted above by Lord Milverton:

'. . . a large proportion of the so-called under-developed countries are in total stagnation . . . in some of them the rate of economic growth is lower than that of population increase . . .

These characteristics are not fortuitous; they correspond strictly to the nature of the Capitalist system in full expansion, which transfers to the dependent countries the most abusive and bare-faced forms of exploitation.'

12.3 North Africa and Southwest Asia

Introduction

Although the region of North Africa and Southwest Asia extends from the Atlantic in Morocco to the Indian Ocean in Oman it has some characteristic features throughout. Most of the region is desert, semi-desert or at best subhumid. About 60 percent of the oil reserves and 35 percent of the natural gas reserves of the world are located there. Culturally considerable homogeneity is given to the region by the almost universal presence of the Muslim religion, while the Arab League (*see* chapter 8) includes most of the countries.

For the purposes of this section four groups of countries may be distinguished within the region, with the 1981 population in millions:

(1) North Africa (Morocco, Algeria, Tunisia, Libya, Egypt) 94 million (*see Figure 12.6*)
(2) Eastern Mediterranean (Cyprus, Syria, Israel, Lebanon, Jordan, Gaza) 21 million
(3) Northern group (Turkey, Iran, Afghanistan) 102 million
(4) Arabian peninsula (Iraq, Saudi Arabia, Yemen and smaller units) 35 million

The whole region had about 252 million inhabitants, 5.6 percent of the total population of the world.

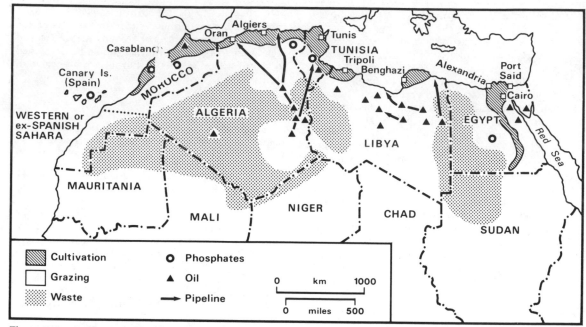

Figure 12.6 *Economic aspects of North Africa*

The present countries of North Africa and Southwest Asia are mostly of recent origin though several are formed round national groups that have been in existence for a very long time. In the Ancient World much of the region was in or on the margins of the Roman Empire and later Arabia, Persia and Turkey have influenced various parts.

In the 19th century France colonized Northwest Africa and many French settlers went into Algeria. Britain and France were interested in the construction and protection of the Suez Canal and had a strategic interest in the whole Mediterranean and Red Sea areas. Italy occupied the unpromising, largely desert area now Libya, while Germany, through the Balkans and Turkey, sought economic influence in Southwest Asia. From the north, Russian conquest of what is now Soviet Middle Asia pushed the Russian Empire to the borders of Persia (now Iran) and Afghanistan where Britain, based in India, was also an influence.

After the First World War modern Turkey was reduced largely to areas inhabited by Turkish speaking people and Britain and France took over or 'protected' areas in Southwest Asia. Yet another outside force in the 20th century were the Jews, resettling their 'homeland' of Palestine and forming the state of Israel which since its creation in 1948 has been strongly supported economically and militarily by the USA.

The boundaries of the present countries of Southwest Asia (*see Figure 12.7*) do not everywhere coincide with clearly defined national groups. Thus the Kurdish people, numbering several million, are divided between Turkey, Syria, Iran and Iraq (*see Figure 3.4*) and their aspirations to form a separate sovereign state have led to numerous conflicts. The Palestinians have been either absorbed into Israel or dispersed into neighbouring countries.

The great extent of the oil and natural gas reserves of Southwest Asia and North Africa was not fully appreciated until after the Second World War. The region now has exceptional significance in a world in which development has for three decades depended so much on the use of cheaply extracted and easily transported fossil fuels.

Natural resources and production

Only a few percent of the land area of North Africa and South Asia is cultivated (column 5 in *Table 12.5*). In some favoured irrigated areas, notably in the Nile valley and delta of Egypt, yields are high. Elsewhere much of the land under crops is in areas where rainfall is low and erratic in occurrence, and evaporation great. The data in column 4 of *Table 12.5* show that the number of people per square kilometre of arable land varies greatly in the region. The economically active population in agriculture (columns 6 and 7) has declined sharply as a share of total economically active population in many countries since 1960. In Afghanistan and the Yemen Republic agriculture is still the principal activity but at the

217

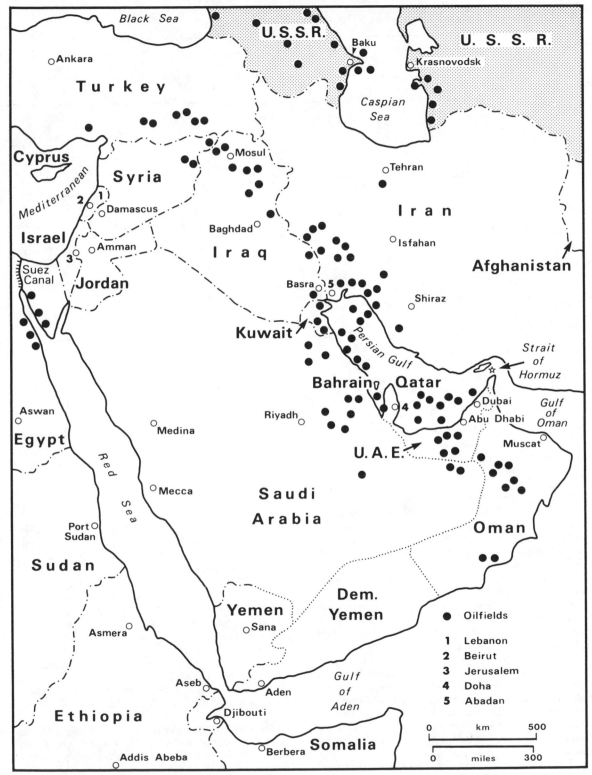

Figure 12.7 *The countries and oilfields of Southwest Asia*

Table 12.5 *Data for selected countries of North Africa and Southwest Asia*

Country	Area 1	Popn 2	Popn change 3	Popn/ arable 4	% arable 5	Non- agric 1960 6	Non- agric 1979 7	Energy cons 8	Infant mort 9	Pupils/ teacher 10	GNP 11
North Africa											
Morocco	447	21.8	3.0	107	18	36	48	290	133	115	740
Algeria	2382	19.3	3.2	44	3	33	49	690	127	67	1580
Tunisia	164	6.6	2.5	87	27	43	59	540	123	72	1120
Libya	1760	3.1	3.5	33	1	44	83	1890	130	24	8210
Egypt	1001	43.5	3.0	1533	3	41	49	270	90	80	460
Southwest Asia											
Turkey	781	46.2	2.2	83	33	22	44	790	125	68	1330
Syria	185	9.3	3.4	66	30	46	52	970	81	49	1070
Lebanon	10	3.2	2.6	904	33	47	89	940	45	24	370[a]
Israel	21	3.9	1.8	321	20	86	93	2360	16	18	4170
Jordan	98	3.3	3.3	223	14	56	73	540	97	51	1180
Iraq	435	13.6	3.4	145	13	47	59	630	92	49	2410
Saudi Arabia	2150	10.4	3.0	12	1	28	39	1310	118	66	7370
Yemen R	195	5.4	2.3	63	8	17	25	50	160	303	420
Iran	1648	39.8	3.0	66	10	46	61	1810	112	56	2340[b]
Afghanistan	647	16.4	2.7	28	12	15	22	50	185	258	170

Sources and definition of variables: *see* Table 12.2

[a] low on account of conflicts
[b] reduced after 1978–79 revolution and war

other extreme, in Libya, the Lebanon and Israel it is of relatively minor importance.

In addition to its oil and natural gas reserves the region of North Africa and Southwest Asia has some non-fuel minerals. Morocco is estimated to have about two-thirds of the world's phosphate rock and Tunisia also has reserves, while a few percent of the world's reserves of potash are in Jordan. Turkey has tungsten deposits and Algeria has mercury. Iron ore (North Africa) and coal and lignite (Turkey) of local importance also occur.

Southwest Asia has about 55 percent of the world's oil reserves and North Africa about six percent. *Table 12.6* shows the reserves in major regions of the world for 1981. In *Table 12.6*, reserves are given by country according to UN 1978 data, against 1981 production. It is therefore possible to calculate the approximate life of existing reserves at 1981 rates of production. New reserves will no doubt be found and on recent experience yearly production rates will also fluctuate greatly. For the whole of Southwest Asia (the 'Middle East'), the reserves of roughly 50 000 million tonnes of oil were being used on average at the rate of about 1000 million tonnes a year in the 1970s.

The following features of the fossil fuel situation in North Africa and Southwest Asia should be noted:

(1) Some countries have virtually no oil or natural gas at all.
(2) Even among the largest producers the quantity of reserves *per inhabitant* varies greatly. There are about 7000 tonnes of oil in the ground for each inhabitant of Kuwait. Each Iranian has only about 160 tonnes of oil to his name.
(3) At production rates of 1980 there are great differences in the life of reserves, ranging from about 150 years in Kuwait to about 30 in Saudi Arabia and Algeria and ten in Egypt.
(4) In 1970 the Middle East consumed only about one-tenth of all energy it produced and in 1980 only about one-eighth. Production and consumption of energy per inhabitant vary greatly, however, between countries as does gross national product per inhabitant (*see Table 12.5* columns *8* and *11*). Compare 1580 dollars per inhabitant in Algeria with 8210 in Libya and only 460 in Egypt. With 17 270, Kuwait is the 'richest' country in the world (USA 10 820).

While the oil industry overshadows other activities in some of the countries of North Africa and Southwest Asia its direct impact on employment is limited and local. The oilfields of the region cover a very small area and some of them are in virtually

Table 12.6 *Oil and gas reserves and production in selected regions*

'Published proved' reserves at end of 1981 by major world regions

Region	Oil		Gas	
	Thousand million tonnes	% share of world total	Trillion cubic metres	% share of world total
Mainly developing countries				
Middle East	49.3	53.5	21.6	26.2
All Africa	7.5	8.3	6.0	7.3
China	2.7	2.9	0.7	0.8
Rest of Eastern hemisphere[a]	2.6	2.8	3.6	4.4
Latin America	11.9	12.5	5.0	6.1
Mainly developed countries				
North America	5.7	6.6	8.2	9.9
USSR and East Europe	9.0	9.7	33.0	40.1
West Europe	3.4	3.7	4.3	5.2
Developing total	74.0	80.0	36.9	44.8
Developed total	18.1	20.0	45.5	55.2
World total	92.1	100.0	82.4	100.0

Oil reserves 1978 and production 1981 in North Africa and Southwest Asia in millions of tonnes

Country	Reserves	Annual prodn	Life (years)	Country	Reserves	Annual prodn	Life (years)
Algeria	1309	46	28	Iraq	4 702	44	107
Tunisia	304	5	61	Kuwait	10 184	48	212
Libya	3719	54	69	Saudi Arabia	15 911	492	32
Egypt	292	32	9	Bahrain	37	3	12
Syria	237	10	24	Qatar	515	20	26
Turkey	30	3	10	UAE[b]	4 318	73	59
Iran	6148	66	93	Oman	447	16	28

Sources: BP Statistical Review of World Energy 1981; UNSYB 79/80

[a] includes Indonesia and Australia

[b] Abu Dhabi, Dubai, Sharjah

uninhabited places. Most of the oil is exported crude and is merely piped to suitable coastal terminals. Only since the dramatic rises in the price of oil starting in 1973 have the exporting countries acquired the potential to invest heavily in the development of other sectors, but falling prices in 1982 could indicate that the oil-dependent developed countries are beginning to need oil less than they did a decade ago.

Agricultural production in North Africa and Southwest Asia is largely geared to local or national needs. Cereals, especially wheat, barley and rice (where irrigation is possible) occupy a large proportion of the cultivated area in North Africa, Turkey and Iran. The livestock sector is directed mainly to the raising of sheep (and goats) rather than cattle. Animals are still moved seasonally between different areas of pasture. Permanent pastures, however, are mostly of poor quality and stands of good quality

timber are also very limited. The region exports a wide variety of agricultural products, but mainly in limited quantities: dates, olives, early vegetables, cotton. Since the population of North Africa and Southwest Asia is expected to grow from about 250 million in 1980 to about 410 million by the year 2000 any increases in agricultural production will undoubtedly be used to feed the population of the region. In some parts of the region, notably Algeria, Afghanistan and Yemen, food supply is already well below theoretical minimum requirements and some countries are already net importers of food.

The impact of modern industry in North Africa and Southwest Asia has been limited and local. A few countries produce limited quantities of pig iron and steel (Turkey, Egypt, Algeria, Tunisia) but metal-working and engineering are almost non-existent. In the 1970s the Shah of Iran had ambitious plans to

Table 12.7 *World oil summary*

The sixteen largest oil producing countries in 1980 in millions of tonnes

Country	1970	1980	Country	1970	1980
USSR	353	603	Saudi Arabia	178	493
USA	538	484	Iraq	77	130
Venezuela	195	116	Libya	160	86
Mexico	24	106	Abu Dhabi and Dubai	38	83
China	28	106	Iran	191	74
Nigeria	53	102	Kuwait	139	71
UK	0	80	Algeria	49	48
Canada	72	80			
Indonesia	42	79	World	2363	3074

Oil production by major world regions in 1980

Developing countries	Total	%	Developed countries	Total	%
Middle East	927	30.2	USSR and East Europe	622	20.2
Latin America	297	9.7	USA and Canada	564	18.3
North Africa	171	5.6	West Europe	126	4.1
South and Southeast Asia	127	4.1	Australasia	19	0.6
Other Africa	126	4.1			
China	106	3.4	World	3074	100.0

develop heavy industry in his country and to become one of the world's major industrial powers. Such development depended on a fast rate of depletion of oil reserves, but whatever its prospects for success it has been shelved if not abandoned in view of internal conflicts, policy changes and a drastic decline in oil exports. Saudi Arabia seems to be the country best placed to industrialize rapidly. Its oilfields on the Gulf are to supply local iron and steel, petrochemicals and fertilizer plants and also via new trans-Arabia pipelines to supply industrial centres on the Red Sea coast.

The main oil producing countries of North Africa and Southwest Asia contain only about three percent of the population of the developing world. With some difficulty they, like some developing countries in East Asia and Latin America, seem to have both the financial resources and government policy needed to become 'developed'. In the case of the major oil producers it is a question of compromise. They have gained greater control over oil resources and production than they had a few decades ago and they get more generous remuneration for exports. They need to use their oil and natural gas as fuel and raw materials for industry and other sectors of production. They must, however, satisfy their customers, several of which have suddenly become interested in alternative sources of energy to oil, notably coal

(USA), nuclear power (France, Japan), solar and wind power or, like the UK and Norway, have themselves found oil.

The countries of North Africa and Southwest Asia have some raw materials other than oil and natural gas for development but locally they lack water. Almost everywhere technical skills are limited. It would be surprising however if the region did not achieve considerable industrial growth even if this were concentrated in only a few places. The necessary rises in the price of oil, ironically, have made development and industrialization even more difficult than before in the many other developing countries both inside and outside the region that depend on imported oil.

In spite of a number of crises and conflicts in North Africa and Southwest Asia the region has survived the shocks. Some may be noted:

Conflict in Palestine and formation of Israel, 1948
   (*see Figure 12.8* for Israel and its neighbours)
Iran takes over foreign oil companies, 1950
France's involvement in Algeria, 1950s to 1962
Suez crisis, Israel, France and UK attack Egypt, 1956
US and UK send forces to Lebanon and Jordan, 1958
Israel occupies lands in Egypt, Jordan and Syria, 1967
Suez Canal closed 1956–57, 1967–75

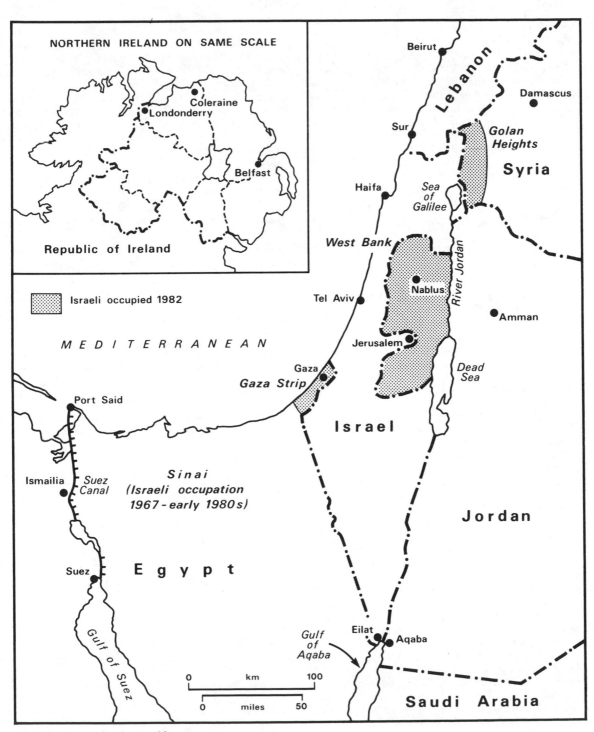

Figure 12.8 *Israel and its neighbours*

Israel survives war with neighbours, 1973
Algeria and Morocco in dispute over Spanish Sahara
    late 1970s
Revolution in Iran, 1979
Soviet forces enter Afghanistan, 1979
Iraq attacks Iran, 1980

Increasingly the USA and the USSR appear to be involved in Southwest Asia. Almost all the oil exported from the region goes either to the USA or to its industrialized allies in West Europe and Japan. If somehow the supply of oil from Southwest Asia to the Western industrial countries could be cut off then economic disaster would follow because at short notice other sources of oil would not be adequate to replace it. 55 percent of the world's oil exports originated in the Middle East in 1980. The following percentages of oil imports were supplied by the Middle East:

North America   25%
West Europe   61%
Japan   72%

Soviet interest in Middle East oil may have been growing in the 1970s for two reasons. Its own reserves of 8600 million tonnes which may, however, understate the true reserves, yielded a production in 1980 of 603 million tonnes. The Soviet 1981–85 Five-Year Plan allows for an even greater output in the future. The life of the official reserves would therefore be only 14 years or less. In the early 1980s the USSR was a net exporter of oil but in future it might import oil itself or force its CMEA partners to do so by reducing exports to them. Soviet trade figures with Middle East countries in *Table 12.6* are an indication of Soviet involvement in the region.

The Soviet Union shares common boundaries with Turkey, Iran and Afghanistan. The data in *Table 12.8* show a marked increase in trade with its three neighbours in the late 1970s. By treaty it has some right or obligation to give military help to Iran should that country require it. It did not feel the need for a treaty to invade Afghanistan. The USA 'defends' the Middle East (with or without treaties) from a great distance. In recent years a sharpening of awareness in the Middle East subgame of gaining allies and establishing bases can be noted. *Figure 1.13* is an attempt to show the situation and some prospects. In 1981 the US military policy was to think up deterrents and tripwires to prevent a direct military invasion by Soviet land forces.

Table 12.8 *Soviet trade with Middle East countries in selected years in millions of roubles*

Country	1975	1976	1977	1978	1979	1980
Iraq	600	715	602	1084	1182	732
Iran	510	445	708	671	409	334
Turkey	95	115	139	158	379	443
Afghanistan	132	154	190	215	324	507
Syria	168	235	207	205	199	321
Yemen PDR	14	20	36	27	65	61
Yemen AR	6	9	21	33	46	48
Cyprus	17	27	27	27	33	43
Saudi Arabia	6	13	14	8	25	31
Kuwait	4	10	24	37	8	17

Source: Various years of Vneshnyaya torgovlya SSSR

Introduction

In a traditional society there would be little point in speculating extensively about the future because it should be broadly similar to the recent past. For several centuries now the whole complex of world affairs has been changing. It has even been argued that at least in the last two centuries the rate of change has itself been accelerating. Speculation about the future is, therefore, of more practical relevance now than the study of the past. Although the past can be reinterpreted by historians it cannot be changed. In contrast the future can (at least apparently) be planned and to some extent controlled.

Until the 1960s speculation about the future tended to consist of attempts to predict a particular event with accuracy (will it rain the day after tomorrow?), to project past trends into the future (as with population) or to formulate loosely defined stages of development of societies or economies on the lines proposed by K. Marx and modified by his followers, or by the US economic historian W. W. Rostow, with his model of stages of growth. H. Kahn's work *The Year 2000* was one of several attempts in the 1960s to look at the future in a more professional way. Speculation about the future requires adherence to some sound practices and now includes several distinct methods of reasonable repute. Electronic computers can help by handling much larger sets of data than could be handled before the 1960s and they allow many alternative futures to be tested.

The following features of modern forecasting make it different from more traditional approaches:

(1) Forecasts about the future are usually based on a number of assumptions. One should not, therefore, attempt to construct an exact prediction of events, but to examine alternative futures based on various combinations of assumptions. The British oil industry may be used as an example. For simplicity one could say that consumption of oil in Britain could decline, stay the same or rise. With regard to oil reserves there could be no new discoveries, some new discoveries or many new discoveries. There are nine possible different combinations each with one consumption future and with one discovery future. Consumption and discovery could to some extent be influenced or controlled by pricing policy and by investment in exploration.

(2) A forecast should be monitored as time passes and, if necessary, revised in the light of new information. It is self-defeating to stick obstinately to one's original forecast if things are not going the way they 'should'. Marxism has only survived because the original model of socio-economic change in the world has been adapted and modified to allow for developments not anticipated a century ago.

(3) Compatibility of projections is vital if the projections are to be valid. An example will illustrate the problem. It has been proposed that as fossil fuel supplies become more limited (and prices rise excessively high) it will be worth obtaining fuel (wood, alcohol, diesel fuel) from the cultivation of fast growing trees or fieldcrops such as sugar cane and the sunflower. It has also been argued that the world's supply of food for humans is likely to become inadequate and that human food could be obtained from the fossil fuel petroleum. The two future developments go in exactly the opposite directions.

(4) In speculating about the future all the latest available information about actual plans, intentions and decisions should be taken into account.

The above are examples of a new concern about good practice in speculating about the future. The new view is nicely expressed by Ayres (1979):

'In my forecasting work, I think of the future as one might think of a play sketched in outline, with the characters and their initial relationships well defined, but the script as yet unwritten. (Indeed, the final script is *never* entirely written before the performance takes place.) The characters live out the play and create it as they go along.'

Several ways of speculating about the future are now accepted and practised. Five possible not entirely mutually exclusive ways will be noted briefly:

(1) The projection of past trends into the future is a traditional method of forecasting. It is generally accepted that the further the trend is projected into the future the wider is the divergence between extreme limits, for example upper and lower limits. It must also be appreciated that some phenomena are more reliable and change more smoothly than others. The total population of a sizeable country can be projected more accurately than the amount of rain that will fall in the coming years in sub-Saharan Africa. *Figure 13.1* shows the remarkable fortunes of the Peruvian fish catch. The reader may visualize the forecasts that could have been based on past trends at different moments in time. Soviet grain yields show a gradual upward trend with year to year fluctuations. How long can the upward trend go on?

(2) The establishment of goals is a projection that takes some target that has been established for the future and examines the feasibility of attaining it. Around 1980, for example, China had about 1000 million inhabitants. At that time an ultimate stable population of 1 200 millions was proposed. Could this target be achieved or would it be exceeded? Knowing the number of females alive in 1980 it is possible to calculate how many children each female should on average to guide the total towards the target. The number would for a time average considerably under two.

A frequent goal in many developing countries a decade or two ago was to reach the level (and 'stage') of a developed country. What would Colombia or Ivory Coast, with a gross national product per inhabitant of about 1000 dollars in the late 1970s, have to do in the next few decades to reach the present gross national product per inhabitant of the USA or France of around 10 000? The population of the two poorer countries is expected to double in two or three decades. Somehow the volume (or value) of total

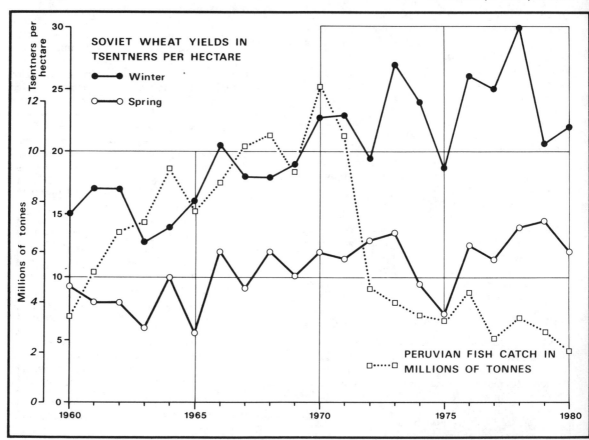

Figure 13.1   *Changes through time exemplified by* (a) *Soviet wheat yields and* (b) *the Peruvian fish catch. Sources: various numbers of Narodnoye khozyaystvo SSSR for (a) and UNSYB for (b)*

production of goods and services would have to be raised not ten but twenty times.

(3) Insurance companies base their premiums for life, health and other insurances on probabilities affecting the lives of large numbers of individuals. In a similar way it is possible to reason out that some futures are more probable than others. If ten sectors of the economy could each have a very good, good, moderate, poor, bad or disastrous future then there are many possible combinations that would give a mix of successes and failures, but only one combination that would give 'very good' for all ten sectors and only one that would give 'disastrous' for all ten. The extreme outcomes are analogous to the rare event of throwing a dice ten times and getting all sixes or all ones.

(4) A system may be defined as a set of interlinked elements, the sum of which is greater than the parts. A change somewhere in the system can have effects throughout the rest of the system. When there are many elements in a system the number of calculations needed to trace changes through time is enormous. Here, once again, electronic computers can help in the application of systems to complex situations. In the early 1970s a world model was developed to trace the interactions between numerous variables according to various assumptions.

(5) The Delphi method can give insight into some aspects of the future. Informed opinion can be collated to give an intelligent consensus about when a particular event might take place. When, for example, does a panel of experts expect nuclear fusion power to be commercially available? Some might estimate 30 years, some 40 years, some never. An informed forecast of this kind at least gives some basis for planning the future.

It is the purpose of the rest of this final chapter to map out a few possible futures and to speculate briefly about them. The more drastic and short-lived ones will be dealt with first. The theme of the chapter may be illustrated by the simple flow chart in *Figure 13.2* in which the question is asked about each future in turn: will it happen or will it not happen? If it does, what will be the result? If it does not, go on to the next prospect. The line of thought might proceed as follows:

(1) Will the human species be eliminated by some natural disaster? If so, might it be the impact of a giant meteorite (or the moon) landing in the Pacific Ocean or sudden unprecedented sunspot activity bringing intolerably hot or cold conditions and atmospheric changes? Could there be changes in the earth's interior upsetting conditions on the crust?

(2) If not the above, will there be an all-out nuclear war or a technological blunder drastically upset-

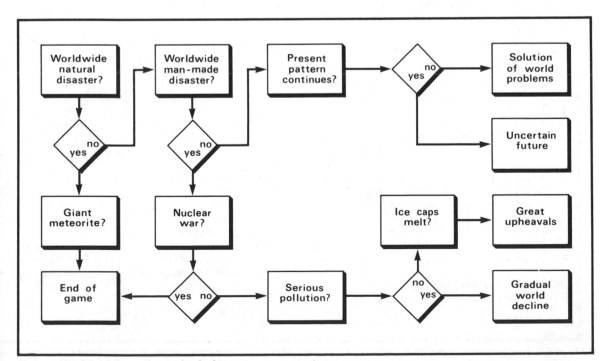

Figure 13.2 *Alternative prospects for the future*

ting man's position in the world by for example causing incurable pollution or melting the ice in Antarctica?

(3) If not, will the human species continue to drift on as now from one conflict to another. If so, what kind of pattern or lack of pattern might we expect?

(4) If not, what would a world be like in which current problems have been solved?

13.2 Natural and manmade disasters

The solar system appears to work smoothly and changes take place on such a slow time scale that the probability of a rare or physically 'impossible' event occurring seems very low indeed. The moon will not collide with the earth, a giant meteorite is not in the offing and the molten material under the earth's crust will not suddenly spill over the surface. Sunspot changes may cause considerable but not drastic variations in climatic conditions.

In the short run it seems much more probable that an end could come to the career of the human species through some manmade catastrophe than as a result of a sudden physical change. A growing problem of world significance in the latter part of the 20th century has been the possibility of either a devastating nuclear or chemical war or of a technological accident or blunder causing some irreversible climatic or biological change detrimental to the further existence of the human species. The effects of a nuclear explosion over a large town were described in chapter 3. A chemical or biological upset will now be discussed.

Early man had few tools, made little impact on the natural environment and like other species depended on hunting or on gathering plant products on a modest scale. Even before the 18th century some features of the natural environment were greatly modified by man through cultivation, grazing and the clearance of forest because agriculture involves limiting and controlling the plants that grow in a given area and keeping out the ones not wanted. Thus the natural vegetation of many parts of the world has been removed. Continued overgrazing and intensive cultivation also changes the physical environment considerably. Until the 19th century, however, human groups existed largely independently, on a local or regional basis, and the chance was remote that some worldwide technological development might change everything suddenly.

With the development of science, the application of technology, the growth of population and a rapid increase in the use of many natural resources, human beings have continued to take over more and more environments and resources. Other species have been pushed out and even eliminated. Resources have been sought in difficult environments: oil in off-shore areas and in Arctic wastes, fish and whales in remote parts of the oceans. It has been argued that in view of its large numbers the human species is reaching a situation of 'swarm'. There will soon be 'too many' people, if there are not already. Moreover, on average every person uses up more non-renewable resources than his predecessor a century or even 30 years ago. Some fairly quick man-induced future paths leading to the complete destruction of, or unacceptable damage to, the human species may be proposed.

It is possible that some civil blunder like the release of a chemical could in some way pollute or upset the structure of the hydrosphere or atmosphere. It has already been suggested that the widespread use of aerosols might damage the present state of the ozone layer of the atmosphere. Atmospheric pollution caused by the burning of fossil fuels and the accumulation of acids, falling as 'acid rain' to threaten plant and animal life, seems a greater problem. Pollution generated in one country (eg the USA, the German FR), may be deposited in another country (eg Canada, Sweden). On the other hand, even a major breakdown at a single nuclear power station, rather than at a large number simultaneously, would probably cause only local or regional damage, not a worldwide disaster. An error of judgement nearly caused an emergency core cooling failure in a nuclear power station in Alabama (*The Times*, 31 March 1975). Released radioactive material could have directly caused 100 000 or more deaths as well as many injuries and much damage. Even more disturbing was the near disaster in the nuclear power station on Three Mile Island in the USA in the late 1970s. There was reputedly an even greater nuclear disaster in the Ural region of the USSR in the late 1950s, an event no doubt known outside the country but hushed up. The contamination of an area around Seveso, near Milan, North Italy, in 1976, illustrates a blunder on a very small scale. It underlines the need for great care and an adequate understanding of processes in the handling of the materials and equipment in the high technology of the modern industrial world.

By excessive use of, or lack of, care for the biological environment, man might act like a cancer, destroying the basis of his own life, other plants and animals. In the end ivy may suffocate the tree it feeds on. The clearance of all the forest in the Amazon region of South America might affect the proportion of oxygen to other gases in the atmosphere. Well-evolved parasites, on the other hand, should not destroy the host animals they live on.

Growing concern is being expressed about the rising content of carbon dioxide in the atmosphere through the burning of fossil fuels and the clearance of forests. The effect is to heat the atmosphere gradually, causing changes in climatic conditions, with a critical lowering of rainfall in many areas cultivated at present. According to Budyko (1979) one area where farming would be adversely affected is the central part of European USSR. In addition the melting of the floating ice in the Arctic Ocean and of part of the Antarctic ice sheet could raise sea level by several metres in the next few decades.

The world may be thought of as a vehicle being taken on a journey through unknown country in the dark by such amateurs as politicians and soldiers, with 'professional' advisers from such fields as economics and religion. Problems and dangers ahead include a conflict between the people driving the machine, hazards for health and genetics, adverse climatic conditions, no food or water somewhere along the route. It is the club of rich countries that largely determines the speed and course of the machine.

The rather nebulous hazards ahead can be tabulated on the basis of a game based on probabilities. The probability of occurrence of each of a number of hazards facing the human species is estimated and held on various imaginary dice, one for each hazard considered. Each dice would have as many outcomes as are required, not necessarily only six faces. It might be estimated that on average once in five years there will be an earthquake serious enough to cause 100 000 deaths. Thus in one year the probability is one-fifth that the event will occur. The probability that a large nuclear power station will give off damaging radioactive material that will kill 100 000 people may be estimated, for example, at once every 20 years in the future when there are many in use. Other events are rarer, such as the melting of all ice rated, perhaps, at one in several thousand.

Table 13.1 illustrates the nature of the probability approach, but no conclusions are to be drawn from the simple example. Many more possible hazards, both natural and manmade, should be included. The probabilities are purely notional. The game, howev-er, *can* be played. It can only show what could happen, not what will happen. If the event occurs then the outcome may be considered 'bad'. If it does not, then the outcome is 'good'. If it is assumed that the events are independent, then the future prospect is described in and/or terms. Next year there could be a 'one-million-deaths' earthquake (one in fifty and/or a nuclear power station emergency (one in fifty) and/or a '10-million deaths' nuclear conflict (one in twenty). The probability that any one of the above will occur is the sum of probabilities minus the occasions when two or more occur together. The probability that all three will occur in the same year is the result of multiplying the three probabilities and is much smaller.

The tabulation of information about future probabilities in *Table 13.1* is thought-provoking. The obvious stands out even more obviously. If you want to remove the risk (dice) of a nuclear war, do not have nuclear weapons. If you think there is a risk (however low) in the use of fluoro-carbons to spray the contents of aerosols, then ban their manufacture. Even on the natural disaster side, preventive action can be taken in a passive way by storing food for bad years and by not removing the Amazon forest. As it is, a fairly sudden and nasty end to the career of *Homo sapiens* could be brought about by man himself.

13.3 World powers and ideologies

Some possible natural and manmade changes that could have disastrous consequences for the human species were considered. One prospect was a conflict between the USA and USSR ending in a large-scale nuclear war. In this and the next section the prospect is considered of continuing friction and conflict between the two super-powers without disastrous consequences for the whole world.

The great range of sizes found among the countries of the world was noted in chapter 2. The size of a

Table 13.1 *Probability of number of deaths resulting from disasters per year*

Disaster	100 000 deaths	1 million deaths	10 million deaths	100 million deaths	1000 million plus deaths
Sudden event					
Serious earthquake	1/5	1/50	1/500	–	–
Nuclear power station	1/20	1/50	1/200	–	1/1000
Nuclear war	–	–	1/20	1/100	1/1000
Change over a period					
Ice melts	–	–	1/500	1/100	1/1000
Forest destroyed, oxygen affected	–	–	–	1/20	1/100
Aerosol on atmosphere	–	–	–	–	1/100
Drought and food supply	1/2	1/5	1/10	1/20	–

Table 13.2 *Shares of world power in per thousands*

Country	1910	1937	1950	1960	1970	1980 I	1980 II
USA	250	250	250	210	180	120	165
USSR (Russia)	80	110	110	140	150	130	185
China	90	80	80	120	150	95	85
UK	90	60	50	40	30	15	15
Germany[a]	80	70	30	30	40	25	25
France	30	30	30	20	30	15	15
Japan	10	30	20	20	40	30	30
Brazil	20	20	20	20	20	35	30
India	–	–	50	50	50	60	50
Canada	20	30	40	30	30	40	35
Australia	10	20	20	20	20	30	25

[a] FR after 1937

given country can be assessed in various ways and the influence of the country on world affairs correlates only roughly with any particular measure of size. While it is reasonably easy to quantify people and products it is unrealistic to give a precise quantitative measure to the overall strength of world powers. Any numerical score depends firstly on what is selected for measurement and secondly on the weight given to each item assessed.

In *Table 13.2* eleven independent countries are assessed on the basis of the combined shares of certain attributes. Calculations of world powers made by the author for the 1972 edition of *Geography of World Affairs* took into account the combined share of area, population and energy consumption of countries or groups of countries in selected years from 1910 to 1970. The shares in *Table 13.2* are expressed in per thousands of total world 'power' held at different times by 11 'major' countries, with the values rounded to eliminate the appearance of spurious precision.

Two assessments are given for 1980. *I* includes area, population, gross national product and natural resources. *II* includes all in *I* plus military strength. The former colonies of the West European countries and of Japan have not been included and groupings of countries such as EEC and CMEA are not shown either.

From the table four types of world power emerge:

(1) All rounders: the USA and USSR, with appreciable scores on all criteria taken.
(2) Large in population: China and India.
(3) Large in area and natural resources: Brazil, Canada, Australia.
(4) Large in industrial capacity: Japan, German FR, France, UK.

The data in *Table 13.2* show that together the USA and USSR have held about one-third of the 'power' in the world throughout this century, a prospect noted by de Tocqueville in the 19th century. Early in the 20th century the USA was largely isolated from world affairs and the USSR was roughly equal in strength to several other European countries. The USSR and USA had become more influential in the world and more equal to each other by 1950 than they had been in 1910. By 1980, if military strength is taken into account, each was still much stronger than any other power.

If the USA and USSR chose to form an alliance they could no doubt make a deal which would enable them to share influence in the world much in the way that Germany and Japan might have done had they been successful in the Second World War. As it is the two superpowers are bitter rivals in an ideological struggle, each seeking the support of other powers to gain additional strength. The break between the USSR and China around 1960 and the discovery of mutual interests between China and the USA since the early 1970s marked a critical shift and realignment of power in the world. The fact that all the other powers in *Table 13.2* except India are in some way allied to the USA makes the USSR appear much more isolated and weak than it is portrayed in the West, even if by 1980 it might have replaced the USA as the strongest single power in the world.

The USSR is regarded by the USA as the last 'predatory' empire. Soviet leaders are seen to have a self-appointed mission to convert the world to socialism and communism. They profess to regard stages of social and political evolution as inevitable. They will take any opportunity to further their cause provided they themselves are not at risk. Conflict in Viet Nam reflected adversely on the USA for its part in the struggle but Soviet arms and advisers helped North Viet Nam to win, though at enormous cost in lives and property. The ends justify the means. Viet Nam was doing the job for the USSR. In military establishments it is regarded as sound practice to reason out how the potential enemy may be viewing a given tactical or strategic situation. The next section is an attempt to view world affairs through Soviet eyes.

13.4 Soviet aims and aspirations

The following passage illustrates a recent official Soviet view of the future of the world. It has been translated from the document provided for discussion with regard to the 1976–80 Five-Year Plan of the USSR and was published widely in various languages. The source used here was *Ekonomicheskaya gazeta*, number 51 December 1975.

'(During 1970–75) the economic power of the USSR has grown considerably. The gradual process continues of overcoming the substantial differences between intellectual and manual work, between town and country life. The ideological and political unity of Soviet society has been further strengthened in the community of peoples of our multinational state . . .

The developed socialist society that has been built in the USSR once again shows the enormous possibilities and advantages of the planned economy which has no unemployment, inflation and crises that shake the capitalist world.

Great steps have been taken in foreign policy, in the realization of the Peace Programme, worked out at the XXIV Congress of the party. The international position of the USSR has been strengthened. *The influence exerted by socialism on world affairs has been intensified.* International tension has been reduced.'

Fidel Castro of Cuba faithfully echoes Soviet views and in 1977 he put even more bluntly the position as follows (*Granma*, Havana, August 9, 1977):

'With the development of socialism and communism, mankind will eventually become one big family and our planet, one single country. The new generations must prepare for that world of the future.'

It is common to find articles in the Soviet press drawing attention to comparisons between the Soviet Union and the Western industrial countries. In the late 1950s Khrushchev pledged that the USSR would overtake the USA in total production, then in *per capita* production, within about 15 years. Sometimes there is a straight, non-controversial comparison of production data. On other occasions completely spurious comparisons mingle with aggressive predictions of doom in the capitalist world. One such

article appeared in *Ekonomicheskaya gazeta* number 14, April 1980. It was produced on the 110th year of Lenin's birth. It came at a time when Soviet industrial production was actually falling in some branches (*see Figure 13.3*). Two poor grain harvests were compounding problems and CMEA was in debt to the West to the tune of some 75 billion US dollars. Notes follow on the article (entitled *Leninskim Kursom*, following the path of Lenin). Observations by the author are in brackets.

Today (it is claimed) the world socialist friendship movement (*sodruzhestvo*) is in the fore of social progress. It is the most dynamic economic and political force in the modern world. The USSR itself is moving into a state of developed socialism.

During 1965–79 the national income of the USSR doubled and real per capita income nearly doubled. Industrial production rose 2.4 times and the agricultural production 1.4 times. Productive funds increased three times and foreign trade five times. Efficiency, quality and planning methods all improved. It took Soviet industrial output ten years to double to 1979, US industrial output 16 years, German FR 18 years and British 26 years (but the *absolute* output per inhabitant was much higher to start with in the three 'capitalist' countries). In 1950 Soviet industrial capacity was only about three-tenths the size of US industrial capacity, but by 1980 it was four-fifths the size.

Table 13.3 shows how a careful choice of branches of production can show Soviet progress in a favourable light. If motor vehicles, maize, meat or many other sectors of production had been included the picture would have been very different.

CMEA itself, initially the USSR and its East European partners, was joined by the Mongolian People's Republic in 1962, Cuba in 1972 and Viet Nam in 1978 and now has 440 million inhabitants or about ten percent of the total population of the world. Other countries defined as socialist in the USSR but not members (or full members) of CMEA are Yugoslavia and Albania, China and North Korea. Other countries strongly influenced by the USSR, but not (yet) defined officially as socialist include Kampuchea, Laos, Angola, Ethiopia, Afghanistan and possibly Nicaragua.

Since the 1950s the Soviet sphere of influence in the world has become 'far-flung' though still fragmented. It is well to remember that Soviet leaders still believe, profess to believe and/or assume that the world will eventually all be socialist, that history is on their side and that they must help history along when there is no risk to themselves in so doing. Nevertheless the article discussed here does confirm the need for continuing peaceful coexistence. Might 'peaceful' coexistence and competition lead to a military encounter involving US and Soviet forces in direct combat? Would an encounter remain local?

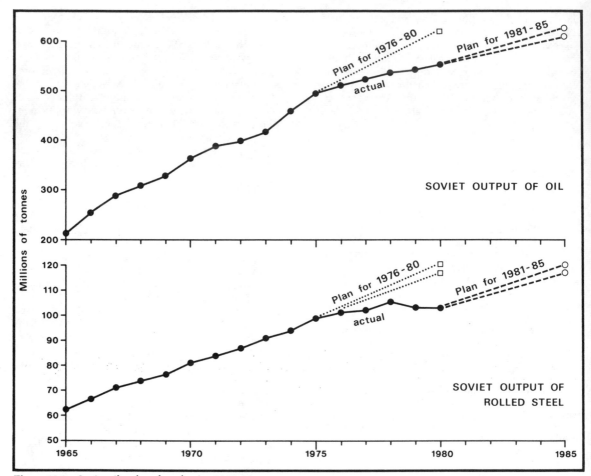

Figure 13.3   *Soviet oil and steel production 1965–85*

It will be assumed that only the USA and the USSR are capable of exerting worldwide influence. An ideal world for the USA would be one in which all countries would be open to 'free' economic transactions and in which all governments would be 'democratically' elected. An ideal world for the USSR would be one in which every country had a Communist Party sympathetic to (and subordinate to) its own Communist Party. The enthusiasm with which each would move towards the risk of a nuclear conflict with the other must to some extent be influenced firstly by the size of a particular country or set of natural resources at risk and secondly by the location of a country or some resources. It makes little difference to the USA whether or not the Mongolian People's Republic is neutral or pro-Soviet. On the other hand should the German FR or Japan be about to come under Soviet control then the USA would react strongly.

When one looks at the positions of the USA and

USSR on the world map the true picture of the situation is difficult to appreciate on account of the problem of projecting the surface of the globe on a flat map. *Figure 13.4* is an attempt to illustrate one aspect of global relations between the two superpowers for the 1980s.

In the game of world affairs the situation in the late 1940s was one in which Soviet influence was almost exclusively in Europe and Asia and was confined by a ring of US allies and bases. If China had become wholeheartedly devoted to the Soviet cause or completely under Soviet control then Soviet influence would have been much greater, but still felt mainly in Asia. Given that the USSR could not take military action far outside its confines, its prospect of weakening the USA and its Western allies was initially through gaining political influence in countries outside its existing sphere. It could attempt to control places of particular strategic and military value or areas with key natural resources. Obvious

Table 13.3a *Comparison of US and Soviet industrial production in 1950 and 1979*

Product	USSR 1950	USA 1950	USSR 1979	USA 1979
Oil (millions of tonnes)	38	267	586	420
Coal (millions of tonnes)	257	508	658	599
Steel (millions of tonnes)	27	90	149	127
Railway locomotives	227	3551	1748	1008
Tractors (thousands)	109	542	557	253
Fertilizers (millions of tonnes of active part)	1.2	4.0	22.1	20.0
Wheat (millions of tonnes)	31	28	90	58
Milk (millions of tonnes)	35	53	93	56
Sugar (millions of tonnes)	2.5	3.0	7.3	4.7

Table 13.3(b) *Growth of the socialist world*

Year when country became socialist		Millions sq km	Millions popn at the time	Millions popn mid-1980s
1917	USSR	21.7	143.5	266
1921	Mongolian PR	1.6	0.7	1.7
1944–45	Albania, Bulgaria, Hungary, German DR, Korean DPR, Poland, Romania, Czechoslovakia*a*, Yugoslavia	1.6	126.8	165
1949	China	9.6	548.8	975
1959	Cuba	0.1	6.9	10
1975–78	Laos and Viet Nam	0.5	53.4	57

Percentage of world total in socialist countries

	1919	1979
Territory	16	26
Population	8	34
Industry	3	40

a Strictly Czechoslovakia was 'captured' by socialism in 1948 (author)

Table 13.4 *Resource rich regions of the world and their distances from European USSR*

Region	Resources	Distance from USSR (low to high)	Quantity of mineral resources (high to low)
Middle East	oil, gas	1	1
Canada	metals	2 (over pole)	2
North Africa	oil, gas	3	6
West Africa	oil	4	8
Southeast Asia	oil, metals	5	9
Southern Africa	metals	6	4
Mexico	oil	7	7
Andean countries	oil, metals	8	5
Australia	metals	9	3

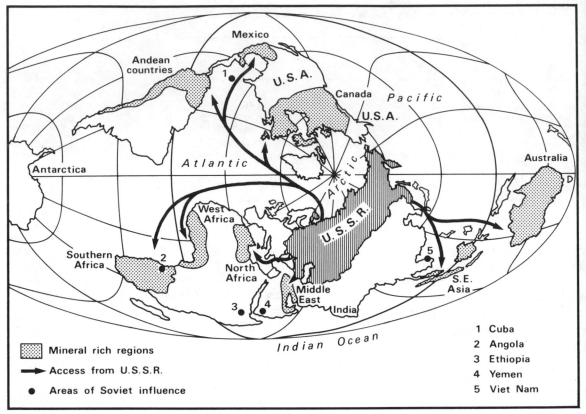

Figure 13.4   *The position of the USSR in relation to the mineral rich regions of the world*

strategic prizes would be the Suez and Panama canals and Singapore. Influence close to the USA in small Caribbean and Central American countries was also attractive. In the event the USSR controls Cuba but has a considerable bill to pay for Cuba's unwanted sugar exports and little strategic reward because Soviet nuclear missiles cannot be based in Cuba.

Some concentrations of minerals of major significance to the Western industrial countries are indicated roughly in *Figure 13.4*. The 'loss' of any of these would be a considerable setback to the industrial countries, always in need of fuel and non-fuel minerals. The nine areas distinguished in *Figure 13.4* are ranked in *Table 13.4* according firstly to their 'distance' from the USSR and secondly approximately to the quantity of mineral resources there. How is the USSR progressing if it is aiming to gain control over these areas? On the way it has picked up as protegés some of the poorest countries in Asia and Africa: Yemen DR, Afghanistan, Viet Nam, Angola, Ethiopia. The reader may imagine the game of GO in *Figure 1.13* extended to the world as a whole.

It would be naive to imagine that the USA was not also putting pressure on the USSR. Indeed, as already implied, the USSR regards itself as the defender of world peace and the promoter of correct political and social principles and practices. It is historically to be expected that the USA would try to thwart the aspirations of its rival. One weapon openly used by the USA is selective trade sanctions. Not only is the USSR likely to be irritated by such a procedure but, as is obvious to anyone who knows a little about Soviet leaders, they can ride out such sanctions, a view expressed in a report (*The Times* 9 September 1982) by the US Trilateral Commission:

'Experience has shown that the Soviet Union is better able than the West to withstand economic strains, while pursuing political ends.'

Surely West Europe is going more and more to need fuel and raw materials from other parts of the world. The United States might help its allies by cutting down on imports of fuel since it has abundant coal reserves of its own.

13.5 World views

In the last section one possible view of world affairs was described, the struggle between the two super-powers. The Soviet Union is after world domination and makes well thought-out but cautious moves, each of which the USA counters. As the game is played it actually seems as if the USSR can only either win a point or stay the same while the USA can only lose a point or stay the same. To fit every event in world affairs since 1945 into such a framework must over-simplify the true situation because other players are playing their smaller games as well. In the 1970s Chinese leaders have publicly denounced both the USSR (reactionary socialism) and the USA (capitalist imperialism) and have not aligned themselves closely with either side. India also keeps an independent line and many developing countries have attempted to exert their modest influence by promising support to and accepting aid from both West and East.

New alignments of whole groups of countries have occurred since 1945, the most significant being 'North–South' or rich versus poor. Countries that are rich in natural resources, whether developed or developing, have also tended to find common interests. Australia and Canada feel that their natural resources are at risk and sympathize with developing countries in the same 'plight'. There is no simple model for interpreting all aspects of world affairs. In this section some alternative views of world affairs will be compared. It will be seen in *Table 13.5* that during the present century different views and interpretations have been fashionable, but that at any given time two or more may be valid simultaneously. Five possible views will now be discussed.

Empires

Figure 13.5 shows the world in which a number of 'metropolitan' countries, mostly European, dominated the world scene. There was little concern over non-religious ideologies, the possible independence of colonies was largely disregarded, and the indus-trialized and rich countries were mostly the metropolitan ones. Colonies held many of the natural resources and provided many of the primary products lacking at home. The Empire-dominated world technically came to a virtual end when Portugal finally gave up its remaining African colonies in the 1970s. Many relics remain, however, not least the intact Russian Empire under the flag of the Soviet Union. Political colonialism has been replaced by economic imperialism, still widely practised by Western industrial countries in their former colonies.

Nationalism

The importance of nationalism in world affairs has fluctuated throughout the present century. Nationalism was strong in Europe in the 1930s (Fascists, Soviet Communists, Nazis). The concept of the sovereign state has only been slightly in question just after the two world wars when the League of Nations and United Nations were formed. The sovereignty of nations was underlined in the United Nations charter. Sovereign states have only reluctantly surrendered parts of their sovereignty to supranational groups since the 1950s.

Ideology

In the 1950s the prime criterion for sub-dividing the world would have been for many the dichotomized world of the Capitalist and Communist blocs. *Figure 2.8* is an interpretation used by the present author in the first edition of *Geography of World Affairs*, written in 1957 and published in 1959. Interest in the development gap was growing at that time but there was little concern about non-renewable natural resources. China's break with the USSR around 1960 reduced the validity of the rigid ideological division of the world and 'non-aligned' countries showed up in increasing numbers. The rivalry between the USSR and USA continued in the 1960s, but was reduced somewhat in the 1970s.

Table 13.5 *Fashions in the interpretation of world affairs*

	1900s	WW I	1920s	1930s	WW II	1950s	1960s	1970s	1980s	1990s
Empires	high	mod	mod	mod	mod	mod	low	low	–	–
Nationalism	mod	high	mod	high	high-mod	high	mod	mod	mod	mod
Ideological	low	low	mod	high	mod	high	high	mod	high	mod
Rich-poor	low	low	low	low	low	mod	high	high	high	high
Natural resources	low	low	low	low-mod	low	low	low	high	high	high

WW = World War; mod = moderate

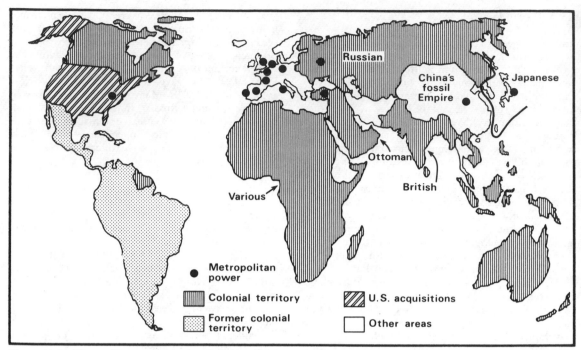

Figure 13.5 *The world around 1900, dominated by European powers and their empires*

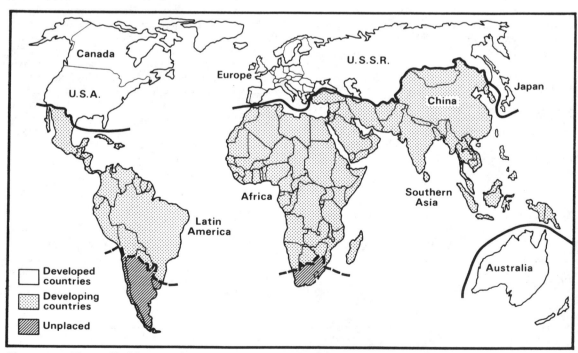

Figure 13.6 *The world of the 1970s characterized by rich and poor countries*

Rich–poor

Meanwhile in the 1960s increasing concern was expressed about the poor or underdeveloped countries, many of them newly established. It was politically and strategically expedient to offer them aid. By the 1970s many would have regarded the division between rich and poor countries of the world to be the prime one. The Brandt Report (1980) certainly made this point. *Figure 13.6* shows their distribution. The huge and growing contrast between rich and poor would remain unless the rich countries did something about it. Very little had been done, however, by the early 1980s.

Natural resources

Meanwhile around 1970 a new concern about an old idea became quickly fashionable as Forrester and Meadows (1971) produced technical and popular reports on the results of long-term projections of population, production, the use of natural resources and the pollution of the environment. Whatever assumptions were made, eventually something would go wrong or run out. The rise in oil prices in 1973–74 (a political move, not directly a commercial necessity) drew attention to the comparatively short prospective life of world oil reserves. In the 1970s it was common practice to stress the fact that minerals are extracted from non-renewable resources and that food is produced from a limited extent of cultivable land, itself at risk to various kinds of degradation. *Figure 13.7* shows a rough division of the world, not fitting actual national boundaries, into regions that are well, moderately and poorly endowed with regard to natural resources in relation to population size.

The most favoured regions are:

(1) Northwestern North America
(2) Siberia
(3) Amazonia
(4) Southwestern Africa
(5) Australia

The first three are tributary to other parts of the countries in which they are located.

The least favoured regions are:

(1) Most of western Europe and the Mediterranean area
(2) Southern Asia
(3) Japan and China

Early in the 20th century the metropolitan countries of Europe and Japan controlled the natural resources of other parts of the world.

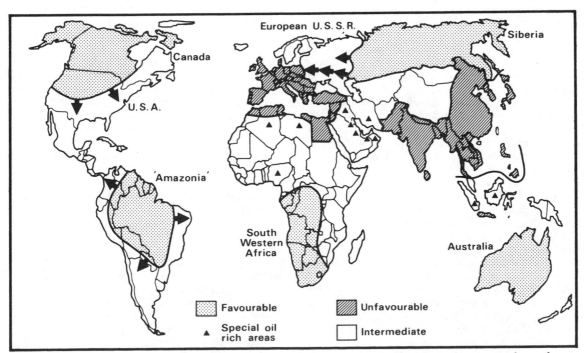

Figure 13.7   *Resource rich and resource poor regions of the world. Arrows indicate selected flows of raw materials out of resource rich regions*

When the availability of natural resources is related to their rate of use then Europe and Japan are seen to be in the most precarious position. In the future they may lose access to natural resources (needed by the host countries) and end up with a reduced share of the world's manufacturing capacity, as developing countries industrialize. The prospects for southern and eastern Asia are also bleak because they can never have access to enough natural resources to lift them far above their present very low levels of consumption of goods and services. The USSR and the USA are industrialized and are comparatively well endowed with natural resources (if Canada is attributed to the USA). Southwestern Africa might eventually have to use some of its mineral resources in the region but nowhere except in South Africa is it capable of doing so at the moment. The top, large country on the scale of natural resources per inhabitant is without any doubt Australia.

13.6 Transfers to close the development gap

Two major types of event could greatly change the course of world affairs in the next few decades: a large-scale natural or manmade disaster or the political reorganization of the world leading to the establishment of a world government. Some possible major disasters have already been discussed in this chapter. In the unlikely event of the sudden emergence of a world government in which laws could be enforced in Castro's 'world like a single country', could world problems be sorted out quickly?

In Cole (1981) the case has been argued that even if the world were run as a single state with a government given complete control over population, natural resources and productive forces, the transfers that would be needed to reduce substantially the disparities between rich and poor regions of the world could only take place gradually. Possible inter-regional net transfers will now be noted.

The interdependence of rich and poor countries results largely from trade. The precise terms of trade at a given time do not satisfy every trading country, perhaps not any. It is unlikely that there will be another shift in the relative value of a particular type of product as drastic as the rises in oil prices in the 1970s. In other words, if world trade continues on the lines it has done for the last three decades there is not likely to be much reduction in the gap between rich and poor countries although small parts of the developing world may make spectacular improvements in their living standards (eg Kuwait, Saudi Arabia, Southeast Brazil, Taiwan, the Korean Republic) and depressed areas in the developed world (old industrial areas, the centres of cities, marginal farming areas) may experience an absolute decline in living standards.

Since the Second World War many countries have attempted to narrow existing internal gaps between rich and poor regions. One way of tackling the problem is to transfer people or products from some regions to others. In Italy, for example, many Italians have moved from the South and Islands into the North. Capital has been moved in the opposite direction to create employment in the poorer regions. The UK, USA, Brazil and India are a few of many countries that have made similar internal transfers. Is it possible to apply the same methods to the world as a whole, as if it were one country? The immediate answer is that with the present system of sovereign states in the world the incentive is not there. If, however, attitudes changed and transfers from rich to poor regions were made compulsory, what could be achieved?

It will be assumed that the following could be transferred or donated from one country to another: people, food, fuel and raw materials, consumer manufactures, services, capital goods to produce goods or services. By definition natural resources could not be themselves shifted except by the transfer of the territory containing them (eg eastern USSR to China). The transfer of 'money' is also a misleading concept because the money is only meaningful when converted into goods or services. The above transferable items will now be considered in turn:

(1) Population could either be moved from countries poor in natural resources to countries rich in them (eg Indians or Japanese to Australia) or from developing to developed countries (eg Algerians to France, Turks to the German FR, Haitians to the USA). According to calculations made by the author (Cole 1981) it would have been necessary in 1975 to shift about half of the total population of the world (some 2000 million people) from some countries to other countries to achieve a reasonable balance between natural resources and population in the world. Even at the height of the flow of emigrants from Europe to the Americas and elsewhere only about ten million people left *per decade*. The large scale transfer of population about the world is not only virtually out of the question politically, but also physically entirely unrealistic and practically impossible.

(2) The richer countries could put aside and forgo part of their production of food and consumer manufactures and send them to poorer countries. Food, blankets, clothing and medical supplies are indeed given generously for brief periods and in emergencies that capture the imagination. Giving such products on a regular basis to raise living standards in poorer countries would be an act of charity that could cause resentment and would not remove the development problem but would

only patch it up artificially while producing a new kind of dependence.

(3) At present there is a net flow of fuel and raw materials from developing and developed countries. The richer countries could only help here by cutting down their imports of these products (while still paying the same) because it would be meaningless to 'give back' what they are already taking from developing countries.

(4) The developed countries could transfer means of production to developing ones. In this way developing countries could increase yields in agriculture, extend their existing industries, and establish new branches, thereby in the future satisfying more fully their own needs of various kinds of product. Eventually, however, they would not need many of the manufactured goods at present obtained from the industrialized countries. The effect would be gradually to reduce the fuel and raw materials available for the developed countries that are poor in natural resources, notably Japan and Europe (excluding the USSR).

(5) The transfer of services has only limited possibilities. It could involve the secondment of such people as doctors, engineers and teachers from developed and developing countries at the expense of the developed ones.

If the limitations attributed by the author to the various transfers described above are valid then one of the main themes of *North-South* (The Brandt Report, 1980) is invalid. If natural resources were for practical purposes unlimited there would be enough for every part of the world to industrialize to a high level. The paradox, however, is that the rich and poor would then be determined according to the natural resources they possessed, not as now according to

their level of industrialization. The regions that are rich in natural resources would no longer need to import manufactured goods and would not necessarily be enthusiastic about losing their fuel and/or raw materials just to keep countries such as the UK or Japan prosperous. If on the other hand natural resources are regarded as limited then the countries possessing plenty would be ever more anxious to retain what they have.

13.7 Future prospects in a 'surprise-free' world

In this section the theme is a future in which things will go on in much the same way in the next three decades as they did in the last three. Already in chapters 3 to 12 the future prospects of various aspects of world affairs have been referred to. This section, then, is a greatly simplified and rather subjective summary of the outlook. In *Table 13.6* the possible score of each region on each of its attributes is on a scale from +2 to −2. The score of +2 is the best possible prospect while −2 is the worst possible prospect.

In determining a score for each entry in the matrix the following features were kept in mind:

(1) Population is likely to grow fast in the six developing regions for some decades to come. Thus some increase is needed in production even to keep the same level per inhabitant.

(2) Existing natural resources are being used up and new material resources are being discovered, but the likely net decrease or increase is difficult to estimate.

(3) The developed countries use up non-renewable natural resources at a much faster rate per inhabitant than the developing ones.

Table 13.6 *Prospects for major regions of the world*

Country	Popn 1	Natural resources			Production and consumption				
		Bio-clim 2	Fossil fuel 3	Non-fuel minl 4	Food 5	Energy 6	Manuf 7	Services 8	Tech-nology 9
Japan	+1	−2	−2	−2	−1	−2	+2	+1	+2
West Europe	+2	−1	−1	−2	0	−1	+1	+1	+2
East Europe	+2	−1	−1	−2	0	−1	0	+1	+1
USA	+2	+1	+1	0	+1	0	+1	+2	+2
Canada-Australia	+2	+2	+2	+2	+2	+2	+1	+2	+1
USSR	+1	+1	+2	+2	0	+2	+1	+1	+2
South Asia	−2	−2	−2	−2	−1	−2	−1	−2	−1
Southeast Asia	−2	−2	−1	−1	−1	−1	−2	−2	−2
China	−1	−2	0	−1	−2	−1	−1	−1	−1
Latin America	−1	+1	−1	+1	0	0	0	−1	−1
Africa (south of the Sahara)	−1	0	−1	+1	−1	−1	−1	−2	−2
North Africa/Southwest Asia	−1	−1	+2	−1	−1	+2	−1	−1	−2

The data in the table can be read across the rows to give an overall impression of the prospects for each region or down the columns to view each sector. The standard of living is generally much higher in the first six regions than in the second six. The reasons why the disparity has come about and the prospects that it will continue are much more complex than is popularly thought.

In *The Wealth of Nations* two centuries ago Adam Smith already noted very big disparities in the world. Since his time access to natural resources and to technology has favoured the economic growth of Europe (including Russia), North America and Japan. The first three regions in *Table 13.6* (West and East Europe and Japan) now depend heavily on the primary products of other parts of the world whereas the next three (USA, Canada-Australia and the USSR) are much more self-sufficient in primary products, though the USA has become a major importer of primary products in recent decades as well as an exporter of them.

In the six developing regions the extensive application of modern technology is limited and local. South Asia (especially India) and China lack natural resources but have some basis of industry and technology. Latin America and Southwest Asia have considerable natural resources and have made some technological progress. Africa also has considerable natural resources but low levels of technology (except in South Africa) while most of Southeast Asia has not much of either.

It is often said that the world situation is unfair or unjust. No region can be blamed for having or not having abundant natural resources. Do some regions exploit others, though, in international trade? The answer is not straightforward. Many of the technological advances in the less developed regions have been made possible through the importation of industrial products from West Europe and later from other now developed regions. Japan and to a lesser extent West Europe and the USA on the other hand 'milk' poorer regions and in so doing perhaps hinder industrialization there.

The argument as to whether the industrialized countries exploit the developing ones can be assessed with the help of a diagram. Each country has within its territory given bioclimatic and mineral resources. In *Figure 13.8* only nine selected countries are

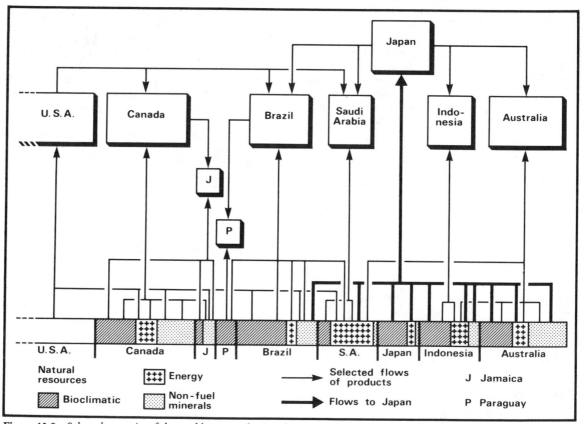

Figure 13.8 *Selected countries of the world portrayed as 'predators' and 'predated'*

represented. They are arranged arbitrarily along a line. The natural resources in the 'patch' of ground belonging to each country are like the plants in the food chain. All countries do some grazing on their own patches, but some extend their roots into the patches of other countries. The 'herbivores', after absorbing materials from their own patch, pass these on, often in processed form, to the 'carnivores' in the food chain above them.

In desperation one might try to give an impartial answer to the question 'who exploits whom?' by speculating what might happen firstly if the six developing regions were to disappear beneath the sea and secondly if the six developed ones were to do so. Who would lose what? The 12 major regions of the world in *Table 13.6* are arranged in groups of three in *Figure 13.9*. Three types of inter-regional flow are noted:

(1) Between developed and developed
(2) Between developing and developing
(3) Between developed and developing

What would happen if the flow of goods between developed and developing regions was cut entirely? The developed regions would feel particularly the lack of fuel and raw material supplies. The three regions rich in natural resources would cope with the situation better than the three lacking in natural resources. Although the developing countries have a large share of the world's oil and natural gas reserves most of the coal and lignite is in the developed ones. Food supplies and technology are available in the developed regions.

If the developed regions disappeared under the sea the non-renewable natural resources of the developing ones would face a longer life. On the other hand

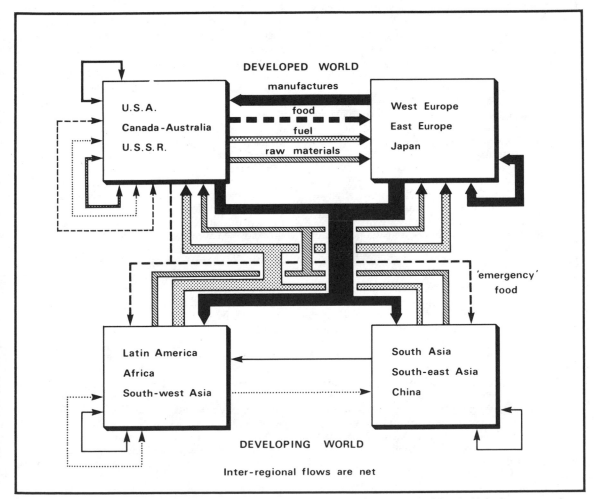

Figure 13.9 *Main types of flow of commodities between developed and developing regions of the world, both resource rich and resource poor*

the sophisticated capital goods obtained from developed countries would no longer be available. One can imagine Brazil, Mexico, South Africa, India and China turning into the industrial leaders of the world. Continuing population growth in the developing world would put increasing pressure on food supplies and emergency famine relief provided especially by the USA would not be forthcoming. The cessation of foreign aid from rich to poor countries would hardly be noticed except locally.

13.8 Concluding views

The contribution of the geographer to the study of world affairs depends particularly on two facts affecting the whole of the earth's surface: places differ from one another in various ways, and each place is uniquely located in relation to all other places. In what respects, to what extent and why place X differs from place Y is a common theme in geography. How X is located in relation to W, Y and Z is central in the spatial tradition of geography. Seven aspects of world affairs will be briefly reviewed with the two approaches noted above in mind: population, the physical environment, natural resources, technology, history, ideology, and location.

Demographic contrasts in the world were emphasized in chapter 3.4. If no drastic armed conflict takes place in the coming decades and world food production continues to rise at the rate it has been doing then it seems inevitable that world population will rise relentlessly from 2500 million in the early 1950s through 4000 million around 1975 to more than 6000 million by the year 2000. Almost all the gain will occur in developing countries. In spite of a drop in fertility in many countries in the 1970s estimates of a reduction in the rate of population increase have lately been modified in view of preliminary census results. Government planners in India had underestimated the census figures of 684 million in 1981 by 12 million. The first census of China since 1953 put the population of that country in 1982 at 1008 million, far more than the 1000 million or so attributed to China by various observers and published in Chinese sources. In many developing countries censuses still apparently undercount the population.

The influence of the natural environment on man's activities has been played down by geographers in recent decades as new techniques have developed in the conquest of nature. Contrasting physical environments still, however, present different sets of problems to man, as vividly exemplified by the ways in which conflicts are conducted in such different places as the forests of Viet Nam and Kampuchea, the arid lands of the Near and Middle East, the mountains of Afghanistan and the seas around the Falkland Islands.

Due to the great difficulty of assessing natural resources geographers have been reluctant to put precise or apparently precise quantitative values to productive land, economic minerals and other sources of food, fuel and raw materials. Chapter 4 was devoted to this neglected, controversial, but important topic and the point was made that in whatever way natural resources are estimated and qualified, each Australian has many more natural resources in his country than each Japanese. It is to be hoped that future military conflicts will not arise over the acquisition and control of natural resources as they have done in the past.

No less marked than the natural resource gap between countries is the gap in levels of consumption of goods and services between rich and poor countries. The present technologically advanced countries seem destined to retain most of the initiative in the development of yet more effective labour saving devices and weapons of destruction. Even so, whatever lies ahead, the decade 1973–83 seems likely to turn out to have been one of diminishing rates of economic growth for most countries. It is tempting to conclude that the fast economic growth in many parts of the world in the 1950s and 1960s may to some extent have been a period of catching up on the four decades 1910–1950, during which economic growth was distorted by two world wars, depression in the Western industrial countries and teething troubles in the newly established Soviet Union.

In chapter 2 the point was made that many problems in world affairs in the 1980s are the result of the particular process by which various European powers extended their influence over much of the rest of the world roughly during the period 1500–1900. It may not be entirely futile to speculate as to how the world would be organized now if world dominance had been achieved by the Chinese rather than the Europeans or by Moslem peoples rather than by Christian peoples. The present system of sovereign states in the world is not the only possible one. How resilient is it? Many assumptions about the future are based on comparatively little change in the present arrangement of some 200 units of greatly varying size and strength. There are however many territories of greatly varying size in the 'wrong hands' in someone's view. If Argentina claims the Islas Malvinas, may not China claim a million square kilometres of the Soviet Far East, and Mexico the Southwest of the USA?

As a result of the concentration of power since the Second World War in the hands of two superpowers the world has been subjected to an ideological battle focussed above all, perhaps on the ownership of means of production in an economy. The 'choice' lies between two extremes, private or public ownership; there is an intermediate type of 'mixed' economy.

The secret of understanding the way the Soviet Union is run seems to be to take what happens in the USA and turn it the other way round or at least go for something different. In the USA the private producer tries to produce (and sell) what the consumer wants. In the USSR the state producer tries to sell as little as possible to the consumer. In the USA contraceptives are widely available. In the USSR abortion is the main weapon of family planning. In the USA people can move freely from one part of the country to another. In the USSR mobility is greatly restricted. In the USA the computer has entered many aspects of life. Such an extension of the powers of the human brain was initially widely regarded with suspicion in the USSR. Unemployment rises to high levels in the USA while in the USSR, theoretically, it does not occur at all.

A geographer, David Smith, said *so* pertinently at a British-Soviet Geographical Seminar in Moscow in 1982: 'the works of Marx help in the understanding of 19th century Western industrial societies but they have no bearing on 20th century Soviet society'. The portraits of Marx, Engels, Stalin and Lenin, prominent still in the main square in Beijing in 1980 had disappeared in 1982. Chinese leaders are no doubt still trying to discover where they fit in the 'East-West' conflict.

Perhaps the haute cuisine of the geographer is the domain of the space of the earth's finite but unbounded surface. Here 200 countries share hundreds of international boundaries of greatly differing lengths and give tens of thousands of potential pairs of transactions. Soviet and US contacts alone may be occurring simultaneously in numerous different corners of the world: a link-up in space, a party in an embassy in Madrid or in Manila. More modest transactions continue between Belgium and Burundi, Czechoslovakia and Chad and so on.

A key feature of the arena of world affairs is that there is no centre and no periphery as there are in individual countries, and in most games of conflict that are carried out on a board or playing field. Command of a particular part of the earth's surface does not apparently give a special advantage in spite of arguments developed earlier in the century that control of East Europe and the interior of Asia would give control of the 'world island'. Centre and periphery refer more to the 'core' of industrial countries and the 'periphery' of developing ones. A Polish friend of the author's, Colonel J. Kowalewski, 30 years ago simplified the interface of the global struggle between the USA and the USSR to the line on a tennis ball separating the two halves. Who, he asked, encircles whom?

An eminent and highly respected British geographer of a few decades ago, S. W. Wooldridge, was credited with saying: 'the eyes of the fool are on the ends of the earth'. Many geographers are happy to work on a small patch to further their research or to exemplify geographical situations and processes in their teaching. In the author's back garden it is possible to identify many geographical situations in miniature. It would be a pity if modern geographers abandoned altogether any one of the regions that for several centuries their predecessors have helped to explore, map and interpret. It is hoped that the present book has served to show that every place in the ultimate region, the whole of the world, can suddenly be vital in world affairs and that there is a continuing need for the study of global features and problems.

References Specifically Used in Text or in Figure Captions

Ayres, R. U. (1979). *Uncertain futures*, Wiley, Chichester

Berry, B. J. L. (1961). In N. Ginsburg (ed.) *Atlas of Economic Development*, University of Chicago Press, Chicago

Budyko, M. I. (1979). Impending Climatic Change, *Soviet Geography, Review and Translation*, Vol. 20, pp. 395–411

Calder, N. (1972). *Restless Earth: A Report on the New Geology*, BBC Publications, London

Callahan, J. J. (1976). The Curvature of Space in a Finite Universe, *Scientific American*, Vol. 235 (No. 2), August, pp. 93–95

Canada Handbook (1979), (48th Annual) Publishing Section, Information Division, Statistics, Canada

Cavalli-Sforza, L. L. (1974). The Genetics of Human Populations, *Scientific American*, Vol. 231 (No. 3), September, pp. 80–89

Chung-Tong Wu in Stöhr, W. B. and Fraser Taylor, D. R. (1981). *Development from Above or Below?* Wiley, Chichester

Cole, J. P. (1980). *China 1980*, Geography Department, Nottingham University

Cole, J. P. (expected 1983). *Geography of the USSR*, Butterworth, London

Crowson, P. (1980). *Non-fuel minerals data base*, Royal Institute of International Affairs, London

Dewey, J. F. (1972). Plate Tectonics, *Scientific American*, Vol. 226, (No. 5), May, pp. 56–68

Eyre, S. R. (1978). *The Real Wealth of Nations*, Arnold, London

Forrester, J. W. (1971). *World Dynamics*, Wright Allen, Cambridge, Mass.

Glassner, M. I. (1978). The Law of the Sea, *Focus*, Vol. XXVIII (No. 4), March–April

Global 2000: Barney G. O. (Study Director), (1982). *The Global Report to the President*, Penguin Books, Harmondsworth

Harris, M. (1952). See Wagley, C. W.

Hauser, P. M. (1971). The Census of 1970, *Scientific American*, Vol. 225, (No. 1), July, pp. 17–25

Hodgson, B. (1978). Natural Gas: the Search Goes on, *National Geographic*, Vol. 154, (No. 5), November, pp. 632–651

Ilich, I. D. (1973). *Deschooling Society*, Penguin, Harmondsworth

Johnson, B. L. C. (1979). *India resources and development*, Heinemann, London

Kahn, H. and Wiener, A. J. (1967). *The Year 2000. A Framework for Speculation on the Next Thirty-Three Years*, Macmillan, New York

Kessler, S. E. (1976). *Our Finite Mineral Resources*, McGraw-Hill Book Company, New York

Lewis, K. N. (1979). The Prompt and Delayed Effects of Nuclear War, *Scientific American*, Vol. 241, (No. 1), July, pp. 27–39

Mackinder, Sir H. J. (1904). The Geographical Pivot of History, *Geographical Journal*, Vol. XXIII, (No. 4), April 1904, pp. 421–2

Mackrell, K. (1980). Oil still vital in 2000, *Chairman's Bulletin for Shareholders*, 'Shell' Transport and Trading Co Ltd, (No. 2), January 1980

Macleish, K. The Tasadays Stone Age Cavemen of Mindanao, *National Geographic*, Vol. 142, (No. 2), August 1972, pp. 219–248

McDonald, J. R. (1978). Europe's Restless Regions, *Focus*, Vol. XXVIII, (No. 5), May–June

McIntyre, L. (1980). Jari: a billion dollar gamble (massive technology transplant in the Brazilian Amazon jungle), *National Geographic*, Vol. 157, (No. 5), May, pp. 686–711

Meadows, D. H., Meadows, D. L., Randers, J. and Behrens III, W. W. (1972). *The Limits to Growth*, (Subtitle: A Report for the Club of Rome's Project on the Predicament of Mankind), Earth Island Ltd., London

Oxford University Press (1965). *The Oxford Annotated Apocrypha*, New York

Parkhurst, R., Ed. (1965). *Travellers in Ethiopia*, Oxford University Press, London, pp. 21–2

Pattison, W. D. (1964). The Four Traditions of Geography, *Journal of Geography*, Vol. LXIII

Riley, D. and Young, A. (1977). *World Vegetation*, Cambridge University Press, p. 95

Rodney, W. (1974). *How Europe Under-developed Africa*, Tanzania Publishing House, Dar-es-Salaam

Rona, P. A. (1973). Plate Tectonics and Mineral Resources, *Scientific American*, Vol. 229 (No. 1), July, pp. 86–95

Rostow, W. W. (1960). *The Stages of Economic Growth*, Cambridge University Press

Russett, B. M. (1965). *World Handbook of Political and Social Indicators*, Yale University Press, New Haven

Sipri, *World Armaments and Disarmament SIPRI Yearbook 1981* (1981). Taylor and Francis Ltd., London

Smith, A. (1776). *The Wealth of Nations*, Penguin Books (1970), Harmondsworth

Toynbee, A. (1953). *The World and the West*, BBC Reith Lectures, Oxford University Press, London

US Bureau of Mines (1970). *Mineral Facts and Problems, 1970* (Bulletin 650), US Dept. of the Interior, Bureau of Mines, Washington

Wagley, C. W. ed. (1952). *Race and Class in Rural Brazil*, UNESCO, Paris

Watson, J. W. (1967). *Mental Images and Geographical Reality in the Settlement of North America*, University of Nottingham

Williamson, J. G. (1968). Regional inequality and the process of national development: a description of the patterns, *Regional Analysis, Selected Readings*, ed Needleman, L. Penguin Books Ltd, Harmondsworth

Wilson, A. (1968). *The Bomb and the Computer*, Barrie & Rockliff, London

Weaver, K. F. (1979). The Promise and Peril of Nuclear Energy, *National Geographic*, Vol. 155, (No. 4), April, pp. 459–493

Zeitschrift für Geopolitik, (No. 11), 1934, pp. 710–712

Further Reading, by Chapters

Chapters 1 and 2

Chisholm, M. (1982). *Modern World Development*, Hutchinson University Library, London

Kahn, H. (1979). *World Economic Development, 1979 and Beyond*, Croom Helm, London

Kidron, M. and Segal, R. (1981). *The State of the World Atlas*, Pan Books, London

Rostow, W. W. (1978). *The World Economy: History and Prospects*, Macmillan

Sheffield, C. (1981). *Earth Watch, A Survey of the World from Space*, Sidgwick and Jackson, London

Short, J. R. (1982). *An Introduction to Political Geography*, Routledge and Kegan Paul, Henley-on-Thames

Times Historical Atlas of the World, (1980)

Chapter 3

Epstein, W. (1980). A Ban on the Production of Fissionable Material for Weapons, *Scientific American*, July, Vol. 243, (No. 1), pp. 31–39

Gwatkin, D. R. and Brandel, S. K. (1982). Life Expectancy and Population Growth in the Third World, *Scientific American*, Vol. 246, (No. 5), May, pp. 33–41

Lewis, H. W. (1980). The Safety of Fission Reactors, *Scientific American*, Vol. 242, (No. 3), March, pp. 33–45

Lewis, K. N. (1980). Intermediate-range Nuclear Weapons, *Scientific American*, Vol. 243, (No. 6), December, pp. 41–51

Meselson, M. and Robinson, J. P. (1980). Chemical Warfare and Chemical Disarmament, *Scientific American*, Vol. 242, (No. 4), April, pp. 34–43

Tsipis , K., (1981). Laser Weapons, *Scientific American*, Vol. 245, (No. 6), December, pp. 35–41

Walker, P. F. (1981). Precision-guided Weapons, *Scientific American*, Vol. 245, (No. 2), August, pp. 20–29

Wit, J. S. (1981). Advances in Antisubmarine Warfare, *Scientific American*, Vol. 244, (No. 2), February, pp. 27–37

Chapter 4

Brown, L. C. (1978). The Worldwide Loss of Cropland, *Worldwatch Paper* 24, October, Worldwatch Institute, Washington DC

Brown, L. R. (1981). World Food Resources and Population: The Narrowing Margin, *Population Bulletin*, Vol. 36, (No. 3), September, Population Reference Bureau

Deffeyes, K. S. and MacGregor, I. D. (1980). 'World Uranium Resources, *Scientific American*, Vol. 242, (No. 1), January, pp. 50–60

Doornkamp, J. C. (1982). The Physical Basis for Planning in the Third World, *World Planning Review*, Vol. 4, (Nos. 1–3)

Gold, T. and Soter S. (1980). The Deep-Earth-Gas Hypothesis, *Scientific American*, Vol. 242, (No. 6), June, pp. 130–137

Gore, R. (1979). The Desert: An Age-old Challenge Grows, *National Geographic*, Vol. 156, (No. 5), November, pp. 586–639

Griffith, E. D. and Clarke, A. W. (1979). World Coal Production, *Scientific American*, Vol. 240, (No. 1), January, pp. 28–37

Keyfitz, N. (1976). World Resources and the World Middle Class, *Scientific American*, Vol. 235, (No. 1), July, pp. 28–35

Manners, G. (1981). Our Planet's resources, *Geographical Journal*, Vol. 147, pp. 1–22

National Geographic, February 1981. Energy, a 'special report in the public interest'

Chapter 5

Alexandersson, G. and Klevebring, B.-I. (1978). *World Resources, Energy, Metals, Minerals*, Walter de Gruyter, Berlin, New York

Economic Development (1980). Special number of *Scientific American*, September Vol. 243, (No. 3), (includes general topics and China, India, Tanzania, Mexico)

Flower, A. R. (1978). World Oil Production, *Scientific American*, Vol. 238 (No. 3), March, pp. 42–49

Mitchell Beazley (1979). *Atlas of Earth Resources*, London

National Research Council (US National Academy of Sciences) (1979). *Science and Technology*, W. H. Freeman and Co, San Francisco

Sheldon, R. P. (1982). Phosphate Rock, *Scientific American*, Vol. 246, (No. 6), June, pp. 31–38

Weaver, K. F. (1979). The Promise and Peril of Nuclear Energy, *National Geographic*, Vol. 155, (No. 4), April, pp. 459–493

World Bank, *World Bank Atlas* (various years), Washington DC

Chapter 6

Coates, B. E., Johnston, R. J. and Knox, P. L. (1977). *Geography and Inequality*, Oxford University Press

Stöhr, W. B. and Fraser Taylor, D. R. (1981). *Development from Above or Below?* Wiley, Chichester

Wriggins, W. H. and Adler-Karlsson, G. (1978). *Reducing Global Inequalities*, McGraw-Hill, New York

Chapters 7 and 8

Cole, J. P. (1981). *The Development Gap*, Wiley, Chichester

House of Commons (1981). *The Mexico Summit*, HMSO, London

Chapter 9

European Economic Community, various publications in *European Documentation*, Office for Official Publications of the European Communities, Boite postale 1003 – Luxembourg

Japan of Today (1980). Ministry of Foreign Affairs, Japan

Kishimoto, M. (1980). Concentrated Japanese, *Geographical Magazine*, Vol. LII, (No. 7), pp. 477–482 and subsequent numbers in series Determined Japan

Nowak, S. (1981). Values and Attitudes of the Polish People, *Scientific American*, Vol. 245, (No. 1), July, pp. 23–31

Vesilind, P. J. (1982). Two Berlins – A Generation Apart, *National Geographic*, Vol. 161, (No. 1), January, pp. 2–51

Chapter 10

Australia: a survey, *The Economist*, 31 October, 1981

Ginzberg, E. and Vojta, G. J. (1981). The Service Sector of the US Economy, *Scientific American*, Vol. 244, (No. 3), March, pp. 32–39

Hauser, P. M. (1981). The Census of 1980, *Scientific American*, Vol. 245, (No. 5), November, pp. 37–45

Keely, C. B. (1982). Illegal Migration, *Scientific American*, Vol. 246, (No. 3), March, pp. 31–37

Leydet, F. (1980). Coal vs Parklands, *National Geographic*, Vol. 158, (No. 6), December, pp. 776–803

Pallot, J. and Shaw, D. J. B. (1981). *Planning in the Soviet Union*, Croom Helm, London

Chapter 11

China Reconstructs, monthly by China Publications Center, P.O. Box 399, Beijing, China

Edwards, M. (1980). China's Born-again Giant (Shanghai), *National Geographic*, Vol. 158, (No. 1), July, pp. 15–43

Ellis, W. S. (1981). Pakistan Under Pressure, *National Geographic*, Vol. 159, (No. 5), May, pp. 668–701

Garrett, W. E. (1980). Thailand: Refuge from Terror, *National Geographic*, Vol. 157, (No. 5), May, pp. 633–641

Gore, R., (1980). Journey to China's Far West, *National Geographic*, Vol. 157, (No. 3), March, pp. 292–331

Grove, N. (1982). Taiwan Confronts a New Era, *National Geographic*, Vol. 161, (No. 1), January, pp. 92–119

Husain, S. (1979). Strategy for the North-West Frontier, *Geographical Magazine*, Vol. LI, (No. 12), September, pp. 816–821

Johnson, B. L. C. (1979). *India resources and development*, Heinemann, London

Lynch, B. (1977). *Indonesia. Problems and prospects*, Sorrett Publishing, Malvern (Victoria, Australia)

Pannell, C. W. (1982). Less land for Chinese farmers, *Geographical Magazine*, Vol. LIV, (No. 6), June, pp. 324–329

Visaria, P. and Visaria, L. (1981). India's Population: Second and Growing, *Population Bulletin*, Population Reference Bureau, Vol. 36, (No. 4), October

Zheng, L. and others (1981). *China's Population: Problems and Prospects*, New World Press, Beijing

Chapter 12

Arden, H. (1982). The Two Souls of Peru, *National Geographic*, Vol. 161, (No. 3), March, pp. 284–321

Azzi, R. (1980). Saudi Arabia: The Kingdom and Its Power, *National Geographic*, Vol. 158, (No. 3), September, pp. 286–332

Birks, J. S. and Sinclair, C. A. (1980). Well Lubricated Emirate Economy, *Geographical Magazine*, Vol. LII, (No. 7), April, pp. 470–476

Blake, G. (1981). Turkish Guard on Russian Waters, *Geographical Magazine*, Vol. LIII, (No. 15), December, pp. 950–955

Cobb, C. E. (1981). A Nation Named Zimbabwe, *National Geographic*, Vol. 160, (No. 5), November, pp. 616–651

Edwards, M. (1978). Mexico, *National Geographic*, Vol. 153, (No. 5), May, pp. 612–647

Edwards, M. (1980). Tunisia: Sea, Sand, Success, *National Geographic*, Vol. 157, (No. 2), February, pp. 184–217

Grove, N. (1981). The Caribbean: Sun, Sea and Seething, *National Geographic*, Vol. 159, (No. 2), February, pp. 244–271

Grove, N. (1979). Nigeria Struggles with Boom Times, *National Geographic*, Vol. 155, (No. 3), March, pp. 413–444

Pelissier, R. (1980). Africa without the Portuguese, *Geographical Magazine*, Vol. LII, (No. 12), September, pp. 793–802

Chapter 13

Allen, R. (1980). *How to save the world*, Kogan Page, London

Freeman, C. and Jahoda, M. (1979). *World Futures*, Martin Robertson, Oxford

Kahn, H., Brown, W. and Martel, L. (1977). *The next 200 years*, Associated Business Programmes, London

Mesarovic, M. and Pestel, E. (1975). *Mankind at the Turning Point*, (The Second Report of the Club of Rome), Hutchinson, London

North-South: A programme for survival, ('The Brandt Report'), Pan Books, London, 1980

Revelle, R. (1982). Carbon Dioxide and World Climate, *Scientific American*, Vol. 247, (No. 2), August, pp. 33–41

Rostow, W. W. (1975). *How it all began*, Methuen, London

Index

Note: Italic figures denote reference to maps, charts or tables